Lecture Notes in Mathematics

2010

Editors:
J.-M. Morel, Cachan
F. Takens, Groningen
B. Teissier, Paris

FONDAZIONE
CIME
ROBERTO CONTI
CENTRO INTERNAZIONALE MATEMATICO ESTIVO
INTERNATIONAL MATHEMATICAL SUMMER CENTER

Fondazione C.I.M.E., Firenze

C.I.M.E. stands for *Centro Internazionale Matematico Estivo*, that is, International Mathematical Summer Centre. Conceived in the early fifties, it was born in 1954 in Florence, Italy, and welcomed by the world mathematical community: it continues successfully, year for year, to this day.

Many mathematicians from all over the world have been involved in a way or another in C.I.M.E.'s activities over the years. The main purpose and mode of functioning of the Centre may be summarised as follows: every year, during the summer, sessions on different themes from pure and applied mathematics are offered by application to mathematicians from all countries. A Session is generally based on three or four main courses given by specialists of international renown, plus a certain number of seminars, and is held in an attractive rural location in Italy.

The aim of a C.I.M.E. session is to bring to the attention of younger researchers the origins, development, and perspectives of some very active branch of mathematical research. The topics of the courses are generally of international resonance. The full immersion atmosphere of the courses and the daily exchange among participants are thus an initiation to international collaboration in mathematical research.

C.I.M.E. Director
Pietro ZECCA
Dipartimento di Energetica "S. Stecco"
Università di Firenze
Via S. Marta, 3
50139 Florence
Italy
e-mail: zecca@unifi.it

C.I.M.E. Secretary
Elvira MASCOLO
Dipartimento di Matematica "U. Dini"
Università di Firenze
viale G.B. Morgagni 67/A
50134 Florence
Italy
e-mail: mascolo@math.unifi.it

For more information see CIME's homepage: http://www.cime.unifi.it

CIME activity is carried out with the collaboration and financial support of:
– EMS - European Mathematical Society

Angiolo Farina · Axel Klar · Robert M.M. Mattheij
Andro Mikelić · Norbert Siedow

Mathematical Models in the Manufacturing of Glass

C.I.M.E. Summer School,
Montecatini Terme, Italy, 2008

Editor:
Antonio Fasano

Editor
Antonio Fasano
Università degli Studi di Firenze,
Dipartimento di Matematica "Ulisse Dini"
Viale Morgagni 67/A
I-50134 Firenze
Italy
fasano@math.unifi.it

Authors: see List of Contributors

ISBN: 978-3-642-15966-4   e-ISBN: 978-3-642-15967-1
DOI: 10.1007/978-3-642-15967-1
Springer Heidelberg Dordrecht London New York

Lecture Notes in Mathematics ISSN print edition: 0075-8434
ISSN electronic edition: 1617-9692

Mathematics Subject Classification (2010): 76D05, 76D07, 76B10, 76B45, 76M10, 80A20, 80A23, 35K05, 35K60, 35R35

Cover design: SPi Publisher Services

Printed on acid-free paper

Springer is part of Springer Science+Business Media (www.springer.com)

Preface

The EMS-CIME Course on Mathematical Models in the Manufacturing of Glass, Polymers and Textiles was held in Montecatini Terme (Italy) from September 8 to September 19, 2008. The course was co-directed by John Ockendon (OCIAM, Oxford, UK) and myself. The following topics were treated:

(1) *Nonisothermal flows and fibres drawing* (Angiolo Farina and Antonio Fasano, Univ. Firenze, Italy, Andro Mikelic, Univ. Lyon, France) (*)
(2) *The mathematics of glass sheets and fibres* (Peter Howell, OCIAM, Oxford, UK)
(3) *Radiative heat transfer in glass industry: modelling, simulation and optimisation* (Axel Klar and Norbert Siedow, ITWM – Fraunhofer, Kaiserslautern Germany) (*)
(4) *Modelling and simulation of glass forming processes* (Robert Mattheij, TU Eindhoven, The Netherlands) (*)
(5) *Injection moulding* (Hilary Ockendon, OCIAM, Oxford, UK)
(6) *Fibre assembly modelling* (Hilary Ockendon, OCIAM, Oxford, UK)
(7) *The mathematics of the windscreen sag process* (John Ockendon, OCIAM, Oxford, UK)

The focus was largely on glass manufacturing processes, with some digression to polymers and textile fibres in a context very close to the area of glass manufacturing. This volume collects the lecture notes of the courses marked with (*), all devoted to problems in glass industry. It is regrettable that the other lecturers could not provide a chapter, because the subjects they illustrated were extremely interesting.

John Ockendon presented a fascinating and quite difficult problem: the production of a windscreen by the natural bending under gravity of a still soft glass layer clumped at the boundary. The audience was very excited by his colourful explanation of the underlying mechanics, making use of any deformable object he had at hand.

Hilary Ockendon posed stimulating questions about injection moulding and the "flow" of fibres in a fluffy tuft subject to traction. We had exciting afternoon sessions discussing such problems.

Peter Howell gave a series of lectures on the manufacturing of glass sheets and fibres which provided an excellent complementary view of the subjects treated by Farina, Fasano, Mikelic.

Indeed he addressed different problems in the same area (e.g.: how to get a fibre of a desired cross section), each with a different mathematical approach.

Fortunately most of the material not included in this volume is retrievable on the CIME web site, either in the form of slides or of excerpts from books.

Altogether the Course presented a remarkable review of quite advanced technological problems in the glass industry and of the mathematics involved. It was quite amazing to realize that such a seemingly small research area is on the contrary extremely rich and it calls for an impressively large variety of mathematical methods.

Despite the fact that the volume is not collecting all the material presented at the Course, it deals with a number of problems which are very typical in the field of glass manufacturing and it can certainly be useful not only to applied mathematicians, but also to physicists and engineers, who can find in it an overview of the most advanced models and methods.

The Chapter by J.A.W.M. Groot, R.M.M. Mattheij, and K.Y. Laevsky illustrates the various processes of glass forming, starting from the basic physical information, developing the mathematical models for each process, and analyzing the procedures of numerical computations.

Then we have two Chapters on radiative heat transfer in glass. The first one is by M. Frank and A. Klar, treating in detail the physics of radiation in glass and various approximated methods to model it, with an eye to numerical complexity. This is a quite substantial piece of work, due to the extension and the intrinsic difficulty of the problem. It is followed by the contribution of N. Siedow, who, after continuing the investigation of numerical methods for heat transfer problems including radiation and convection, passes to a question of great importance: the measurement of glass temperature from the observation of the spectrum of emitted optical radiation. From the mathematical point of view this is formulated as an inverse problem, which is typically ill posed.

The way of circumventing this difficulty is explained in detail and examples are provided.

The last Chapter, by A. Farina, A. Fasano, A. Mikelic, deals with the industrial process of glass fibre drawing, which goes through several stages having different thermal and mechanical characterizations, and analyzes in general non-isothermal motions of viscous fluids which are mechanically incompressible and thermally expansible.

I must abstain from commenting the scientific level of the present volume, since I am among the contributors, but at least I wish to express my deep gratitude to the Authors for their valuable work. Finally, also on behalf of John Ockendon, I wish to thank EMS and CIME for having made this Course possible. A particular thank to the Secretary of CIME, Prof. Elvira Mascolo, who took care of so many details.

Antonio Fasano

Contents

Contributors

Angiolo Farina Università degli Studi di Firenze, Dipartimento di Matematica "Ulisse Dini", Viale Morgagni 67/A, I-50134 Firenze, Italy farina@math.unifi.it

Antonio Fasano Università degli Studi di Firenze, Dipartimento di Matematica "Ulisse Dini", Viale Morgagni 67/A, I-50134 Firenze, Italy fasano@math.unifi.it

Martin Frank University of Kaiserslautern, Erwin-Schrödinger-Strasse, 67663 Kaiserslautern, Germany frank@mathematik.uni-kl.de

J.A.W.M. Groot Department of Mathematics and Computer Science, Eindhoven University of Technology, PO Box 513, 5600 MB Eindhoven, The Netherlands j.a.w.m.groot@tue.nl

Axel Klar University of Kaiserslautern, Erwin-Schrödinger-Strasse, 67663 Kaiserslautern, Germany and Fraunhofer ITWM, Fraunhofer Platz 1, 67663 Kaiserslautern, Germany klar@itwm.fhg.deá

K.Y. Laevsky konstantin.laevsky@asml.com

Robert M.M. Mattheij Department of Mathematics and Computer Science, Eindhoven University of Technology, PO Box 513, 5600 MB Eindhoven, The Netherlands r.m.m.mattheij@tue.nl

Andro Mikelić Université de Lyon, Lyon, F-69003, FRANCE; Université Lyon 1, Institut Camille Jordan, UMR 5208 CNRS, Bât. Braconnier, 43, Bd du onze novembre 1918 69622 Villeurbanne Cedex, FRANCE Andro.Mikelic@univ-lyon1.fr

Norbert Siedow Fraunhofer-Institut für Techno- und Wirtschaftsmathematik Kaiserslautern, Germany norbert.siedow@itwm.fraunhofer.de

Mathematical Modelling of Glass Forming Processes

J.A.W.M. Groot, Robert M.M. Mattheij, and K.Y. Laevsky

Abstract An important process in glass manufacture is the forming of the product. The forming process takes place at high rate, involves extreme temperatures and is characterised by large deformations. The process can be modelled as a coupled thermodynamical/mechanical problem including the interaction between glass, air and equipment. In this paper a general mathematical model for glass forming is derived, which is specified for different forming processes, in particular pressing and blowing. The model should be able to correctly represent the flow of the glass and the energy exchange during the process. Various modelling aspects are discussed for each process, while several key issues, such as the motion of the plunger and the evolution of the glass-air interfaces, are examined thoroughly. Finally, some examples of process simulations for existing simulation tools are provided.

Nomenclature

Br	Brinkman number
Fr	Froude number
Nu	Nusselt number
Pe	Péclet number
Re	Reynolds number
F_e [N]	External force on plunger

J.A.W.M. Groot (✉)
Department of Mathematics and Computer Science, Eindhoven University of Technology, PO Box 513, 5600 MB Eindhoven, The Netherlands
e-mail: j.a.w.m.groot@tue.nl

R.M.M. Mattheij
Department of Mathematics and Computer Science, Eindhoven University of Technology, PO Box 513, 5600 MB Eindhoven, The Netherlands
e-mail: r.m.m.mattheij@tue.nl

K.Y. Laevsky
e-mail: konstantin.laevsky@asml.com

A. Fasano (ed.), *Mathematical Models in the Manufacturing of Glass*,
Lecture Notes in Mathematics 2010, DOI 10.1007/978-3-642-15967-1_1,
© Springer-Verlag Berlin Heidelberg 2011

1

F_g [N]   Force of glass on plunger
L [m]   Characteristic length
T [K]   Temperature
T_g [K]   Glass temperature
T_m [K]   Mould temperature
V [m s^{-1}]   Characteristic flow velocity
V_p [m s^{-1}]   Plunger velocity
c_p [J kg^{-1} K^{-1}]   Specific heat
g [m s^{-2}]   Gravitational acceleration
p [Pa]   Pressure
r_p [m]   Radius of plunger
t [s]   Time
$\bar{z}_p$ [m]   Vertical plunger position
α [W m^{-2} K^{-1}]   Heat transfer coefficient
β [N m^{-3} s]   Friction coefficient
λ [W m^{-1} K^{-1}] Effective conductivity
μ [kg m^{-1} s^{-1}]   Dynamic viscosity
ρ [kg m^{-3}]   Density
n   Unit normal
t   Unit tangent
u [m s^{-1}]   Flow velocity
u$_w$ [m s^{-1}]   Wall velocity
x [m]   Position
I   Unit tensor
$\dot{\mathsf{E}}$ [s^{-1}]   Strain rate tensor
T [Pa]   Stress tensor

1 Introduction

Nowadays glass has a wide range of uses. By nature glass has some special characteristics, including shock-resistance, soundproofing, transparency and reflecting properties, which makes it particularly suitable for a wide range of applications, such as windows, television screens, bottles, drinking glasses, lenses, fibre optic cables, sound barriers and many other applications. It is therefore not surprising that glass manufacture is an extensive branch of industry.

1.1 Glass Forming

The manufacture of a glass product can be subdivided into several processes. Below a common series of glass manufacturing processes are described in the order of their application.

1.1.1 Melting

In industry the vast majority of glass products is manufactured by melting raw materials and recycled glass in tank furnaces at an elevated temperature [33, 61]. Examples of raw materials include silica, boric oxide, phosphoric oxide, soda and lead oxide. The temperature of the molten glass in the furnace usually ranges from $1,200$ to $1,600\,^\circ$C. A slow formation of the liquid is required to avoid bubble forming [61].

1.1.2 Forming

The glass melt is cut into uniform *gobs*, which are gathered in a forming machine. In the forming machine the molten gobs are successively forced into the desired shape. The forming technique used depends on the type of product. Forming techniques include pressing, blowing and combinations of both, and are discussed further on. During the formation the glass slightly cools down to below $1,200\,^\circ$C. After the formation the glass objects are rapidly cooled down as to take a solid form.

1.1.3 Annealing

Development of stresses during the formation of glass may lead to static fatigue of the product, or even to dimensional changes due to relaxation or optical refraction. The process of reduction and removal of stresses due to relaxation is called *annealing* [61].

In an annealing process the glass objects are positioned in a so-called *Annealing Lehr*, where they are reheated to a uniform temperature region, and then again gradually cooled down. The rate of cooling is determined by the allowable final permanent stresses and property variations throughout the glass [61].

1.1.4 Surface Treatment

An exterior surface treatment is applied to reduce surface defects. Flaws in the glass surface are removed by chemical etching or polishing. Subsequent flaw formation may be prevented by applying a lubricating coating to the glass surface. Crack growth is prevented by chemical tempering (ion exchange strengthening), thermal tempering or formation of a compressive coating. For more information about flaw removal and strengthening of the glass surface the reader is referred to [61].

The process step of interest in this paper is the actual glass formation. Below three widely used forming techniques are discussed. See [22, 61, 73] for further details on glass forming.

Fig. 1 Pressing machine

1.1.5 Press Process

Commercial glass pressing is a continuous process, where relatively flat products (e.g. lenses, TV screens) are manufactured by pressing a gob that comes directly from the melt [61]. This process is usually referred to as the *press* process. In this paper also the term *direct press* is used to distinguish the process from the parison press, which is explained further on.

The direct press is depicted in Fig. 1. Initially, the glass gob is positioned in the centre of a mould. Over the mould a plunger is situated. In order to enclose the space between the mould and the plunger, so that the glass cannot flow out during the process, a *ring* is positioned on top of the mould. During the direct press the plunger moves down through the ring so that the gob is pressed into the desired shape.

1.1.6 Press-Blow Process

A hollow glass object is formed by inflating a glass preform with pressurised air. This is called the *final blow*. The preform is also called *parison*.

In a *press-blow* process first a preform is constructed by a press stage, to avoid confusion with the direct press here referred to as the *parison press*. Figure 2 shows a schematic drawing of a press-blow process. In the press stage a glass gob is dropped down into a mould, called the *blank mould*, and then pressed from below by a plunger (see Fig. 2a). Once the gob is inside the blank mould, the *baffle* (upper part of the mould) closes and the plunger moves gradually up. When the plunger is at its highest position, the ring closes itself around the plunger, so that the mould-plunger construction is closed from below. Finally, when the plunger is lowered, the ring is decoupled from the blank mould and the glass preform is carried by means of a robotic arm to another mould for the blow stage (final blow). In the blow stage the preform is usually first left to sag due to gravity for a short period. Then pressurised

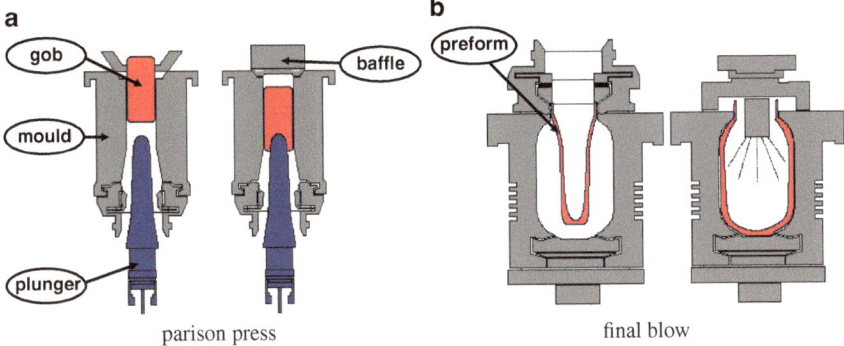

Fig. 2 Schematic drawing of a press-blow process

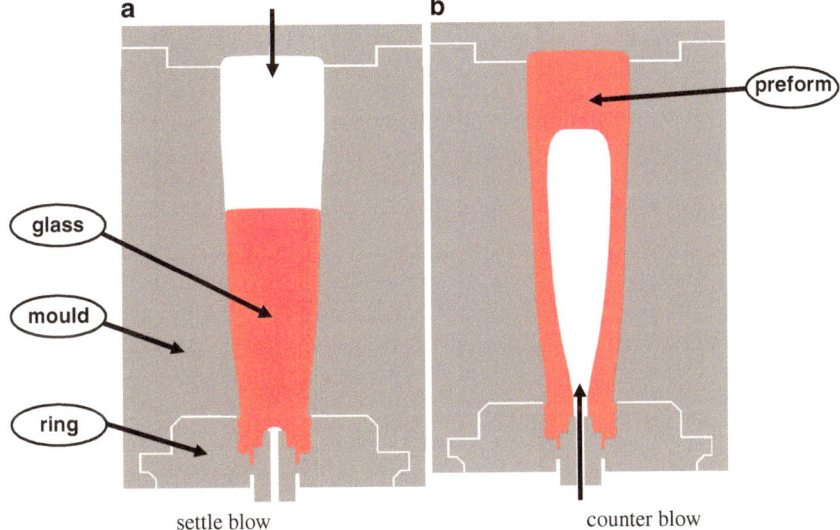

Fig. 3 Schematic drawing of the first blow stage of a blow–blow process

air is blown inside to force the glass in a mould shape (see Fig. 2b). It is important to know the right shape of the preform beforehand for an appropriate distribution of glass over the mould wall.

1.1.7 Blow–Blow Process

A *blow-blow* process is based on the same principle as a press-blow process, but here the preform is produced by a blow stage. By means of a blow stage a hollow preform can be formed, which is required for the production of narrow-mouth containers. In practice the glass gob is blown twice to create the preform. Figure 3 shows a schematic drawing of the first blow stage of a blow-blow process. First in the *settle blow* the glass gob is blown from above to form the neck of the container (see Fig. 3a), then in the *counter blow* from below to form the preform (see Fig. 3b). After the counter blow the preform is carried to the mould for the second blow stage, the final blow. The final blow is basically the same as in the press-blow process, but the preform is typically different.

The temperature of the material of the forming machine is typically around 500 °C. Because of the high temperature of the gob, the surface temperature of the material will increase. To keep the temperature of the material within acceptable bounds the mould and plunger are thermally stabilised by means of water-cooled channels.

1.2 Process Simulation

In the recent past glass forming techniques were still based on empirical knowledge and hand on experience. It was difficult to gain a clear insight into forming processes. Experiments with glass forming machines were in general rather expensive and time consuming, whereas the majority had to be performed in closed constructions under complicated circumstances, such as high temperatures.

Over the last few decades numerical process simulation models have become increasingly important in understanding, controlling and optimising the process [33, 71]. Since measurements are often complicated, simulation models may offer a good alternative or can be used for comparison of results.

The growing interest of glass industry for computer simulation models has been a motivation for a fair number of publications on this subject. The earliest papers dealt with computer simulation of glass container blowing [12, 52]. The first publication in which both stages of the press-blow process were modelled was presented in [74], although the model did not include aspects such as the drop of the gob into the blank mould and the transfer of the preform to the mould for the final blow [28].

Shortly afterward also the first PhD theses on the modelling of glass forming processes appeared. A numerical model for blowing was published in [54]. Not much later a simplified mathematical model for pressing was presented in [27]. Both models assumed axial symmetry of the forming process.

Subsequent publications reported the development of more advanced models. A fully three-dimensional model for pressing TV panels was addressed in [35]. A complete model for the three-dimensional simulation of TV panel forming and conditioning, including gob forming, pressing cooling and annealing was developed by TNO [7]. A simulation model for the complete press-blow process, from gob forming until the final blow, was presented in [28]. The model also included effects of viscoelasticity and surface tension. Finally, an extensive work on the mathematical modelling of glass manufacturing was published in the book entitled 'Mathematical Simulation in Glass Technology' [33].

More recent papers focus on optimisation of glass forming processes [25]. For instance a numerical optimisation method to find the optimum tool geometry in a model for glass pressing was introduced in [42]. Optimisation methods have also been developed to estimate heat transfer coefficients [9] or the initial temperature distribution in glass forming simulation models [41]. An engineering approach to find the optimum preform shape for glass blowing was addressed in [39]. Their algorithm attempts to optimise the geometry of the blank mould in the blow–blow process, given the mould shape at the end of the second blow stage. More recently, an optimisation algorithm for predictive control over a class of rheological forming processes was presented in [4].

By far most papers on modelling glass forming processes use FEM (Finite Element Methods) for the numerical simulation. Exceptions are [75], in which a Finite Volume Method is used, and [14], in which Boundary Element Methods are used. FEM in models for forming processes usually go together with re-meshing techniques, sometimes combined with an Eulerian formulation or a Lagrangian method. In [7] remeshing was completely avoided by using an arbitrary Euler–Lagrangian approach to compute the mesh for the changing computational domain due to the motion of the plunger and by using a Pseudo-Concentration Method to track the glass-air interfaces. In [20] a Level Set Method was used to track the glass-air interfaces in a FEM based model for glass blowing.

In this paper a general mathematical model for the aforementioned forming processes is derived. Subsequently, the model is specified for different forming processes, thereby discussing diverse modelling aspects. The paper focusses on the most relevant aspects of the forming process, rather than supplying a model that is as complete as possible. As discussed previously, a considerable amount of work on modelling glass forming processes has been done in the recent past. For completeness, comparison or reviewing various references to different simulation models are included. Finally, some examples of process simulations for existing simulation tools are provided. Most of the work in this paper has been done by CASA[1] (Centre for Analysis, Scientific computing and Applications).

1.3 Outline

The paper is structured as follows. First in Sect. 2 the physical aspects of glass forming are described and a general mathematical model for glass forming is derived. Then the mathematical model is specified for the parison press in Sect. 3, for both the counter blow and the final blow in Sect. 4 and for the direct press in Sect. 5. In each of these sections also a computer simulation model for the forming process in question is described and some results are presented.

---

[1] A research group in the Department of Mathematics and Computer Science of Eindhoven University of Technology.

2 Mathematical Model

This section presents a mathematical model for glass forming in general. The section has the following structure. First Sect. 2.1 defines the physical domains into which a glass forming machine can be subdivided and in which the boundary value problems for glass forming are defined. Then Sect. 2.2 describes the physical aspects of glass forming, sets up the resulting balance laws and derives the corresponding boundary value problems. Sect. 2.2.1 is concerned with the thermodynamics and Sect. 2.2.2 with the mechanics.

2.1 Geometry, Problem Domains and Boundaries

In order to formulate a mathematical model for glass forming, the forming machine is subdivided into subdomains. First the space enclosed by the equipment is subdivided into a glass domain and an air domain. Then separate domains for components of the equipment (e.g. mould, plunger) are considered. Subdomains of the equipment are of interest when modelling the heat exchange between glass, air and equipment. On the other hand, for less advanced heat transfer modelling it can be assumed that the equipment has constant temperature, so that the mathematical model can be restricted to the glass and air domain. In this case the equipment domains are disregarded.

Figure 4 illustrates the domain decomposition of 2D axi-symmetrical forming machines for the direct press, parison press, counter blow and final blow, respectively. The entire open domain of the forming machine, consisting of equipment, glass melt and air, is denoted by Σ. The 'flow' domain Ω consists of the open glass domain Ω_g, the open air domain Ω_a and the glass-air interface(s) Γ_i. For blowing $\Omega := \Omega_g \cup \Omega_a \cup \Gamma_i$ is fixed, while for pressing Ω changes in time. Furthermore, Ω_g and Ω_a are variable in time for any forming process. The boundaries of the domains are:

Γ_b : Baffle boundary	Γ_i : Glass-air interface	Γ_m : Mould boundary
Γ_o : Outer boundary	Γ_p : Plunger boundary	Γ_r : Ring boundary
Γ_s : Symmetry axis		

Note that not necessarily all boundaries exist for each forming machine. Domain Σ is enclosed by $\Gamma_o \cup \Gamma_s$. Domain Ω is enclosed by $\partial\Omega := \overline{\Omega} \cap \left(\Gamma_e \cup \Gamma_o \cup \Gamma_s \right)$, with

$$\Gamma_e := \Gamma_b \cup \Gamma_m \cup \Gamma_p \cup \Gamma_r \qquad \text{(equipment boundary)}. \qquad (1)$$

In addition, define $\partial\Omega_a := \overline{\Omega}_a \cap \partial\Omega$ and $\partial\Omega_g := \overline{\Omega}_g \cap \partial\Omega$. Finally, the boundaries for the glass domain and the air domain are distinguished:

$$\Gamma_{a,e} = \partial\Omega_a \cap \Gamma_e, \quad \Gamma_{g,e} = \partial\Omega_g \cap \Gamma_e, \quad \Gamma_{a,o} = \partial\Omega_a \cap \Gamma_o, \quad \Gamma_{g,o} = \partial\Omega_g \cap \Gamma_o.$$

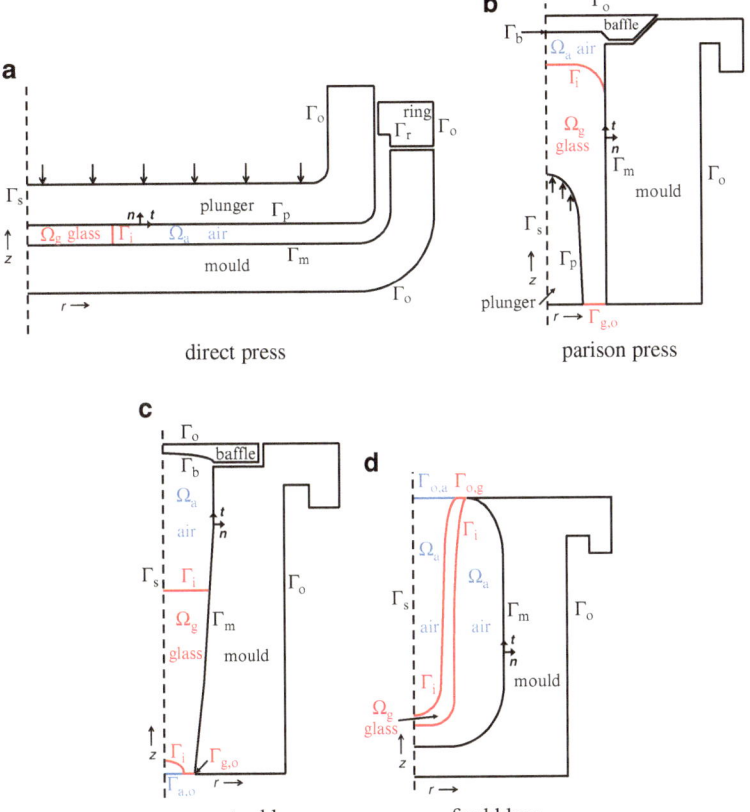

Fig. 4 2D axi-symmetrical problem domains of glass forming models

Remark 2.1. In pressing sometimes only the forces acting on the glass are of interest. If the density and viscosity of air are negligible compared to the density and the viscosity of glass, the mathematical model may be restricted to the glass domain Ω_g. In this case the interface Γ_i is replaced by Γ_o.

2.2 Balance Laws

In this section the physical aspects of glass forming are described and the resulting balance laws are formulated in the form of boundary value problems. Subsequently a dimensional analysis of the balance laws is applied. Section 2.2.1 is concerned with the thermodynamics and Sect. 2.2.2 with the mechanics. In subsequent sections these balance laws are specified for different forming processes.

2.2.1 Thermodynamics

Glass forming involves high temperatures within a typical range of 800–$1,400\,^\circ$C. Temperature variations within this range may cause significant changes in the mechanical properties of the glass.

- The range of viscosity for varying temperature is relatively large for glass: it amounts from 10 Pa s at the melting temperature (about $1,500\,^\circ$C) to 10^{20} Pa s at room temperature. The viscosity increases rapidly as a glass melt is cooled, so that the glass will retain its shape after the forming process. Typical values for the viscosity in glass forming processes lie between 10^2 and 10^5Pa s [61,73].
  The temperature dependence for the viscosity of glass within the forming temperature range is given by the VFT-relation, due to Vogel, Fulcher and Tamman [61,73]:

$$\mu(T)\backslash[\mathrm{Pa\,s}] = \exp\left(-A + B/(T - T_L)\right). \qquad (2)$$

Quantities A $[-]$, B $[^\circ\mathrm{C}]$ and T_L $[^\circ\mathrm{C}]$ represent the Lakatos coefficients, which depend on the composition of the glass melt. Figure 5 shows how strongly the viscosity depends on temperature for soda-lime-silica glass with Lakatos coefficients $A = 3.551$, $B = 8575\,^\circ$C and $T_L = 259\,^\circ$C [49].

- The following density-temperature relation can be deduced

$$\rho(T) = \rho_0\left(1 - \alpha_V\left(T - T_{\mathrm{ref}}\right)\right), \qquad (3)$$

where

- α_V $[^\circ\mathrm{C}^{-1}]$ is the volumetric thermal expansion coefficient
- T_{ref} $[^\circ\mathrm{C}]$ is a reference temperature
- ρ_0 $[\mathrm{kg\,m}^{-3}]$ is the density at the reference temperature

The volumetric thermal expansion coefficient is often assumed constant; for molten glass it is typically ranged from $5 \cdot 10^{-5}$ to $8 \cdot 10^{-5}\,^\circ\mathrm{C}^{-1}$. The density

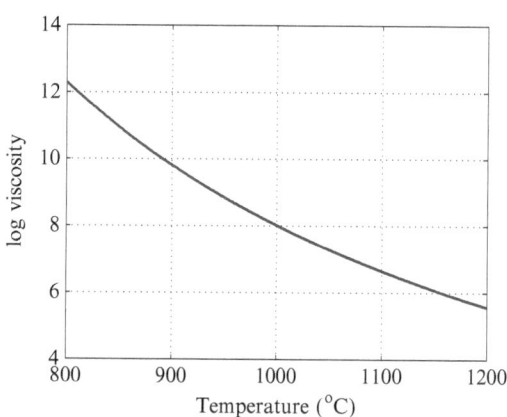

Fig. 5 VFT-relation

of molten glass is of the order 2,300–2,500 kg m^{-3} and is 5–8% lower than at room temperature. Thus it is quite reasonable to assume incompressibility for molten glass [73].

Clearly, the mechanics of glass forming is related to the heat transfer in the glass. The heat transfer in glass, air and equipment is described by the heat equation for incompressible continua:

$$\underbrace{\rho c_p\left(\frac{\partial T}{\partial t}+\mathbf{u}\cdot\nabla T\right)}_{\text{advection}} = -\underbrace{\nabla\cdot\mathbf{q}}_{\substack{\text{conduction}\\\text{radiation}}} + \underbrace{2\mu\left(\dot{\mathbf{E}}:\nabla\otimes\mathbf{u}\right)}_{\text{dissipation}}, \quad \text{in } \overset{\circ}{\Sigma}, \tag{4}$$

where the temperature distribution T [K] is unknown. Here $\overset{\circ}{\Sigma}$ denotes domain Σ minus the interfaces between the continua. On the interfaces between different continua a steady state temperature transition is imposed,

$$\left[\!\left[\lambda\mathbf{n}\cdot\nabla T\right]\!\right] = 0. \tag{5}$$

The heat flux $\mathbf{q}$ [W m^{-2}] is the result of the contribution of both conduction and radiation,

$$\mathbf{q} = -\lambda\cdot\nabla T, \tag{6}$$

where λ is the effective conductivity [W m^{-1}K^{-1}], given by

$$\lambda = \lambda_c + \lambda_r. \tag{7}$$

Here λ_c is the thermal conductivity and λ_r is the radiative conductivity. The thermal conductivity measures 1.0 W m^{-1}K^{-1} at room temperature for soda-lime glass and increases with approximately 0.1 W m^{-1}K^{-1} per 100 K. In this paper it is assumed to be constant. The calculation of the radiative conductivity λ_r is often a complicated process. However, for non-transparent glasses it can be simplified by the Rosseland approximation [16,51,73]

$$\lambda_r(T) = \frac{16\,n^2\sigma T^3}{3}\frac{}{\alpha}, \tag{8}$$

where

- σ is the Stefan Boltzmann radiation constant [W m^{-2}K^{-4}]
- n is the average refractive index [−]
- α is the absorption coefficient [m^{-1}]

The radiative conductivity λ_r in the sense of (8) is called the Rosseland parameter. Relation (8) cannot be applied for highly transparent glasses, since in this case not all radiation is absorbed by the glass melt [73]. A more simple approach is to omit the radiative term, which is often reasonable for clear glass [34, 37].

For more information on heat transfer in glass by radiation the reader is referred to [33, 37, 38, 43, 73]. The specific heat c_p is in general slightly temperature dependent. In [3] an increase in the specific heat of less than 10% in a temperature range of 900–1,300 K for soda-lime-silica glass is reported. For various specific heat capacity models the reader is referred to [3, 60].

In order to analyse the energy exchange problem quantitatively the heat equation is written in dimensionless form. Define a typical: velocity V, length scale L, viscosity $\bar{\mu}$, specific heat $\bar{c}_p$, effective conductivity $\bar{\lambda}$, glass temperature T_g and mould temperature T_m. Then introduce the dimensionless variables

$$t^* := \frac{Vt}{L}, \qquad \mathbf{x}^* := \frac{\mathbf{x}}{L}, \qquad \mathbf{u}^* := \frac{\mathbf{u}}{V}, \qquad T^* := \frac{T - T_m}{T_g - T_m},$$

$$c_p^* := \frac{c_p}{\bar{c}_p}, \qquad \lambda^* := \frac{\lambda}{\bar{\lambda}}, \qquad \mu^* = \frac{\mu}{\bar{\mu}}, \tag{9}$$

For convenience all dimensionless variables, spaces and operators with respect to the dimensionless variables are denoted with superscript *. Substitution of the dimensionless variables (9) in the heat equation and splitting up the effective conductivity into (7) lead to the dimensionless form,

$$\mathrm{Pe}\left(\frac{\partial T^*}{\partial t^*} + \mathbf{u}^* \cdot \nabla^* T^*\right) = \nabla^* \cdot \left(\lambda^* \nabla^* T^*\right) + 2\mu^* \mathrm{Br}\left(\dot{\mathbf{E}}^* : \nabla^* \otimes \mathbf{u}^*\right), \quad \text{in } \overset{\circ}{\Sigma}^*, \tag{10}$$

where

$$\mathrm{Pe} = \frac{\rho \bar{c}_p V L}{\bar{\lambda}}, \tag{11}$$

$$\mathrm{Br} = \frac{\bar{\mu} V^2}{\bar{\lambda}\left(T_g - T_m\right)} \tag{12}$$

are the Péclet number and the Brinkman number, respectively. The Péclet number represents the ratio of the advection rate to the diffusion rate. On the other hand, the Brinkman number relates the dissipation rate to the conduction rate. The dimensionless numbers are useful for assessing the order of magnitude of the different terms in (10).

The energy BCs follow from symmetry and heat exchange with the surroundings:

$$\left(\lambda \nabla T\right) \cdot \mathbf{n} = 0, \qquad \text{on } \Gamma_s$$
$$\left(\lambda \nabla T\right) \cdot \mathbf{n} = \alpha\left(T - T_\infty\right), \quad \text{on } \Gamma_o,$$

where T_∞ is the temperature of the surroundings. The heat transfer coefficient α [W m^{-2}K^{-1}] can differ for separate equipment domains, such as the mould and the plunger. Let $\bar{\alpha}$ be a typical value for the heat transfer coefficient, then the Nusselt number is defined by

$$\mathrm{Nu} = \frac{\bar{\alpha} L}{\bar{\lambda}}. \tag{13}$$

The dimensionless boundary conditions become

$$\begin{aligned}
\left(\lambda^* \nabla T^*\right) \cdot \mathbf{n} &= 0, & &\text{on } \Gamma_s \\
\left(\lambda^* \nabla T^*\right) \cdot \mathbf{n} &= \mathrm{Nu}\, \alpha^* \left(T^* - T_\infty^*\right), & &\text{on } \Gamma_o,
\end{aligned} \tag{14}$$

with $\alpha^* = \alpha/\bar{\alpha}$. On the other hand, if the heat transfer in the equipment domain is not of interest, the boundary condition on Γ_o in (14) is imposed on Γ_e, with T_∞ the surface temperature of the mould.

2.2.2 Mechanics

A balance law for the mechanics of the glass melt is formulated. In general, glass can be treated as an isotropic viscoelastic Maxwell material [3,53,54,61], that is the strain rate tensor can be split up into an elastic and a viscous part:

$$\dot{\mathbf{E}} = \dot{\mathbf{E}}_e + \dot{\mathbf{E}}_v, \tag{15}$$

where the elastic and viscous strain rate tensors, $\dot{\mathbf{E}}_e$ and $\dot{\mathbf{E}}_v$ respectively, are given by [54]

$$\dot{\mathbf{E}}_e = \frac{1-2v}{E}\left(\alpha_V \frac{\partial T}{\partial t} + \frac{1}{3}\mathrm{tr}(\dot{\mathbf{T}})\right) + \frac{1+v}{E}\mathrm{dev}(\dot{\mathbf{T}}) \tag{16}$$

$$\dot{\mathbf{E}}_v = \frac{1}{2\mu}\mathrm{dev}(\mathbf{T}). \tag{17}$$

Here α_V [°C^{-1}] is the volumetric thermal expansion coefficient, E [Pa] is the Young's modulus, v [-] is the Poisson's ratio and $\dot{\mathbf{T}}$ [Pa s^{-1}] denotes the stress rate tensor. However, at relatively low viscosities the relation between shear stress and viscosity becomes approximately linear. For example, for soda lime silica glasses the viscosity as a function of the strain rate and the temperature becomes [6,63]

$$\mu(\dot{\mathbf{E}}, T) = \frac{\mu_0(T)}{1 + 3.5 \cdot 10^{-6}\dot{\mathbf{E}}\, \mu_0^{0.76}(T)}, \tag{18}$$

where μ_0 is the Newtonian viscosity. Consequently, the motion of glass is dominated by viscous flow and the influence of elastic effects can be neglected [3, 53, 61]. Moreover, as verified in Sect. 2.2.1, glass is practically incompressible in the forming temperature range, from which it follows that

$$\mathrm{tr}(\dot{\mathbf{E}}) = 0. \tag{19}$$

It can be concluded that glass in the forming temperature range behaves as an incompressible Newtonian fluid [3,53,73], i.e.

$$\mathbf{T} = -p\mathbf{I} + 2\mu\dot{\mathbf{E}}, \tag{20}$$

with

$$\dot{\mathsf{E}} = \frac{1}{2}\left(\nabla \otimes \mathbf{u} + (\nabla \otimes \mathbf{u})^T\right). \tag{21}$$

For simplicity the pressurised air is considered as an incompressible, viscous fluid with uniform viscosity. Thus the motion of glass melt and pressurised air is described by the Navier–Stokes equations for incompressible fluids. These involve the momentum equations,

$$\underbrace{\rho\left(\frac{\partial \mathbf{u}}{\partial t} + \mathbf{u}\cdot(\nabla \otimes \mathbf{u})\right)}_{\text{inertia}} = \underbrace{-\nabla p}_{\text{pressure}} + \underbrace{\rho\mathbf{g}}_{\text{gravity}} + \underbrace{2\nabla\cdot(\mu\dot{\mathsf{E}})}_{\text{viscosity}}, \quad \text{in } \Omega\setminus\Gamma_i, \tag{22}$$

and the continuity equation, which follows directly from (19),

$$\nabla\cdot\mathbf{u} = 0, \quad \text{in } \Omega\setminus\Gamma_i. \tag{23}$$

The unknowns are the flow velocity $\mathbf{u}$ [m s^{-1}] and the pressure p [Pa].

Flow problem (22)–(23) is coupled to the energy problem (10) in two ways: firstly the viscosity is temperature dependent and secondly the heat transfer is partly described by convection and diffusion.

In order to apply a quantitative analysis the Navier–Stokes equations are written in dimensionless form. First a dimensionless pressure is defined by

$$p^* := \frac{Lp}{\bar{\mu}V}. \tag{24}$$

The gravity force can be written as $-\rho g \mathbf{e}_z$, where $\mathbf{e}_z$ is the unit vector in z-direction. Substitution of the dimensionless variables (9) and (24) into the Navier–Stokes equations (22)–(23) and division by the order of magnitude of the diffusion term, $\frac{\bar{\mu}V}{L^2}$, lead to the dimensionless Navier–Stokes equations

$$\text{Re}\left(\frac{\partial \mathbf{u}^*}{\partial t^*} + \mathbf{u}^*\cdot(\nabla^* \otimes \mathbf{u}^*)\right) = -\nabla^* p^* - \frac{\text{Re}}{\text{Fr}}\mathbf{e}_z + 2\nabla^*\cdot\left(\mu^*\dot{\mathsf{E}}^*\right), \quad \text{in } \Omega^*\setminus\Gamma_i^*,$$

$$\nabla^*\cdot\mathbf{u}^* = 0, \quad \text{in } \Omega^*\setminus\Gamma_i^*, \tag{25}$$

where

$$\text{Re} = \frac{\rho VL}{\bar{\mu}}, \tag{26}$$

$$\text{Fr} = \frac{V^2}{gL} \tag{27}$$

are the Reynolds number and the Froude number, respectively. The Reynolds number measures the ratio of inertial forces to viscous forces, while the froude number measures the ratio of inertial forces to gravitational forces.

The jump conditions between two immiscible viscous fluids are the continuity of the flow velocity,

$$[[\mathbf{u}]] = \mathbf{0}, \tag{28}$$

as well as the continuity of its tangential derivative,

$$[[(\nabla \otimes \mathbf{u}) \cdot \mathbf{t}]] = \mathbf{0}, \tag{29}$$

and a dynamic jump condition stating the balance of stress across the fluid interface [31, 45],

$$[[\mathbf{Tn}]] = \mathbf{0}. \tag{30}$$

Remark 2.2. If the influence of surface tension is taken into account, the dynamic jump condition is

$$[[\mathbf{Tn}]] = -\gamma\kappa\mathbf{n}, \tag{31}$$

where $\mathbf{n}$ points in the air domain. Here $\gamma[\text{N m}^{-1}]$ denotes the surface tension and $\kappa[\text{m}^{-1}]$ denotes the curvature. The influence of surface tension highly depends on the glass composition [61]. In this paper the influence of surface tension is simply disregarded, although for some glasses a dimensional analysis can point out that the surface tension term is not negligible.

Boundary conditions for the flow problem can be determined as follows. On Γ_s symmetry conditions are imposed. On $\Gamma_{g,e}$ a suitable slip condition for the glass should be adopted. The air can escape through small cavities in (part of) the mould wall. This aspect can be modelled by allowing air to flow freely through the mould wall. Thus free-stress conditions are proposed on this part of the equipment boundary, which is referred to as $\Gamma_{a,e}^{(1)}$. A free-slip condition is prescribed on the remaining part, $\Gamma_{a,e}^{(2)}$. On Γ_o the normal stress should be equal to the external pressure.

A commonly used boundary condition to describe fluid flow at an impenetrable wall [13, 19, 29, 50] is Navier's slip condition:

$$(\mathbf{Tn} + \beta(\mathbf{u} - \mathbf{u}_w)) \cdot \mathbf{t} = 0, \tag{32}$$

where β is the friction coefficient $[\text{N m}^{-3}\text{s}]$ and $\mathbf{u}_w$ is the velocity of the wall $[\text{m s}^{-1}]$. A similar condition can be obtained by using the Tresca model [16]. Introduce a dimensionless friction coefficient,

$$\beta^* := \frac{L\beta}{\bar{\mu}}, \tag{33}$$

then the dimensionless Navier's slip condition reads:

$$(\mathbf{T}^*\mathbf{n} + \beta^*(\mathbf{u}^* - \mathbf{u}_w^*)) \cdot \mathbf{t} = 0. \tag{34}$$

The order of magnitude of the friction coefficient depends on many parameters, such as the type of glass, temperature, pressure or presence of a lubricant [15, 18, 50].

Remark 2.3. For $\beta^* \to \infty$ Navier's slip condition together with the boundary condition for an impenetrable wall, $(\mathbf{u}^* - \mathbf{u}_w^*) \cdot \mathbf{n} = 0$, can be reformulated as a no slip condition:

$$\mathbf{u}^* = \mathbf{u}_w^*. \tag{35}$$

In summary, the boundary conditions for the flow problem can be formulated as:

$$
\begin{array}{llll}
\mathbf{u}^* \cdot \mathbf{n} = 0, & & \mathbf{T}^*\mathbf{n} \cdot \mathbf{t} = 0, & \text{on } \Gamma_s, \\
(\mathbf{u}^* - \mathbf{u}_w^*) \cdot \mathbf{n} = 0, & & \left(\mathbf{T}^*\mathbf{n} + \beta^*(\mathbf{u}^* - \mathbf{u}_w^*)\right) \cdot \mathbf{t} = 0, & \text{on } \Gamma_{g,e}, \\
\mathbf{T}^*\mathbf{n} \cdot \mathbf{n} = 0, & & \mathbf{T}^*\mathbf{n} \cdot \mathbf{t} = 0, & \text{on } \Gamma_{a,e}^{(1)}, \\
(\mathbf{u}^* - \mathbf{u}_w^*) \cdot \mathbf{n} = 0, & & \mathbf{T}^*\mathbf{n} \cdot \mathbf{t} = 0, & \text{on } \Gamma_{a,e}^{(2)}, \\
\mathbf{T}^*\mathbf{n} \cdot \mathbf{n} = p_0, & & \mathbf{T}^*\mathbf{n} \cdot \mathbf{t} = 0, & \text{on } \Gamma_o,
\end{array} \tag{36}
$$

where p_0 is the external pressure.

3 Parison Press Model

This section presents a mathematical model for the parison press. Section 3.1 specifies the mathematical model described in Sect. 2 for the parison press. By restricting the analysis to a narrow channel between the plunger and the mould, an analytical approximation of the flow can be derived. Section 3.2 explains this concept, known as the slender geometry approximation [50]. Section 3.3 describes the motion of the plunger by an ordinary differential equation. It appears that the motion of the plunger is coupled to the flow, which considerably complicates the parison press model. Section 3.4 presents a numerical simulation model for the parison press. The motion of the free boundaries is emphasised. Finally, Sect. 3.5 shows some examples of parison press simulations. The simulation tool used for these results is presented in [34].

3.1 Mathematical Model

In Sect. 2.2 the balance laws for glass forming were formulated. In this section these balance laws are further specified for the parison press process. Typical values for the parison press are:

Glass density	:	$\rho_g = 2.5 \cdot 10^3 \text{ kg m}^{-3}$
Glass viscosity	:	$\bar{\mu}_g = 10^4 \text{ kg m}^{-1}\text{s}^{-1}$
Gravitational acceleration	:	$g = 9.8 \text{ m s}^{-2}$
Flow velocity	:	$V = 10^{-1} \text{ m s}^{-1}$
Length scale of the parison	:	$L = 10^{-2} \text{ m}$
Glass temperature	:	$T_g = 1{,}000\,^\circ\text{C}$
Mould temperature	:	$T_m = 500\,^\circ\text{C}$
Specific heat of glass	:	$\bar{c}_p = 1.5 \cdot 10^3 \text{ J kg}^{-1}\text{K}^{-1}$
Effective conductivity of glass	:	$\bar{\lambda} = 5 \text{ W m}^{-1}\text{K}^{-1}$,

As a result the following dimensionless numbers are found:

$$Pe_{glass} \approx 7.6 \cdot 10^2, \quad Br_{glass} \approx 4.0 \cdot 10^{-3}, \quad Re_{glass} \approx 2.5 \cdot 10^{-4}, \quad Fr \approx 1.0 \cdot 10^{-1}. \quad (37)$$

Apparently, the heat transport in glass is dominated by thermal advection. As a result the heat equation (10) simplifies to

$$\frac{dT^*}{dt^*} = \frac{\partial T^*}{\partial t^*} + \mathbf{u}^* \cdot \nabla^* T^* = 0, \qquad in\ \Omega_g^*, \qquad (38)$$

Thus the temperature remains constant along streamlines. Consequently, if the initial temperature distribution in the glass is (approximately) uniform, the glass viscosity can be considered constant. From the small Reynolds number for glass it can be concluded that the inertia forces can be neglected with respect to the viscous forces. Furthermore,

$$\frac{Re_{glass}}{Fr} \approx 2.5 \cdot 10^{-3},$$

which means that also the contribution of gravitational forces is rather small. In conclusion, the glass flow can be described by the Stokes flow equations:

$$\nabla^* \cdot T^* = \mathbf{0}, \qquad \nabla^* \cdot \mathbf{u}^* = 0, \qquad in\ \Omega_g^*, \qquad (39)$$

where T^* is the dimensionless stress tensor, which satisfies (20) in terms of the dimensionless variables. The air domain is ignored for the following reasons. Firstly, the force of the plunger acts directly on the glass domain. Secondly, the density and viscosity of air are negligible compared to those of glass, so that air hardly forms any obstacle for the glass flow. Thirdly, the simplification of the heat equation to convection equation (38) gives reason to restrict the energy exchange problem to the glass domain, or all together ignore the heat transfer in case of an uniform initial glass temperature. Subsequently, the glass-air interfaces are treated as an outer boundary of the flow domain, Γ_o, on which free-stress conditions are imposed. Note that although the problem can be considered as a free-boundary problem, the geometry is constrained by the mould and the plunger.

Remark 3.1. Close to the equipment wall extreme temperature variations occur over a small length scale. Therefore, the conductive heat flux close to the equipment wall should, strictly speaking, not be disregarded [50,64]. Moreover, the viscosity in this region may increase by several orders, so that the fluid friction may not be negligible and the influence of heat generation by dissipation should be taken into account for optimal accuracy [50]. Although the reader should take notice of these boundary layer effects, in this paper it is simply assumed that they are small enough to be ignored. For more advanced heat modelling during the parison press the reader is referred to [23, 38, 64].

The boundary conditions for the flow problem (2.2.2) can be specified for the parison press:

$$
\begin{aligned}
\mathbf{u}^* \cdot \mathbf{n} &= 0, & \mathbf{T}^*\mathbf{n} \cdot \mathbf{t} &= 0, & &\text{on } \Gamma_s, \\
(\mathbf{u}^* - V_p\mathbf{e}_z) \cdot \mathbf{n} &= 0, & \left(\mathbf{T}^*\mathbf{n} + \beta_p^*(\mathbf{u}^* - V_p\mathbf{e}_z)\right) \cdot \mathbf{t} &= 0, & &\text{on } \Gamma_{g,p}, \\
\mathbf{u}^* \cdot \mathbf{n} &= 0, & \left(\mathbf{T}^*\mathbf{n} + \beta_m^*\mathbf{u}^*\right) \cdot \mathbf{t} &= 0, & &\text{on } \Gamma_{g,m}, \\
\mathbf{T}^*\mathbf{n} \cdot \mathbf{n} &= 0, & \mathbf{T}^*\mathbf{n} \cdot \mathbf{t} &= 0, & &\text{on } \Gamma_o.
\end{aligned}
\tag{40}
$$

In the remainder of this section it is assumed that the glass gob initially has an uniform temperature distribution, so that with (38) it follows that $\mu = \bar{\mu}$ is constant.

3.2 Slender-Geometry Approximation

In the model for glass pressing the analysis can be restricted to the flow in a narrow channel between plunger and mould. In other words, the analysis focusses on the flow in the slender geometry around the plunger, while the flow between the plunger top and the baffle (Fig. 4b) is considered practically stagnant. For a more complete analysis the reader is referred to [50]. In the slender-geometry approximation of the flow two typical length scales ℓ and L are considered, with $\ell \ll L$, where ℓ is the length scale for the width of the channel and L is the length scale for the length of the channel. Thus variations in r-direction are scaled by ℓ and variations in z-direction are scaled by L. By means of this scaling the following dimensionless variables can be defined:

$$
t^* := \frac{Vt}{L}, \qquad \varepsilon := \frac{\ell}{L}, \qquad r^* := \frac{r}{\varepsilon L}, \qquad z^* := \frac{z}{L},
$$

$$
u_r^* := \frac{u_r}{\varepsilon V}, \qquad u_z^* := \frac{u_z}{V}, \qquad p^* := \frac{\varepsilon^2 L p}{\mu V}.
\tag{41}
$$

In the remainder of this section all variables, spaces and operators are dimensionless and the superscript $*$ is ignored. Substitution of the dimensionless variables (41) into the Navier–Stokes equations (22)–(23) leads to the dimensionless 2D axisymmetrical Navier–Stokes equations

$$
\varepsilon^3 \mathrm{Re}_\ell \left(\frac{\partial u_r}{\partial t} + u_r \frac{\partial u_r}{\partial r} + u_z \frac{\partial u_r}{\partial z} \right) = -\frac{\partial p}{\partial r} + \varepsilon^2 \frac{\partial}{\partial r}\left(\frac{1}{r} \frac{\partial}{\partial r}(r u_r) \right) + \varepsilon^4 \frac{\partial^2 u_r}{\partial z^2}, \qquad \text{in } \Omega_g,
$$

$$
\varepsilon \mathrm{Re}_\ell \left(\frac{\partial u_z}{\partial t} + u_r \frac{\partial u_z}{\partial r} + u_z \frac{\partial u_z}{\partial z} \right) = -\frac{\partial p}{\partial z} + \frac{\partial}{\partial r}\left(\frac{1}{r} \frac{\partial}{\partial r}(r u_z) \right) + \varepsilon^2 \frac{\partial^2 u_z}{\partial z^2} + \varepsilon \frac{\mathrm{Re}_\ell}{\mathrm{Fr}}, \quad \text{in } \Omega_g,
$$

$$
\frac{1}{r} \frac{\partial}{\partial r}(r u_r) + \frac{\partial u_z}{\partial z} = 0, \qquad \qquad \text{in } \Omega_g, \tag{42}
$$

where Re$_\ell$ is the Reynolds number with respect to length scale ℓ, i.e.

$$\text{Re}_\ell = \frac{\rho V \ell}{\bar{\mu}} = \varepsilon \text{Re}_L. \tag{43}$$

Thus, for small ε, (42) can be simplified to

$$\frac{\partial p}{\partial r} = O(\varepsilon^2), \quad \text{in} \, \Omega_g,$$

$$\frac{\partial p}{\partial z} = \frac{\partial}{\partial r}\left(\frac{1}{r}\frac{\partial}{\partial r}(ru_z)\right) + O(\varepsilon^2), \quad \text{in} \, \Omega_g,$$

$$\frac{1}{r}\frac{\partial}{\partial r}(ru_r) + \frac{\partial u_z}{\partial z} = 0, \quad \text{in} \, \Omega_g. \tag{44}$$

If the $O(\varepsilon^2)$ terms are neglected, system of (44) can be recognised as Reynolds' 2D axial-symmetrical lubrication flow [19, 50].

The equipment boundary conditions in the slender-geometry approximation can be simplified accordingly [50]. Define the plunger surface and the mould surface by:

$$r = r_p\big(z - \bar{z}_p(t)\big), \qquad r = r_m(z), \tag{45}$$

respectively, where $z = \bar{z}_p(t)$ is the top of the plunger at time t (see Fig. 6). Consider the plunger position $\bar{z}_p \equiv \bar{z}_p(t)$ at a fixed time t. Then the outward unit normal $\mathbf{n}_p$ and the counterclockwise unit tangent $\mathbf{t}_p$ on the plunger wall are given by

$$\mathbf{n}_p = \frac{\varepsilon r_p' \mathbf{e}_z - \mathbf{e}_r}{\sqrt{1 + \varepsilon^2 r_p'^2}}, \qquad \mathbf{t}_p = \frac{-\mathbf{e}_z - \varepsilon r_p' \mathbf{e}_r}{\sqrt{1 + \varepsilon^2 r_p'^2}}. \tag{46}$$

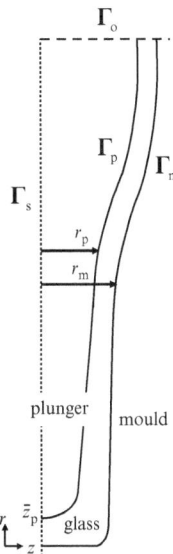

Fig. 6 Geometry of the plunger

Analogously, the outward unit normal $\mathbf{n}_m$ and the counterclockwise unit tangent $\mathbf{t}_m$ on the plunger wall on the mould wall are given by

$$\mathbf{n}_m = \frac{-\varepsilon r'_m \mathbf{e}_z + \mathbf{e}_r}{\sqrt{1+\varepsilon^2 r'^2_m}}, \qquad \mathbf{t}_m = \frac{\mathbf{e}_z + \varepsilon r'_m \mathbf{e}_r}{\sqrt{1+\varepsilon^2 r'^2_m}}. \tag{47}$$

By substituting (46)–(47) into Navier's slip condition (32) and scaling the friction coefficient β by a factor ε the following boundary conditions are obtained

$$\beta_p \left((u_z - V_p) + \varepsilon^2 u_r r'_p \right) = \frac{\left(1 - \varepsilon^2 r'^2_p\right)\left(\frac{\partial u_z}{\partial r} + \varepsilon^2 \frac{\partial u_r}{\partial z}\right) - 2\varepsilon^2 r'_p \left(\frac{\partial u_z}{\partial z} - \frac{\partial u_r}{\partial r}\right)}{\sqrt{1+\varepsilon^2 r'^2_p}},$$

$$\text{on } r = r_p(z - \bar{z}_p) \tag{48}$$

$$-\beta_m \left(u_z + \varepsilon^2 u_r r'_m \right) = \frac{\left(1 - \varepsilon^2 r'^2_m\right)\left(\frac{\partial u_z}{\partial r} + \varepsilon^2 \frac{\partial u_r}{\partial z}\right) - 2\varepsilon^2 r'_m \left(\frac{\partial u_z}{\partial z} - \frac{\partial u_r}{\partial r}\right)}{\sqrt{1+\varepsilon^2 r'^2_m}},$$

$$\text{on } r = r_m(z), \tag{49}$$

where V_p is the velocity of the plunger and β_p, β_m are the friction coefficients corresponding to the plunger and the mould, respectively. For small ε the Navier's slip condition on the plunger and mould surfaces can be written as

$$\beta_p(u_z - V_p) = \frac{\partial u_z}{\partial r} + O(\varepsilon^2), \qquad \text{on } r = r_p(z - \bar{z}_p), \tag{50}$$

$$-\beta_m u_z = \frac{\partial u_z}{\partial r} + O(\varepsilon^2), \qquad \text{on } r = r_m(z). \tag{51}$$

The componentwise boundary condition for the impenetrable wall is

$$u_r = (u_z - V_p) r'_p, \qquad \text{on } r = r_p(z - \bar{z}_p), \tag{52}$$

$$u_r = u_z r'_m, \qquad \text{on } r = r_m(z). \tag{53}$$

Note that for $\beta_{m,p} \to \infty$ the error in boundary condition (50)–(51) due to the slender-geometry approximation vanishes.

Following [40, 50] the analytical solution to system of (44) with set of boundary conditions (50)–(53) on the equipment boundary and free-stress conditions on the other boundaries can be obtained. Neglecting the $O(\varepsilon^2)$ terms system of equations (44) becomes

$$\frac{\partial p}{\partial r} = 0, \quad \text{in } \Omega_g, \tag{54}$$

$$\frac{\partial p}{\partial z} = \frac{\partial}{\partial r}\left(\frac{1}{r}\frac{\partial}{\partial r}(r u_z)\right), \quad \text{in } \Omega_g, \tag{55}$$

$$\frac{1}{r}\frac{\partial}{\partial r}(ru_r) + \frac{\partial u_z}{\partial z} = 0, \quad \text{in } \Omega_g. \tag{56}$$

Firstly, from (54) it follows that p is a function of z only. Secondly, from (55) and the boundary conditions it follows that

$$u_z(r,z) =$$
$$\frac{1}{4}\frac{dp}{dz}(z)\left(r^2 + \frac{\psi_m(z)\chi_p(r,z) - \psi_p(z)\chi_m(r,z)}{\chi_m(r,z) - \chi_p(r,z)}\right) + V_p\frac{\chi_m(r,z)}{\chi_m(r,z) - \chi_p(r,z)}, \tag{57}$$

with

$$\chi_p(r,z) = \log\left(rr_p(z-\bar{z}_p)\right) - \frac{1}{\beta_p r_p(z-\bar{z}_p)}, \quad \psi_p(z) = r_p(z-\bar{z}_p)\left(r_p(z-\bar{z}_p) - \frac{2}{\beta_p}\right),$$
$$\chi_m(r,z) = \log\left(rr_m(z)\right) + \frac{1}{\beta_m r_m(z)}, \quad \psi_m(z) = r_m(z)\left(r_m(z) + \frac{2}{\beta_m}\right). \tag{58}$$

Thirdly, from (56) and the boundary conditions it follows that

$$u_r(r,z) = \frac{1}{r}\left(r_m r_m' u_z(r_m,z) + \int_r^{r_m} s\frac{\partial u_z}{\partial z}(s,z)ds\right) = \frac{1}{r}\frac{d}{dz}\int_r^{r_m} s\,u_z(s,z)ds. \tag{59}$$

Finally, to find the pressure gradient, Gauss' divergence theorem is applied to the continuity equation,

$$0 = \int_{\Omega_g} \nabla\cdot\mathbf{u}d\Omega = \int_{\Gamma_{g,o}} \mathbf{u}\cdot\mathbf{n}d\Gamma + \int_{\Gamma_p} \mathbf{u}\cdot\mathbf{n}d\Gamma + \int_{\Omega_g\cap\Gamma_s} \mathbf{u}\cdot\mathbf{n}d\Gamma + \int_{\Gamma_m} \mathbf{u}\cdot\mathbf{n}d\Gamma. \tag{60}$$

The fluxes through the symmetry axis and the mould wall are zero. Since the continuity equation also holds in the plunger domain Σ_p, it holds that

$$\int_{\Gamma_p} \mathbf{u}\cdot\mathbf{n}d\Gamma = -\int_{\Sigma_p\cap\Gamma_o} \mathbf{u}\cdot\mathbf{n}d\Gamma = \pi V_p r_p^2. \tag{61}$$

As a result,

$$\int_{\Gamma_{g,o}} \mathbf{u}\cdot\mathbf{n}d\Gamma = 2\pi\int_{r_p}^{r_m} ru_z dr = -\pi V_p r_p^2. \tag{62}$$

Substitution of (57) into (62) yields

$$\left[r^2\left(\frac{1}{8}\frac{dp}{dz}\left(r^2 - \frac{\psi_m(2\chi_p+1) - \psi_p(2\chi_m+1)}{\chi_m - \chi_p}\right) + V_p\frac{\chi_m + \frac{1}{2}}{\chi_m - \chi_p}\right)\right]_{r=r_p}^{r=r_m} = -V_p r_p^2. \tag{63}$$

By solving (63) the following solution for the pressure gradient is found:

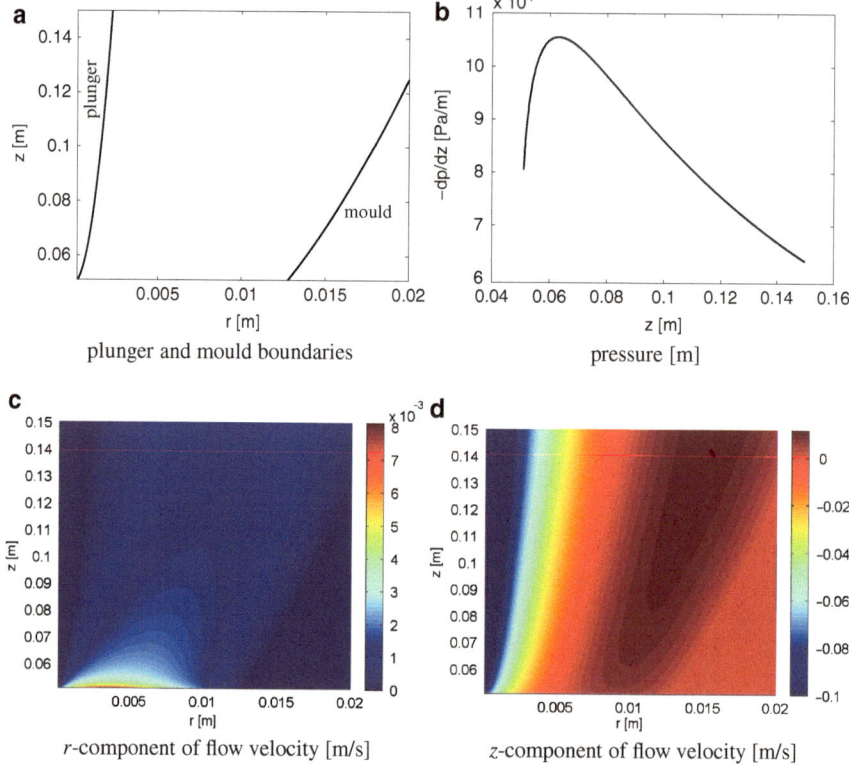

Fig. 7 Analytical solution of slender-geometry approximation for $\bar{z}_p = 0.5$

$$\frac{dp}{dz}(z) = 4V_p \frac{\psi_p(z) - \psi_m(z)}{\left(\psi_p(z) - \psi_m(z)\right)^2 - (\chi_p - \chi_m)(z)\left(\omega_p(z) - \omega_m(z)\right)}, \qquad (64)$$

with

$$\omega_p(z) = \psi_p(z)^2 - \frac{4r_p(z - \bar{z}_p)^2}{\beta_p^2}, \qquad \omega_m(z) = \psi_m(z)^2 - \frac{4r_m(z)^2}{\beta_m^2}. \qquad (65)$$

Figure 7–8 plot the solution at $\bar{z}_p = 0.05\,\mathrm{m}$ and $\bar{z}_p = 0.01\,\mathrm{m}$, respectively, for $\varepsilon = 0.1$ and $V_p = -0.1\,\mathrm{m\,s}^{-1}$. The (dimensionless) geometries of the plunger and mould have been taken from [50]:

$$r_p(z) = -0.1\sqrt{5z}, \qquad r_m(z) = 0.8\sqrt{5z}.$$

For simplicity no-slip boundary conditions at the mould and plunger wall have been used. In [50] it is reported that the results are in good agreement with the numerical solution obtained by using FEM. For more results the reader is referred to [40, 50].

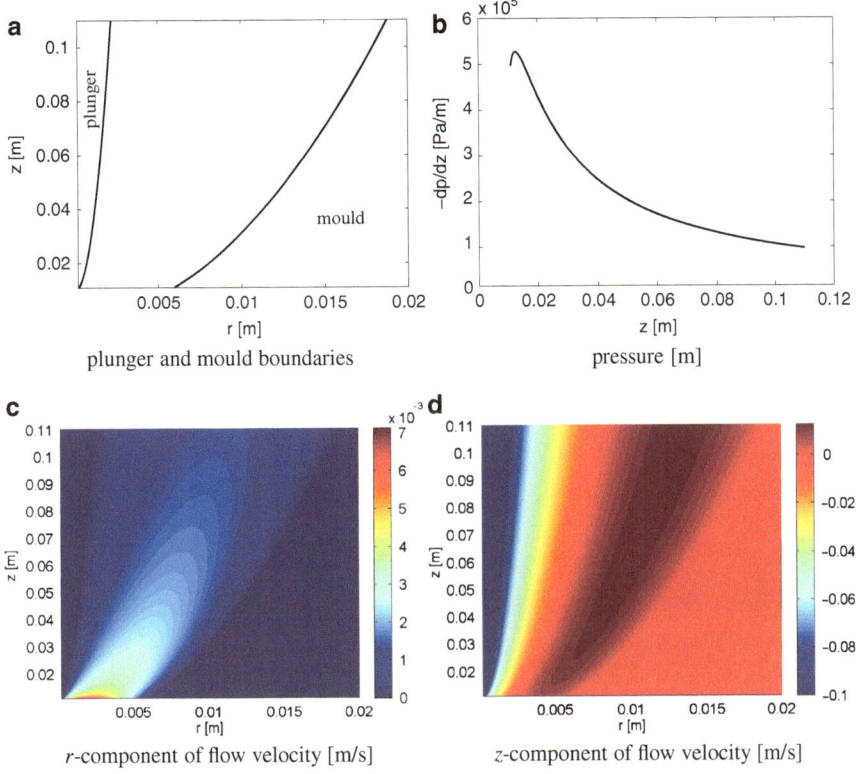

Fig. 8 Analytical solution of slender-geometry approximation for $\bar{z}_p = 0.1$

3.3 Motion of the Plunger

In the previous section it was assumed that the plunger moves with a constant flow velocity. However, in practice the plunger is pushed by a piston. This means that the flow velocity is the result of an external force applied to the plunger. In this section it can be seen that the plunger velocity is coupled to the glass flow, which considerably complicates the parison press model.

The press process is initiated by applying an external force F_e to the plunger. This causes the plunger to move with velocity $V_p(t)$. This plunger velocity is the result of the total force F on the plunger, which is the sum of the external force F_e and the force of the glass on the plunger F_g:

$$\frac{dV_p}{dt}(t) = \frac{F(t)}{m_p} = \frac{F_e + F_g(t)}{m_p}, \tag{66}$$

where m_p is the mass of the plunger [34, 50]. The force F_g is determined by the mechanical forces of the glass acting on the plunger and hence depends on the glass flow. The glass flow at its turn is caused by the plunger motion. Therefore, differential equation (66) is fully coupled to the flow problem. Clearly, if a constant external force is applied to the plunger, the plunger moves until the force of the glass on the plunger is equal in magnitude to the external force.

Below the equation for the motion of the plunger (66) is examined more thoroughly. For a complete analysis the reader is referred to [34, 50]. First the force of the glass on the plunger is analysed. At every time t the force can be fully described by the stress tensor (20) integrated over the plunger surface Γ_p,

$$F_g = \int_{\Gamma_p} (\mathbf{Tn}) \cdot \mathbf{e}_z d\Gamma. \tag{67}$$

Using the definition of the plunger surface (45) the surface element $d\Gamma$ can be written as

$$d\Gamma = 2\pi\sqrt{1 + r_p'^2} r_p dz, \qquad \text{on } \Gamma_p. \tag{68}$$

By means of the definition of the stress tensor (20) and the expressions for the unit normal and tangent (46) and the surface element (68), the expression for the force of the glass on the plunger becomes

$$F_g = 2\pi \int_{\bar{z}_p}^{z_0} \left(\left(p - 2\mu \frac{\partial u_z}{\partial z} \right) r_p' + \mu \left(\frac{\partial u_r}{\partial z} + \frac{\partial u_z}{\partial r} \right) \right) r_p dz. \tag{69}$$

where $z = z_0(t)$ is the bottom glass level at time t. Substitution of the dimensionless variables (41) into (69) yields

$$F_g = 2\pi\mu VL \int_{\bar{z}_p^*}^{z_0^*} \left(\left(p^* - 2\varepsilon^2 \frac{\partial u_z^*}{\partial z^*} \right) r_p^{*\prime} + \left(\varepsilon^2 \frac{\partial u_r}{\partial z^*} + \frac{\partial u_z^*}{\partial r^*} \right) \right) r_p^* dz^*. \tag{70}$$

Considering expression (70) it makes sense to define the dimensionless force of the glass on the plunger as

$$F_g^* := \frac{F_g}{2\pi\mu VL}. \tag{71}$$

For small ε the dimensionless force can be simplified to

$$F_g^* =: \int_{\bar{z}_p^*}^{z_0^*} \left(p^* r_p^{*\prime} + \frac{\partial u_z^*}{\partial r^*} \right) r_p^* dz^* + O(\varepsilon^2). \tag{72}$$

The flow velocity and pressure in (72) are implicit functionals of the plunger velocity V_p. However, the slender-geometry approximation in Sect. 3.2 can be used to find approximate solutions for u_z^* and p^*. By substituting (57) and (65) into (70)–(71) the dimensionless force can be written as

$$F_g^*(t^*) = V_p^*(t^*) I^*(t^*), \tag{73}$$

with an assumed error of $O(\varepsilon^2)$, where the dimensionless function $I^*(t^*)$ only depends on the geometry and the friction coefficient β^* [34,50]. Then the equation for the motion of the plunger (66) in dimensionless form becomes

$$\frac{dV_p^*}{dt^*}(t^*) = \frac{2\pi\mu L^2}{m_p V}\left(V_p^*(t^*)I^*(t^*) + F_e^*\right) + O(\varepsilon^2), \tag{74}$$

where the dimensionless external force F_e^* is defined in the same way as F_g^* in (71). Typical values in the slender-geometry approximation are:

Viscosity : $\bar{\mu}_g = 10^4 \text{ kg m}^{-1}\text{s}^{-1}$
Flow velocity : $V = 10^{-1} \text{ m s}^{-1}$
Length scale of the parison : $L = 10^{-1} \text{ m}$
Mass of the plunger : $m_p = 1 \text{ kg}$

As a result the dimensionless coefficient in (74) is typically

$$\frac{2\pi\mu L^2}{m_p V} \approx 10^4.$$

This value is rather large, which indicates that (74) is a stiffness equation. According to [34] this phenomenon can also be observed if (74) is solved numerically using the Euler forward scheme for time integration. This means that one would have to resort to an implicit time integration scheme in order to solve (74). Unfortunately, a fully implicit scheme is practically impossible, since the plunger velocity is not known explicitly. See [34] for further details on this stiffness phenomenon.

The stiffness phenomenon previously described indicates that the coupling of the equation for the plunger motion (66) to the boundary conditions on the plunger for the Stokes flow problem is undesirable. Therefore, in the following the plunger velocity $V_p(t)$ is decoupled from the parameter V_p in the boundary conditions for the flow problem. For a more detailed analysis the reader is referred to [34]. The following lemma is used:

Lemma 3.2. *Let $(\mathbf{u}_v, p_v)$ be the family of solutions of Stokes flow problem (39)–(40) with plunger velocity $V_p = v$. Define $(\mathbf{u}, p)_v := (\mathbf{u}_v, p_v)$. Then*

$$\left(\mathbf{u}, p\right)_{k_1 v_1 + k_2 v_2} = \left(k_1 \mathbf{u}_{v_1} + k_2 \mathbf{u}_{v_2}, p_0 + k_1(p_{v_1} - p_0) + k_2(p_{v_2} - p_0)\right)$$

for arbitrary constants k_1, k_2.

The lemma can be proven by direct substitution of the solution $\left(k_1 \mathbf{u}_{v_1} + k_2 \mathbf{u}_{v_2}, p_0 + k_1(p_{v_1} - p_0) + k_2(p_{v_2} - p_0)\right)$ into the Stokes flow problem with plunger velocity $V_p = k_1 v_1 + k_2 v_2$ [34]. In the remainder of this section all variables, spaces and operators are dimensionless and the superscript * is ignored. From Lemma 3.2 it immediately follows that

$$\left(\mathbf{u}, p\right)_v = \left(v\mathbf{u}_1, p_0 + v(p_1 - p_0)\right), \tag{75}$$

and as a result,

$$F_{g,\upsilon} = F_{g,0} + \upsilon \left(F_{g,1} - F_{g,0} \right). \tag{76}$$

Substitution of (76) with $\upsilon = V_p(t)$ into the dimensionless form of (66) yields

$$\frac{dV_p}{dt}(t) = \frac{2\pi\mu L^2}{m_p V} \left(V_p(t) \left(F_{g,1}(t) + F_{g,0}(t) \right) + F_{g,0}(t) + F_e \right). \tag{77}$$

The force $F_{g,1}(t)$ can be calculated by solving the Stokes flow problem in $\Omega_g(t)$ with plunger velocity $V_p = 1$. Thus the equation for the motion of the plunger is decoupled from the boundary conditions for the Stokes flow problem. Note that the Stokes flow problem is still coupled to differential equation (77) as the geometry of the glass domain is determined by the plunger velocity.

The time dependency of the plunger velocity seems a bit awkward as the Stokes flow problem is not explicitly time dependent; the flow merely changes in time through the changing geometry. It seems more convenient to define the force of the glass on the plunger and the plunger velocity as functions of the plunger position $z := \bar{z}_p$:

$$F_g := F_g(z), \qquad V_p := V_p(z). \tag{78}$$

As a result also the motion of the plunger can be described by the plunger position. By the chain rule of differentiation,

$$\frac{dV_p}{dt}(t) = \frac{dV_p}{dz}(z)V_p(z). \tag{79}$$

As for the initial condition, let $z = 0$ and $V_p = V_0$ at $t = 0$. Then the motion of the plunger involves the following initial value problem:

$$\begin{cases} \dfrac{1}{2}\dfrac{dV_p^2}{dz}(z) = \dfrac{2\pi\mu L^2}{m_p V} \left(V_p(z) \left(F_{g,1}(z) + F_{g,0}(z) \right) + F_{g,0}(z) + F_e \right) \\ V_p(0) = V_0 \end{cases} \tag{80}$$

Since the plunger velocity does not need to be known to determine the glass domain at given plunger position z, an implicit time integration scheme can be used to solve (80), thus overcoming the stiffness problem [34].

3.4 Simulation Model

A simulation tool for the parison press process was designed [34]. The tool is able to compute and visualise the velocity field, the pressure and also the temperature in the parison during the press stage. The input parameters for the simulations include the 2D axi-symmetrical parison geometry, i.e. the description of the mould and plunger surfaces, the initial positions of the plunger and the glass domain, as well as the

Fig. 9 Mesh for press
simulation of a jar
parison [34]

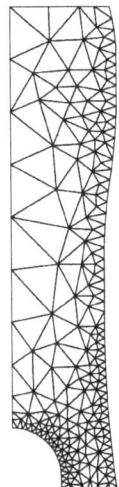

physical properties of the glass. The geometry of the initial computational domain
is defined by that of the glass gob (see Sect. 2.1). At the beginning of each time step
the computational domain is discretised by means of FEM. A typical mesh for the
initial glass domain in a press simulation of the parison for a jar is depicted in Fig. 9.
The mesh distribution depends on the geometries of the mould and the plunger. For
example, the mesh in the ring domain requires a relatively small scale in comparison
with the mould domain and the plunger domain.

The implicit Euler method is used to solve the initial value problem for the
motion of the plunger (80):

$$\begin{cases} \dfrac{1}{2} \dfrac{V_p^{k+1^2} - V_p^{k^2}}{z^{k+1} - z^k} = \dfrac{2\pi\mu L^2}{m_p V} \left(V_p^{k+1} \left(F_{g,1}(z^{k+1}) + F_{g,0}(z^{k+1}) \right) + F_{g,0}(z^{k+1}) + F_e \right) \\ V_p^0 = V_0 \end{cases}$$

$$(81)$$

The plunger position and hence the geometry of the glass domain can be updated
using the approximation

$$z^{k+1} = z^k + \Delta t^k V_p(z^k), \qquad t^{k+1} = t^k + \Delta t^k, \tag{82}$$

where Δt^k denotes the kth time step.

The motion of the free boundaries Γ_o is described by the ordinary differential
equation

$$\frac{d\mathbf{x}}{dt} = \mathbf{u}(\mathbf{x}), \qquad \text{for } t \in [0, t_{\text{end}}). \tag{83}$$

Let $\mathbf{x}_i$, $i = 1, \ldots, N$ be the nodes on a free boundary $\Gamma_f \subset \Gamma_o$ and let $\mathbf{x}_i^k$ be an approximation of $\mathbf{x}(t_k)$ (see Fig. 10). Then the new position of the ith node can be obtained
by the explicit scheme

$$\mathbf{x}_i^{k+1} = \mathbf{x}_i^k + \Delta t \, \mathbf{u}_i^k, \tag{84}$$

Fig. 10 Time integration on
the free boundary [34]

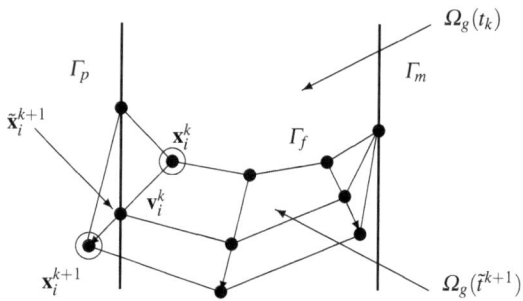

Fig. 11 Clip algorithm

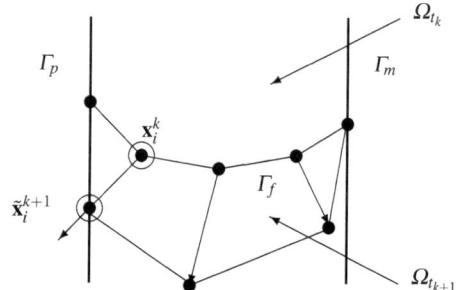

with $\mathbf{u}_i^k := \mathbf{u}(\mathbf{x}_i^k)$. A particular question is how to deal with this moving boundary. Depending on the velocity a situation may be encountered where the obtained position $\mathbf{x}_i^{k+1}$ lies outside Ω_g (see Fig. 10).

In this situation one of the strategies described below may be used. For simplicity only explicit integration is considered.

One approach to deal with the moving boundary is to decrease the time step:

$$\tilde{\mathbf{x}}_i^{k+1} = \mathbf{x}_i^k + \alpha_i^k \Delta t \, \mathbf{u}_i^k, \qquad \alpha_i^k \in (0,1], \tag{85}$$

where $\tilde{\mathbf{x}}_i^{k+1}$ is the new position of the ith node obtained with the decreased time step. The kth time step can be defined by $\Delta \tilde{t}_i^k := \min_i \alpha_i^k \Delta t$, such that the nodes are situated inside the glass domain at time $\tilde{t}^{k+1} := \tilde{t}^k + \Delta \tilde{t}_i^k$, as depicted in Fig. 10. Unfortunately, this algorithm introduces a variable time step that turns out to be too irregular in practice and can be excessively small. In order to have consistency in the topology of the computational domain the time step should be constant [34].

An alternative is illustrated in Fig. 11; the so-called *clip algorithm* leads the nodes along a discrete 'solution curve' until they reach the boundary, thereby *clipping* the trajectories on the boundary. Thus node i, which would originally leave the physical domain at time t^{k+1}, is clipped on the boundary by (85), but with $\alpha_i^k = 1$ for a

Fig. 12 Modified clip
algorithm

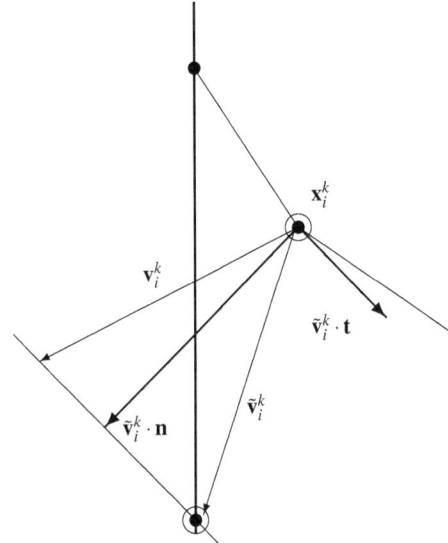

non-clipped node; the time step only differs for clipped nodes. Regrettably, also this algorithm has an evident drawback: the clipping influences the mass conservation property of the glass domain [34].

The clip algorithm can be modified to enforce better mass conservation [34]. To this end alter the velocities at the nodes that would otherwise end up outside the glass domain, such that their normal component stays the same, i.e.

$$\tilde{\mathbf{u}}_i^k \cdot \mathbf{n} = \mathbf{u}_i^k \cdot \mathbf{n}, \tag{86}$$

while the tangential component $\mathbf{u}_i^k \cdot \mathbf{t}$ is obtained by rotating the velocity vector, such that $\mathbf{x}_i^k$ ends up on the equipment boundary (see Fig. 12). This can be formulated as

$$\tilde{\mathbf{u}}_i^k := \alpha_i^k \mathbf{R}_i^k \mathbf{u}_i^k, \tag{87}$$

where α_i^k is the scaling parameter and $\mathbf{R}_i^k$ is the 2×2 rotation matrix,

$$\mathbf{R}_i^k := \begin{pmatrix} \cos \gamma_i^k & -\sin \gamma_i^k \\ \sin \gamma_i^k & \cos \gamma_i^k \end{pmatrix}. \tag{88}$$

The net outflow for the modified velocity field remains zero and hence the algorithm should give better mass conservation, i.e.

$$\int_{\Gamma_f} \mathbf{u} \cdot \mathbf{n} \, d\Gamma = \int_{\Gamma_f} \tilde{\mathbf{u}} \cdot \mathbf{n} \, d\Gamma = 0. \tag{89}$$

An approximate position of a node i that would be outside the glass domain at time t^{k+1} is obtained using the modified velocity field,

$$\tilde{\mathbf{x}}_i^{k+1} = \mathbf{x}_i^k + \Delta t\ \tilde{\mathbf{u}}_i^k. \tag{90}$$

Instead of (84) also an implicit time integration scheme can be used:

$$\mathbf{x}_i^{k+1} = \mathbf{x}_i^k + \Delta t\ \mathbf{u}_i^{k+1}. \tag{91}$$

In order to use (91) it is necessary to compute the flow velocity in $\Omega_g(t_{k+1})$. However, even though the plunger position is known in the sense of (82), the free boundary of $\Omega_g(t_{k+1})$ is still unknown. This difficulty can be overcome by employing an algorithm that iterates on $\mathbf{x}_i^{k+1}$. Unfortunately, this straightforward approach requires each iteration the solution of a Stokes flow problem, which is computationally too costly. In [34] a numerical tool is introduced that overcomes the essential difficulty of the implicitness of the scheme by using the fact that the flow velocity is autonomous. In [69] this matter is examined into more detail.

3.5 Results

In this section simulations of the pressing of a jar and a bottle parison are shown. This includes the visualisation of the velocity and pressure fields, the tracking of the free boundaries of the glass flow domain and the computation of the motion of the plunger. For more results of the simulation tool used the reader is referred to [34]. Furthermore, see [28] for results of a different parison press model.

First consider the simulation of the pressing of a jar parison. The simulation starts at the moment the gob of glass has entered the mould and the baffle has closed. Figure 13 visualises the flow velocity field in the glass domain during pressing. Here full slip of glass at the equipment boundary is assumed, that is $\beta = 0$. The plunger moves upward forcing the glass to fill the space between the mould and the plunger. When the glass hits the mould the pressure increases. Figure 14 depicts the flow velocity of the glass in the neck part ring of the jar in the final stage of the parison press. Figure 15 shows the pressure field. It is important to know the pressure of the glass onto the mould during the process, so that a similar pressure can be applied from the outside in order to keep the separate parts of the mould together [34].

Next consider the simulation of the pressing of a bottle parison. The initial position of the top of the plunger in the simulation is close to the attachment of the ring to the mould. When the glass gob is dropped into the mould it almost reaches the ring before the plunger starts moving (see Fig. 16a). Figure 16 visualises the flow velocity of the glass during pressing. Figure 17 depicts the flow velocity in the bottle neck in the final stage of the parison press. Figure 18 shows the pressure field. The results are similar to the jar parison simulation. Again full slip of glass at the equipment boundary is assumed.

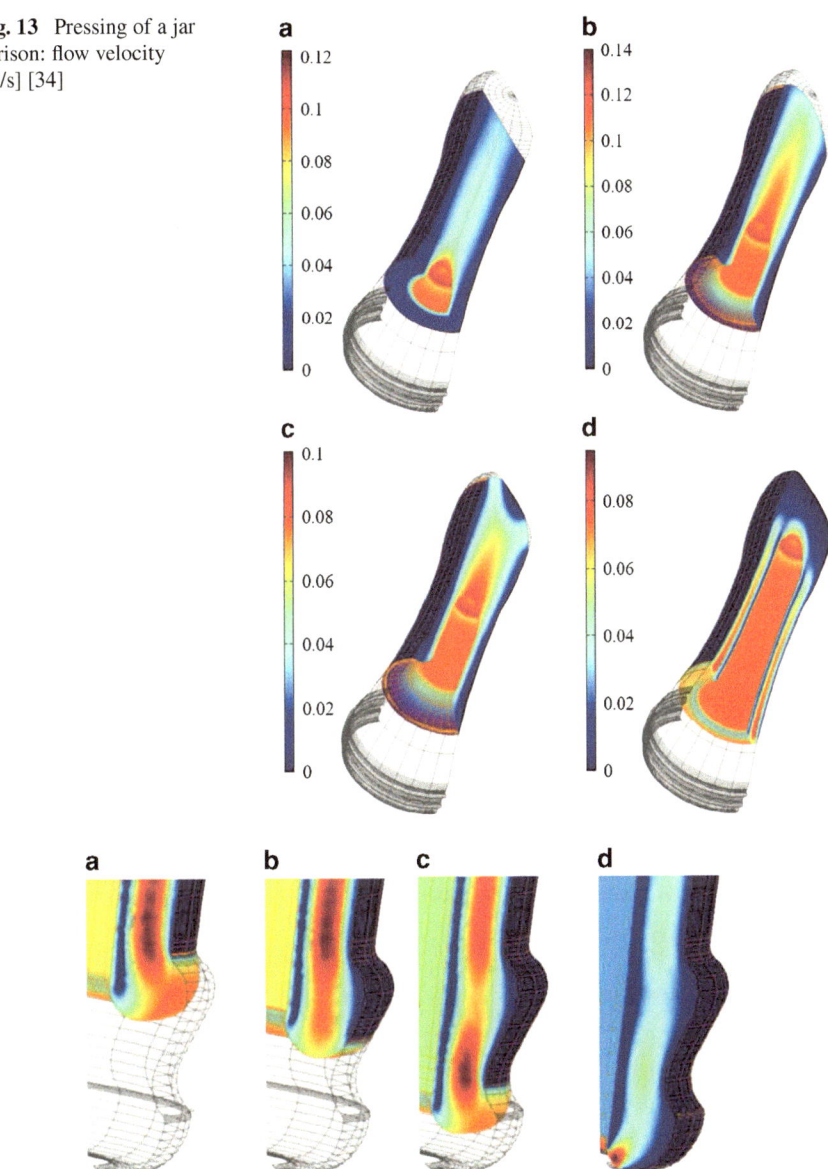

Fig. 13 Pressing of a jar parison: flow velocity [m/s] [34]

Fig. 14 Pressing of a jar parison (neckring part): flow velocity [34]

4 Blow Model

This section presents a mathematical model for blowing. Section 4.1 specifies the mathematical model described in Sect. 2 for the counter blow and the final blow. In addition to the flow of glass, also the flow of air is modelled. A particular question is

Fig. 15 Pressing of a jar
parison: pressure [bar] [34]

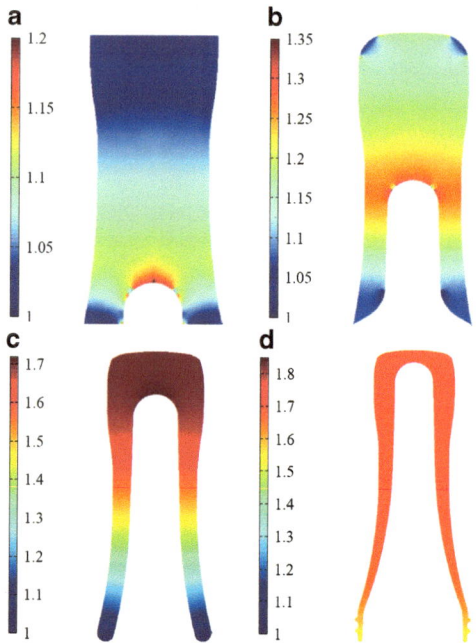

how to deal with the glass-air interfaces. Section 4.2 discusses several techniques. Subsequently, Sect. 4.3 explains how a variational formulation is used to combine the physical problems for glass and air as well as the jump conditions on the interfaces into one statement. Section 4.4 presents a numerical simulation model for the blow–blow process. Finally, Sect. 4.5 shows some examples of process simulations. The simulation tool used for the results is presented in [21].

4.1 Mathematical Model

In Sect. 2.2 the balance laws for glass forming were formulated. In this section these balance laws are further specified for glass blowing. Since the final blow stage starts with the preform obtained in either the parison press stage of the counter blow stage, the orders of magnitude of most physical parameters for these forming processes are typically the same. The temperature of the glass melt in the final blow is usually slightly lower than in the preceding stage, but this does not lead to a significant difference in the order of magnitude of the physical parameters. Therefore, the same typical values for both the counter blow and the final blow are considered:

$$\begin{aligned}
\text{Glass density} &: \rho_g = 2.5 \cdot 10^3 \text{ kg m}^{-3} \\
\text{Glass viscosity} &: \bar{\mu}_g = 10^4 \text{ kg m}^{-1}\text{s}^{-1} \\
\text{Gravitational acceleration} &: g = 9.8 \text{ m s}^{-2} \\
\text{Flow velocity} &: V = 10^{-2} \text{ m s}^{-1}
\end{aligned}$$

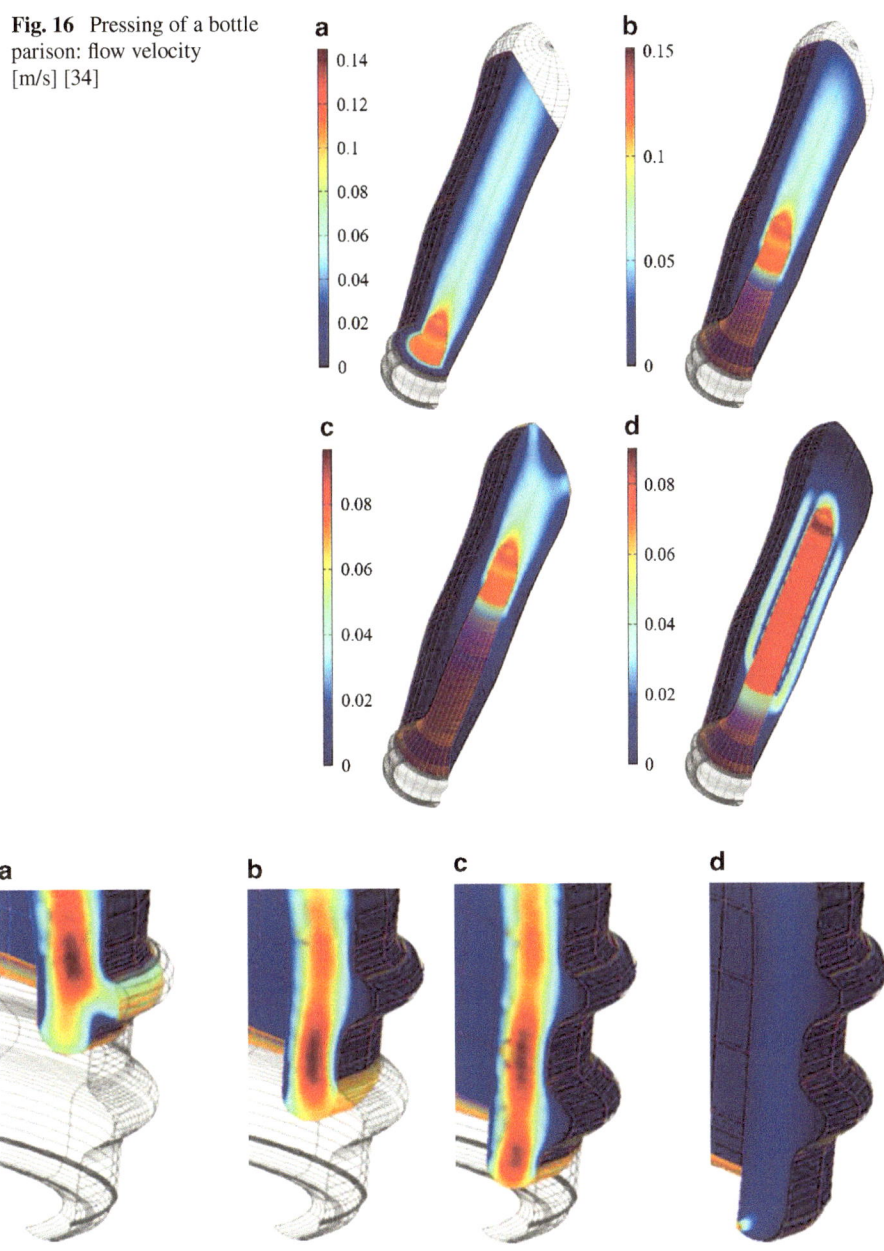

Fig. 16 Pressing of a bottle parison: flow velocity [m/s] [34]

Fig. 17 Pressing of a bottle parison (neckring part): flow velocity [34]

Length scale of the parison	:	$L = 10^{-2}$ m
Glass temperature	:	$T_g = 1,000\,^\circ$C
Mould temperature	:	$T_m = 500\,^\circ$C
Specific heat of glass	:	$\bar{c}_p = 1.5 \cdot 10^3$ J kg^{-1}K^{-1}
Effective conductivity of glass	:	$\bar{\lambda} = 5$ W m^{-1}K^{-1},

Fig. 18 Pressing of a bottle parison: pressure [bar] [34]

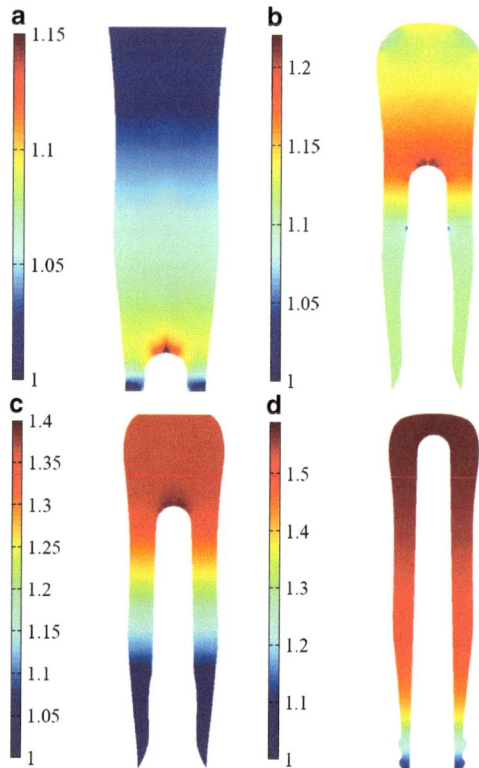

The main difference from the physical parameters in the parison press is that the time duration of a blow stage is typically much larger, hence the flow velocity is much smaller. The dimensionless numbers corresponding to either blow stage are:

$$\text{Pe}_{\text{glass}} \approx 76, \quad \text{Br}_{\text{glass}} \approx 4.0 \cdot 10^{-5}, \quad \text{Re}_{\text{glass}} \approx 2.5 \cdot 10^{-5}, \quad \text{Fr} \approx 1.0 \cdot 10^{-3}. \quad (92)$$

The Péclet number for glass is moderately large, while the Brinkman number is negligibly small. Thus the heat transfer is dominated by advection, convection and radiation:

$$\left(\frac{\partial T^*}{\partial t^*} + \mathbf{u}^* \cdot \nabla^* T^* \right) = \nabla^* \cdot \left(\lambda^* \nabla^* T^* \right), \quad (93)$$

with

$$\lambda^* := \frac{\lambda}{\rho \bar{c}_p V L}, \quad (94)$$

rather than its definition (9) in Sect. 2.2.1. The Reynolds number for glass is sufficiently small to neglect the inertia terms in the Navier–Stokes equations (25). The order of magnitude of the gravity term is given by

$$\frac{\text{Re}_{\text{glass}}}{\text{Fr}} \approx 2.5 \cdot 10^{-2},$$

which is not extremely small. Therefore, the glass flow is described by the Stokes flow equations:

$$\nabla^* \cdot \mathbf{T}^* = \mathbf{g}^*, \qquad \nabla^* \cdot \mathbf{u}^* = 0, \tag{95}$$

where $\mathbf{T}^*$ is the dimensionless stress tensor, which satisfies (20) in terms of the dimensionless variables, and $\mathbf{g}^*$ is the dimensionless gravity force, given by

$$\mathbf{g}^* = \frac{\text{Re}}{\text{Fr}} \mathbf{e}_z. \tag{96}$$

Subsequently, the air domain is considered. Note that the arguments to ignore the air domain in the parison press model do not apply to the blow model. Firstly, the inflow pressure is applied at the mould entrance and not directly on the glass, so that in this case the transport phenomena in air are of higher interest for the blow model. Secondly, because the influence of the heat flux cannot be ignored, the energy exchange between the glass and its surroundings should be taken into account. For hot pressurised air the following typical values are considered[2]:

Initial air temperature : $T_0 = 750\,^\circ\text{C}$,
Specific heat of air : $c_p = 10^3 \text{ J kg}^{-1}\text{K}^{-1}$,
Thermal conductivity of air : $\lambda = 10^{-1} \text{ W m}^{-1}\text{K}^{-1}$.
Air density : $\rho = 1 \text{ kg m}^{-3}$,
Air viscosity : $\mu = 10^{-4} \text{ kg m}^{-1}\text{s}^{-1}$.

The resulting dimensionless numbers are:

$$\text{Pe}_{\text{air}} \approx 10^2, \qquad \text{Br}_{\text{air}} \approx 4 \cdot 10^{-8}, \qquad \text{Re}_{\text{air}} \approx 1. \tag{97}$$

It can be concluded that the heat transfer in air is described by (93), while the flow of air is described by the full Navier–Stokes equations (25). This means that the model for the air flow is more complicated model than for the glass flow, while the motion of glass is most interesting. Therefore, air is replaced by a fictitious fluid with the same physical properties as air, but with a much higher viscosity, e.g. $\mu_a = 1$. Then the Reynolds number of the fictitious fluid $\text{Re} \approx 10^{-4}$ is small enough to reasonably neglect the influence of the inertia forces. On the other hand, the viscosity of the fictitious fluid is still much smaller than the viscosity of glass, so that the pressure drop in the air domain is negligible compared to the pressure drop in the glass domain [2]. Thus the flow of the ficitious fluid can be described by the Stokes flow equations (95).

Remark 4.1. If the glass temperature and the pressure in air can be reasonably assumed to be uniform, the calculations can be restricted to the glass domain and the glass-air interfaces can be treated as free boundaries with a prescribed pressure. In this case it can be recommended to use Boundary Element Methods to solve the flow problem [13, 14].

---

[2] The true orders of magnitude may be slightly different from their rough estimates in the table, but this will not affect the final results.

The boundary conditions for the flow problem are given in Sect. 2.2, but can be further specified for glass blowing. Free-stress conditions are imposed for air at the mould wall. If no lubricate is used, usually a no-slip condition is imposed for glass at the mould wall. For modelling of the slip condition in the presence of a lubricate the reader is referred to [15, 16, 36]. As a result the boundary conditions for the flow problem become:

$$
\begin{aligned}
\mathbf{u}^* \cdot \mathbf{n} &= 0, & \mathsf{T}^* \mathbf{n} \cdot \mathbf{t} &= 0, & &\text{on } \Gamma_s, \\
\mathbf{u}^* \cdot \mathbf{n} &= 0, & \mathbf{u}^* \cdot \mathbf{t} &= 0, & &\text{on } \Gamma_{g,e}, \\
\mathsf{T}^* \mathbf{n} \cdot \mathbf{n} &= 0, & \mathsf{T}^* \mathbf{n} \cdot \mathbf{t} &= 0, & &\text{on } \Gamma_{a,e}, \\
\mathsf{T}^* \mathbf{n} \cdot \mathbf{n} &= p_0, & \mathsf{T}^* \mathbf{n} \cdot \mathbf{t} &= 0, & &\text{on } \Gamma_o,
\end{aligned}
\tag{98}
$$

with

$$
p_0 = \begin{cases} p_{\text{in}} & \text{on } \Gamma_{a,o} \\ 0 & \text{on } \Gamma_{g,o}, \end{cases}
\tag{99}
$$

where p_{in} is the pressure at which air is blown into the mould. Alternatively, one may prefer to introduce the boundary condition

$$
\mathbf{u}^* \cdot \mathbf{n} = 0, \qquad \mathsf{T}^* \mathbf{n} \cdot \mathbf{t} = 0, \qquad \text{on } \Gamma_{g,o}.
\tag{100}
$$

This boundary condition avoids outflow of glass through the mould entrance during blowing. However, this involves the definition of separate boundaries $\Gamma_{a,o}$ and $\Gamma_{g,o}$, rather than the single boundary Γ_o. Moreover, these boundaries can change in time. In the next section it can be seen that this is not always convenient, particularly if a single fixed mesh is used for the discretisation of the flow domain. Instead $\Gamma_{g,o}$ can be conceived as the boundary between the glass and the ring, thereby imposing a no-slip condition,

$$
\mathbf{u}^* = \mathbf{0}, \qquad \text{on } \Gamma_{g,o}.
\tag{101}
$$

Note that in this case $\Gamma_{a,o}$ and $\Gamma_{g,o}$ remain fixed.

The boundary conditions for the energy exchange problem are defined on the boundary of the flow domain:

$$
\begin{aligned}
\left(\lambda^* \nabla T^*\right) \cdot \mathbf{n} &= 0, & &\text{on } \Gamma_s \cup \Gamma_e, \\
\left(\lambda^* \nabla T^*\right) \cdot \mathbf{n} &= \text{Nu}\, \alpha^* \left(T^* - T_\infty^*\right), & &\text{on } \Gamma_o,
\end{aligned}
\tag{102}
$$

4.2 Glass-Air Interfaces

A two-phase fluid flow problem is considered, involving the flow of both glass and air. The flow domain Ω is described by the geometry of the mould and hence fixed. On the other hand, the air domain Ω_a and the glass domain Ω_g are separated by moving interfaces Γ_i, as depicted in Fig. 4c–d, and therefore change in time. In order to model the two-phase fluid flow problem, the glass-air interfaces have to captured. There are different numerical techniques to deal with the moving interfaces in two-phase fluid flow problems. They can be classified in two main categories [20]: *interface-tracking techniques (ITT)* and *interface-capturing techniques (ICT)*.

Interface-tracking techniques (ITT) attempt to find the moving interfaces explicitly. ITT involve separate discretisations of domains Ω_a and Ω_g; the meshes of both domains are updated as the flow evolves by following the flow velocity on the interfaces (see e.g. Sect. 3.4). The major challenge of ITT is the mesh update. The procedure of updating the mesh can become increasingly computationally expensive as the mesh size decreases or the mesh has to be updated more frequently. See [12,28,52] for examples of ITT in mathematical modelling of glass blow processes.

Interface-capturing techniques (ICT) are based on an implicit formulation of the interfaces by means of interface functions, which allow ICT to function on a fixed mesh for domain Ω. An interface function marks the location of the corresponding interface by a given level set. Two widely used ICT are *Volume-Of-Fluid* (VOF) Methods [26,56] and *Level Set Methods* [1,8,58,68].

In VOF methods the interface function denotes the fraction of volume within each element of either fluid. VOF methods are conservative and can deal with topological changes of the interface. However, they are often rather inaccurate; high order of accuracy is hard to achieve because of the discontinuity of the interface function [46]. Furthermore, they can suffer from small remnants of mixed-fluid zones ('flotsam and jetsam') [44,48]. Still VOF methods are attractive because of their rigorous conservation properties.

In Level Set Methods the interface is generally represented by the zero contour of the interface function. Level Set Methods automatically deal with topological changes and it is in general easy to obtain high order of accuracy [46]. In addition, properties of the interfaces, such as the normal and the curvature, are straightforward to calculate. Also Level Set Methods generalise easily to three dimensions [65]. A drawback of Level Set Methods is that they are not conservative. Poor mass conservation of Level Set Methods for incompressible two-phase fluid flow problems is addressed in [17, 46, 72]. A major concern in Level Set Methods is the re-initialisation of the interface function in order to avoid numerical problems. For examples of Level Set Methods in mathematical modelling of glass blow processes the reader is referred to [2, 20].

Of the aforementioned methods Level Set Methods seem to be most attractive for this application. The interfaces are accurately captured, topological changes are naturally dealt with, a generalisation to three dimensions is relatively easy and complicated re-meshing algorithms are avoided. In order to compensate for the mass loss or gain coupled Level Set and VOF Methods have been developed [48,62,67]. However, [20] reports a change in mass of less than 1% during the glass blow process simulations, which can be further improved by using higher order time integration schemes or by taking smaller time steps. This indicates that in this case Level Set Methods are suited to be used as ICT.

The basic idea of Level Set Methods is to embed the moving interfaces as the zero level set of the interface function ϕ, the so-called *level set function* (see Fig. 19):

$$\phi(\mathbf{x},t) = 0, \qquad \mathbf{x} \in \Gamma_i(t). \tag{103}$$

Fig. 19 Level set function
for a glass preform

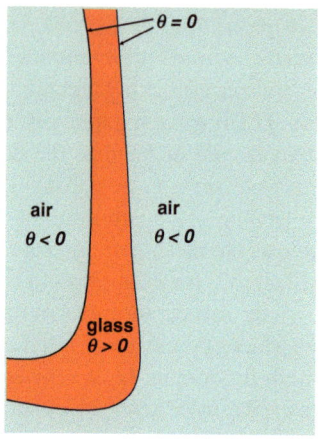

The equation of motion of the interfaces follows by the chain rule,

$$\frac{\partial \phi}{\partial t}(\mathbf{x},t) + \mathbf{u} \cdot \nabla \phi(\mathbf{x},t) = 0, \qquad \mathbf{x} \in \Gamma_i(t) \tag{104}$$

Here the flow velocity is obtained from the Stokes flow problem (95). Initially, the level set function is defined as the signed Euclidean distance function to the interfaces $\Gamma_i(0)$. If furthermore the level set equation (104) is extended to the flow domain the corresponding level set problem becomes

$$\begin{cases} \dfrac{\partial \phi}{\partial t} + \mathbf{u} \cdot \nabla \phi = 0, & \text{in } \Omega, \\[2mm] \dfrac{\partial \phi}{\partial t}(\mathbf{x},0) := \begin{cases} -d(\mathbf{x},\Gamma_i(0)), & \mathbf{x} \in \Omega_a(0) \\ d(\mathbf{x},\Gamma_i(0)), & \mathbf{x} \in \Omega_g(0), \end{cases} \end{cases} \tag{105}$$

where

$$d(\mathbf{x},\Gamma) = \inf_{\mathbf{y} \in \Gamma} \|\mathbf{x} - \mathbf{y}\|_2. \tag{106}$$

Note that since two interfaces are involved ϕ is not everywhere differentiable. This problem can be overcome by defining two level set functions ϕ_1 and ϕ_2, one for each interface, and subsequently solving the two corresponding level set problems. Then the level set function for both interfaces is $\phi = \min\{\phi_1, \phi_2\}$.

One of the difficulties encountered in Level Set Methods is maintaining the desired shape of the level set function. The flow velocity does not preserve the signed distance property, but may instead considerably distort and stretch the shape of the function, which eventually leads to additional numerical difficulties [10, 58]. To avoid this the evolution of the level set function is stopped at a certain point in time to rebuild the signed distance function. This process is referred to as *re-initialisation*. There are several ways to accomplish this. One approach is to solve the partial differential equation [47, 58, 68]

$$\frac{\partial \tilde{\phi}}{\partial \tau} + \text{sign}(\tilde{\phi}_0)\big(\|\nabla \tilde{\phi}\|_2 - 1\big) = 0, \qquad \text{in } \Omega, \tag{107}$$

with $\tilde{\phi}(\mathbf{x},0) = \tilde{\phi}_0 := \phi(\mathbf{x},t)$, to steady state. If properly implemented the function $\tilde{\phi}$ converges rapidly to the signed distance function around the interface [47,68]. However, the technique does not properly preserve the location of the interface [58,66]. This problem was fixed in [66] by adding a constraint to (107) that enforces mass conservation within the grid cells. This re-initialisation technique has appeared to be quite successful in practice. Another approach is to solve the Eikonal equation,

$$\|\nabla \phi\|_2 = 1, \tag{108}$$

given $\phi = 0$ on Γ_i, using *Fast Marching Methods* (FMM) [11,57,58]. FMM build the solution outward starting from a narrow band around the interface and subsequently marching along the grid points. Depending on the implementation FMM can be extremely computationally efficient. A Fast Marching Method is discussed in Sect. 4.4.

4.3   Variational Formulation

The variational formulation combines the physical problems for glass and air as well as the jump conditions on the interfaces into one statement. This results in a single problem formulation for the entire flow domain, while Level Set Methods can be used to deal with the interfaces. Moreover, the variational formulation is used for the discretisation by FEM in the simulation model.

First consider a variational formulation of Stokes flow problem (95), (98). Define vector spaces

$$Q := L^2(\Omega), \tag{109}$$

$$U := \Big\{ \mathbf{u} \in H^1(\Omega)^d \ \Big| \ \mathbf{u} \cdot \mathbf{n} = 0 \text{ on } \Gamma_s, \ \mathbf{u} = \mathbf{0} \text{ on } \Gamma_{g,e} \Big\}, \tag{110}$$

where $L^2(\Omega)$ is the set of Lebesgue 2-integrable functions over Ω and $H^1(\Omega)$ is the first order Hilbert space over Ω. Superscript d denotes the dimension of the flow. Then the variational formulation is to find $(\mathbf{u}, p) \in U \times Q$, such that for all $(\mathbf{v}, q) \in U \times Q$,

$$\begin{cases} a(\mathbf{v},\mathbf{u}) + b(\mathbf{v},p) = c(\mathbf{v}), \\ \qquad\qquad b(\mathbf{u},q) = 0. \end{cases} \tag{111}$$

where $a : U \times U \mapsto \mathbb{R}$ and $b : U \times Q \mapsto \mathbb{R}$ are bilinear forms and $c : U \mapsto \mathbb{R}$ is a linear form, defined by

$$a(\mathbf{v},\mathbf{u}) = \int_\Omega \mu \left(\nabla \otimes \mathbf{v}\right) : \left(\left(\nabla \otimes \mathbf{u}\right)^T + \nabla \otimes \mathbf{u}\right) d\Omega, \tag{112}$$

$$b(\mathbf{v}, q) = -\int_{\Omega} q \nabla \cdot \mathbf{v} \, d\Omega, \tag{113}$$

$$c(\mathbf{v}) = -\int_{\Omega} \mathbf{g} \cdot \mathbf{v} \, d\Omega + \int_{\Gamma_o} p_0 \mathbf{n} \cdot \mathbf{v} \, d\Gamma. \tag{114}$$

It is verified that the variational formulation indeed follows from problem (95), (98) with corresponding jump conditions (28), (30) on Γ_i. First note that

$$a(\mathbf{v}, \mathbf{u}) - b(\mathbf{v}, p) = \int_{\Omega} (\nabla \otimes \mathbf{v}) \vdots \left(\mu \left((\nabla \otimes \mathbf{u})^T + \nabla \otimes \mathbf{u} \right) - p\mathbf{I} \right) d\Omega$$
$$= \int_{\Omega} (\nabla \otimes \mathbf{v}) \vdots \mathbf{T} \, d\Omega. \tag{115}$$

By means of (115) the first equation in (111) can be written as

$$\int_{\Omega} \left((\nabla \otimes \mathbf{v}) \vdots \mathbf{T} + \mathbf{g} \cdot \mathbf{v} \right) d\Omega = \int_{\Gamma_o} p_0 \mathbf{n} \cdot \mathbf{v} \, d\Gamma. \tag{116}$$

The integral on the left side of (116) can be split up into

$$\int_{\Omega} \left((\nabla \otimes \mathbf{v}) \vdots \mathbf{T} + \mathbf{g} \cdot \mathbf{v} \right) d\Omega = \int_{\Omega_g} \left((\nabla \otimes \mathbf{v}) \vdots \mathbf{T} + \mathbf{g} \cdot \mathbf{v} \right) d\Omega$$
$$+ \int_{\Omega_a} \left((\nabla \otimes \mathbf{v}) \vdots \mathbf{T} + \mathbf{g} \cdot \mathbf{v} \right) d\Omega. \tag{117}$$

Successive application of identity

$$\nabla \cdot (\mathbf{T}\mathbf{v}) = \mathbf{v} \cdot (\nabla \cdot \mathbf{T}) + (\nabla \otimes \mathbf{v}) \vdots \mathbf{T}, \tag{118}$$

Gauss' divergence theorem and the boundary conditions yields

$$\int_{\Omega_g} \left((\nabla \otimes \mathbf{v}) \vdots \mathbf{T} + \mathbf{g} \cdot \mathbf{v} \right) d\Omega = \int_{\Omega_g} \nabla \cdot (\mathbf{T}\mathbf{v}) \, d\Omega$$
$$= \int_{\partial\Omega_g} \mathbf{n} \cdot \mathbf{T}\mathbf{v} \, d\Gamma$$
$$= \int_{\Gamma_{i,g}} \mathbf{T}\mathbf{n} \cdot \mathbf{v} \, d\Gamma, \tag{119}$$

since $p_0 = 0$ on $\Gamma_{g,o}$, and

$$\int_{\Omega_a} \left((\nabla \otimes \mathbf{v}) \vdots \mathbf{T} + \mathbf{g} \cdot \mathbf{v} \right) d\Omega = \int_{\Gamma_{i,a}} \mathbf{T}\mathbf{n} \cdot \mathbf{v} \, d\Gamma + \int_{\Gamma_{a,o}} p_0 \mathbf{n} \cdot \mathbf{v} \, d\Gamma. \tag{120}$$

Here $\Gamma_{i,g}$ and $\Gamma_{i,a}$ denote the interfaces construed as part of the boundary of Ω_g and Ω_a, respectively. As a result,

$$\int_{\Gamma_i} [[\mathbf{T}\mathbf{n} \cdot \mathbf{v}]] \, d\Gamma = 0, \tag{121}$$

for all $\mathbf{v} \in U$, which implies condition (30). Analogously, from the second equation in (111) it follows that

$$\int_{\Gamma_i} [[\mathbf{u} \cdot \nabla q]] \, d\Gamma = 0, \tag{122}$$

for all $q \in Q$, which implies condition (28).

In addition, consider a variational formulation of energy exchange problem (93), (4.1). Define vector spaces

$$V := \left\{ T \in H^1(\Omega) \times [0, \infty) \mid T = T_0 \text{ on } \partial\Omega, \ T(\mathbf{x}, 0) = T_0(\mathbf{x}) \text{ for } \mathbf{x} \in \Omega \right\}, \tag{123}$$

$$W := \left\{ \omega \in H^1(\Omega) \times [0, \infty) \mid \omega = 0 \text{ on } \partial\Omega, \ \lim_{t \to \infty} \omega(\mathbf{x}, t) = 0 \text{ for } \mathbf{x} \in \Omega \right\}. \tag{124}$$

Then the variational formulation is to find $T \in V$, such that for all $\omega \in W$,

$$\int_0^\infty \int_\Omega \left(T \frac{\partial \omega}{\partial t} + (T\mathbf{u} - \lambda \nabla T) \cdot \nabla \omega \right) d\Omega dt$$

$$= \int_0^\infty \int_{\Gamma_o} \alpha \omega (T_\infty - T) \, d\Gamma dt + \int_\Omega T_0 \omega_0 \, d\Omega, \tag{125}$$

where $\omega_0(\mathbf{x}) = \omega(\mathbf{x}, 0)$ for $\mathbf{x} \in \Omega$. It is verified that the variational formulation indeed follows from problem (93)–(4.1) with corresponding jump condition (5) on Γ_i. To this end split up the integral on the left side of (125) into

$$\int_0^\infty \int_\Omega \left(T \frac{\partial \omega}{\partial t} + (T\mathbf{u} - \lambda \nabla T) \cdot \nabla \omega \right) d\Omega dt$$

$$= \int_0^\infty \int_{\Omega_g} \left(T \frac{\partial \omega}{\partial t} + (T\mathbf{u} - \lambda \nabla T) \cdot \nabla \omega \right) d\Omega dt$$

$$+ \int_0^\infty \int_{\Omega_a} \left(T \frac{\partial \omega}{\partial t} + (T\mathbf{u} - \lambda \nabla T) \cdot \nabla \omega \right) d\Omega dt. \tag{126}$$

Successive application of the product rule for differentiation, Gauss' divergence theorem and the boundary conditions yields

$$\int_0^\infty \int_{\Omega_g} \left(T \frac{\partial \omega}{\partial t} + (T\mathbf{u} - \lambda \nabla T) \cdot \nabla \omega \right) d\Omega dt$$

$$= \int_0^\infty \int_{\Omega_g} \left(\frac{\partial}{\partial t}(T\omega) + \nabla \cdot (\omega T\mathbf{u} - \lambda \omega \nabla T) \right) d\Omega dt$$

$$= \int_0^\infty \int_{\Gamma_g} \omega (T\mathbf{u} - \lambda \nabla T) \cdot \mathbf{n} \, d\Omega dt$$

$$- \int_0^\infty \int_{\Gamma_g} \omega T\mathbf{u} \cdot \mathbf{n} \, d\Gamma dt + \int_{\Omega_g} T_0 \omega_0 \, d\Omega$$

$$= \int_0^\infty \int_{\Gamma_{g,o}} \alpha \omega (T_\infty - T) \, d\Gamma dt$$

$$- \int_0^\infty \int_{\Gamma_{i,g}} \lambda \omega \nabla T \cdot \mathbf{n} \, d\Gamma dt + \int_{\Omega_g} T_0 \omega_0 \, d\Omega, \tag{127}$$

and

$$\int_0^\infty \int_{\Omega_a} \left(T \frac{\partial \omega}{\partial t} + (T\mathbf{u} - \lambda \nabla T) \cdot \nabla \omega \right) d\Omega dt$$

$$= \int_0^\infty \int_{\Gamma_{a,o}} \alpha \omega (T_\infty - T) \, d\Gamma dt$$

$$- \int_0^\infty \int_{\Gamma_{i,a}} \lambda \omega \nabla T \cdot \mathbf{n} \, d\Gamma dt + \int_{\Omega_a} T_0 \omega_0 d\Omega. \qquad (128)$$

Therefore,

$$\int_{\Gamma_i} [[\lambda \omega \nabla T \cdot \mathbf{n}]] \, d\Gamma = 0, \qquad (129)$$

for all $\omega \in W$, which implies condition (5).

4.4 Simulation Model

A simulation model for the blow–blow process is described in [21]. The model is able to compute and visualise the preform or container shape, the thickness of the final product as well as the stress and thermal deformation the glass and mould undergo during the process. Input parameters for the simulations are the gob volume or the glass preform, the mould shape, the temperature distribution of the glass and a prescribed inlet air pressure, as well as the physical properties of the glass. The 2D axi-symmetrical geometry of the initial computational domain is defined by that of the flow domain (see Sect. 2.1). The model does not take the heat transfer in the mould into account. See [23, 28] for a complete description of the heat transfer between the glass and the mould. FEM are used to solve the problems. Triangular Mini-elements are used for the discretisation of the flow problem. Typical meshes for the moulds for the parison and the bottle are depicted in Fig. 20.

A *streamline-upwind Petrov-Galerkin* (SUPG) method is used to obtain a stabilised formulation of the energy equation [5, 20, 30]. The upwind parameter is defined by

$$\xi = \frac{h\xi}{2} \frac{\mathbf{u} \cdot \nabla \varphi_i}{\|\mathbf{u}\|_2}, \qquad \text{in } e_i. \qquad (130)$$

Here h is the width of the element in flow direction, φ_i is the ith basis function, e_i is the ith element and ξ is a tuning parameter. In the classical case $\xi = 1$ and in the modified case [20, 59]

$$\tilde{\xi} = \frac{1}{2} \sqrt{\Delta t^2 + \left(\frac{h}{\|\mathbf{u}\|_2} \right) + \frac{1}{4} \left(\frac{h}{\mathbf{u} \cdot \mathbf{Au}} \right)^2} \left(\mathbf{u} \cdot \nabla \varphi_i \right), \qquad \text{in } e_i, \qquad (131)$$

for some matrix $\mathbf{A}$.

Fig. 20 Typical meshes
for blow–blow simulation
of a bottle

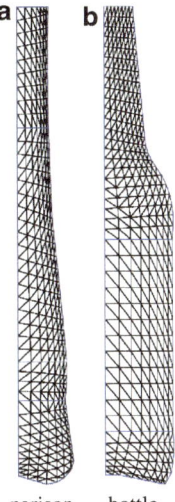

parison   bottle

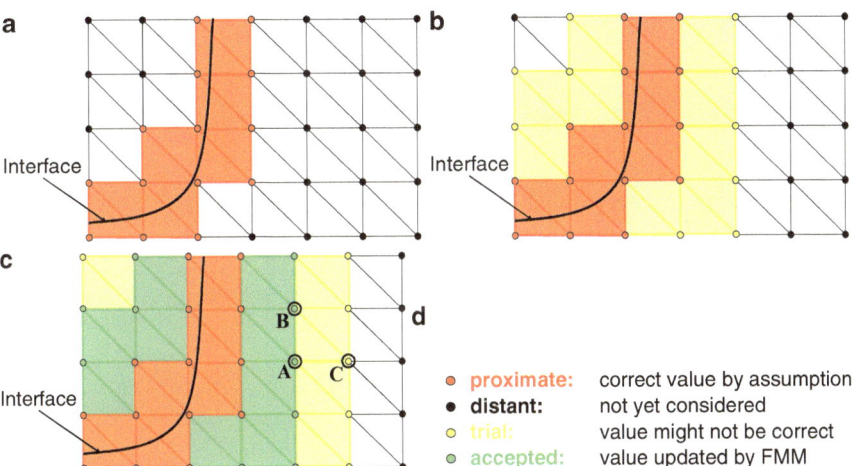

Fig. 21 Fast marching methods

A triangulated Fast Marching Method is used as a re-initialisation algorithm to maintain the level set function as a signed distance function [2, 20, 32, 57]. The method assumes that the level set values at the nodes within a narrow band around the interface are the correct signed distance values and then builds a signed distance function outward by marching along the nodes. Figure 21 shows the algorithmic representation of the Fast Marching Method for a two-dimensional mesh. The curve represents a glass-air interface. The algorithm is initialised by computing initial values for distance function d in all nodes adjacent to the interface $\phi^{-1}(0)$. A second order accurate initialisation procedure is discussed in [11]. The nodes adjacent to the interface, to which the initial values are assigned, are tagged as *proximate* and are colour marked as orange in Fig. 21a. The proximate nodes are the first nodes to be added to the set of *accepted* nodes, which contains all nodes of which the corresponding values are accepted as a distance value. The remaining nodes are initially tagged as *distant*, which are colour marked as black in Fig. 21, and are not (yet) of interest. The nodes adjacent to the set of accepted nodes are candidates to be added to either the set of accepted nodes or the set of distant nodes and are the *trial* nodes, which are colour marked as green in Fig. 21b. These nodes have a trial value assigned to them that might not yet be the correct distance value. When the values of all trial nodes have been updated, the trial nodes are removed from the set of trial nodes and added to the set of accepted nodes (see Fig. 21c). If multiple values are assigned to a trial node, the smallest one holds. This procedure repeats itself until the set of accepted nodes contains all grid points.

An element in Fig. 21c with one trial node C with two accepted nodes A and B is considered. If the mesh is structured as in Fig. 21 and the angle between edges AB and AC is right, then the distance value d_C at node C is the solution of the quadratic equation

$$(d_C - d_A)^2 + (d_B - d_A)^2 = (d_C - d_B)^2. \tag{132}$$

Suppose the mesh is not structured, but the triangles have different edges. Figure 22 shows a triangle with angles α, β, γ and edge lengths a, b, c. The interface is approximated by a line l such that the distance from nodes A and B to l is equal to the approximate distance to the interface, that is d_A and d_B, respectively. Suppose $d_B \leq d_A$. Then the angle δ between AB and l is determined by $\sin(\delta) = \frac{d_A - d_B}{c}$. As a result distance value d_C can be computed by

$$d_C = a \sin(\delta + \beta) + d_B. \tag{133}$$

It should be verified that the shortest distance to l from C intersects the triangle. Therefore the following requirement has to be satisfied:

$$0 \leq a \frac{\cos(\delta + \beta)}{\cos(\delta)} \leq c. \tag{134}$$

If requirement (134) is not fulfilled, the update is performed by taking

$$d_C = \min\left(d_A + b, d_B + a\right). \tag{135}$$

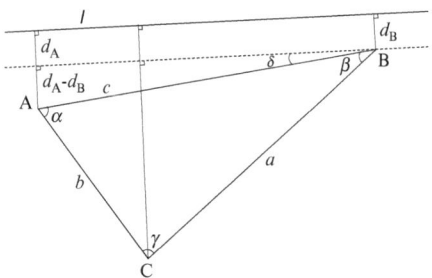

Fig. 22 Triangulated fast marching method

Finally, it is also possible to consider triangles with only one accepted node. If for example only A is accepted, then the values at B and C can be computed as $d_B = d_A + c$ and $d_C = d_A + b$.

4.5 Results

In this section simulations of the blow–blow process for a bottle are shown. The outer surface of the preform in the final blow is used as the mould shape for the counter blow. The propagation of the interfaces and the temperature distribution during the blow stages are visualised. For more results of the simulation model used the reader is referred to [20, 21]. Furthermore, see [12, 28, 53, 55] for results of different simulation models.

First consider the simulation of the counter blow. The initial temperature of the mould and the air in the counter blow are 500 °C. The gob has an uniform initial temperature distribution. The inlet air pressure is 138 kPa. Figure 23 visualises the evolution of the glass domain (red) and the air domain (blue) and Fig. 24 shows the temperature distribution.

Next consider the simulation of the final blow. First the preform is left to sag due to gravity for 0.3 s. Then pressurised air is blown into the mould with an inlet air pressure of 3 kPa. The pressure should be much lower than in the counter blow, because the preform is relatively thin and therefore easily breaks. Figure 25 visualises the evolution of the glass domain (red) and the air domain (blue) and Fig. 26 shows the temperature distribution.

In [21] a change in glass volume of less than 1.5% during the blow–blow simulation is reported. The volume conservation can be further improved by decreasing the time step, improving the mesh quality or by using higher order time discretisation schemes [20, 21].

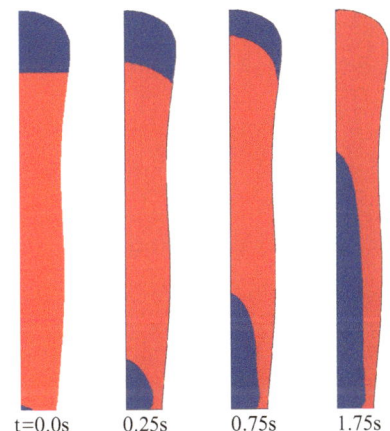

Fig. 23 Evolution of the glass domain during the counter blow [21]. Glass is *red*, air is *blue*

t=0.0s   0.25s   0.75s   1.75s

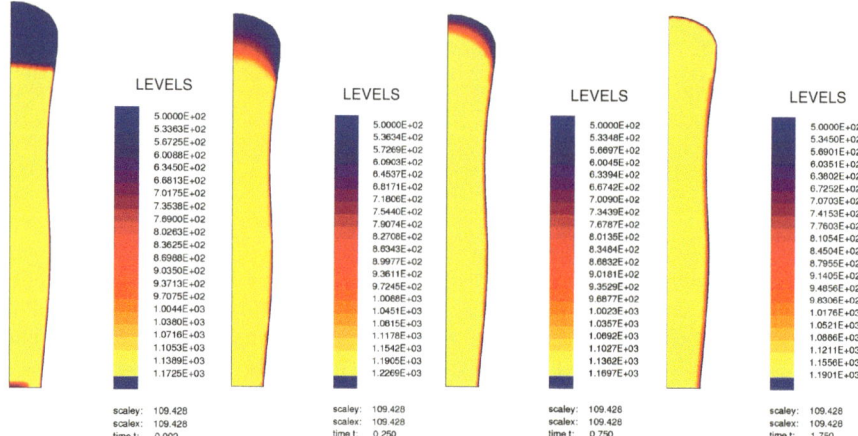

Fig. 24 Temperature distribution during the counter blow [21]

Fig. 25 Evolution
of the glass domain during
the final blow [21]. Glass
is *red*, air is *blue*

t=0.0s   0.3s   0.725s   1.025s

5 Direct Press Model

*This section presents a mathematical model for the direct press. Since physical
problems in the direct press model are described in glass, air and equipment, it
is more complicated than the parison press model described in Sect. 3 and the blow*

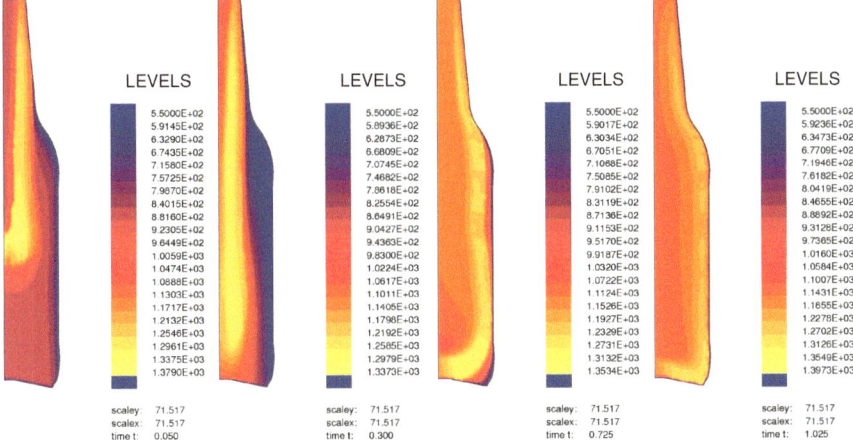

Fig. 26 Temperature distribution during the final blow [21]

model described in Sect. 4. Section 3.1 specifies the mathematical model described in Sect. 2 for the direct press. Features from both the blow model and the parison press model are combined in the direct press model. Section 3.4 presents a numerical simulation model for the direct press. An essential difference with the parison press is the way the mesh is updated. Finally, Sect. 3.5 shows some examples of two-dimensional TV panel press simulations. The simulation tool used for these results is presented in [24].

5.1 Mathematical Model

In Sect. 2.2 the balance laws for glass forming were formulated. In this section these balance laws are further specified for direct pressing. Essentially, the typical values for most physical parameters for direct pressing are the same as for parison pressing; the main difference is the larger time duration of the direct press and hence the smaller velocity:

Glass density	: $\rho_g = 2.5 \cdot 10^3$ kg m^{-3}
Glass viscosity	: $\bar{\mu}_g = 10^4$ kg m^{-1}s^{-1}
Gravitational acceleration	: $g = 9.8$ m s^{-2}
Flow velocity	: $V = 10^{-2}$ m s^{-1}
Length scale of the parison	: $L = 10^{-2}$ m
Glass temperature	: $T_g = 1{,}000\,^\circ$C
Mould temperature	: $T_m = 500\,^\circ$C
Specific heat of glass	: $\bar{c}_p = 1.5 \cdot 10^3$ J kg^{-1}K^{-1}
Effective conductivity of glass	: $\bar{\lambda} = 5$ W m^{-1}K^{-1},

The corresponding dimensionless numbers are:

$$\text{Pe}_{\text{glass}} \approx 76, \quad \text{Br}_{\text{glass}} \approx 4.0 \cdot 10^{-5}, \quad \text{Re}_{\text{glass}} \approx 2.5 \cdot 10^{-5}, \quad \text{Fr} \approx 1.0 \cdot 10^{-3}. \quad (136)$$

In conclusion, the heat transfer in glass is dominated by convection and radiation,

$$\left(\frac{\partial T^*}{\partial t^*} + \mathbf{u}^* \cdot \nabla^* T^* \right) = \nabla^* \cdot \left(\lambda^* \nabla^* T^* \right), \tag{137}$$

where λ^* is defined by (94), and the glass flow is described by the Stokes flow equations,

$$\nabla^* \cdot \mathsf{T}^* = \mathbf{g}^*, \qquad \nabla^* \cdot \mathbf{u}^* = 0, \tag{138}$$

where T^* is the dimensionless stress tensor and $\mathbf{g}^*$ is the dimensionless gravity force, given by (96). As the glass is always in contact with both air and equipment and the influence of radiation and convection cannot be disregarded, it is interesting to model the heat transfer in the whole domain Σ. Consider the typical values for the physical properties of air in Sect. 4.1. Then the following dimensionless numbers can be found:

$$\text{Pe}_{\text{air}} \approx 10^2, \qquad \text{Br}_{\text{air}} \approx 4 \cdot 10^{-8}, \qquad \text{Re}_{\text{air}} \approx 1. \tag{139}$$

Since the Péclet number for air is not extremely large, heat transfer by conduction and radiation should be taken into account. Heat transfer by dissipation can be neglected. The flow of air is described by the full Navier–Stokes flow equations, but these can eventually be simplified to the Stokes flow equations by replacing air by a fictitious fluid (see Sect. 4.4). In the equipment there is no flow, so the heat exchange is described by:

$$\frac{\partial T^*}{\partial t^*} = \nabla^* \cdot \left(\lambda^* \nabla^* T^* \right), \tag{140}$$

Separate domains for the mould, plunger and ring should be defined. The mould and plunger can again consist of parts with different material properties, for which separate subdomains should be defined. For example, if water-cooled channels are used to thermally stabilise the temperature of the plunger and the mould, these should also be taken into account. In this paper the mould, the plunger and the ring are considered as entities with constant physical properties.

    The boundary conditions given in Sect. 2.2 are specified for the direct press. Free-slip conditions are imposed for air at the equipment boundary, except for the furthest end of the ring wall, where free flow of air is allowed. In order to distinguish these boundary conditions the ring boundary Γ_r is subdivided into two parts $\Gamma_r^{(1)}$ and $\Gamma_r^{(1)}$, as it is illustrated in Fig. 27, denoting the inner boundaries of the lower part and the upper part of the ring, respectively. Finally, the boundary conditions for the flow problem can be formulated as:

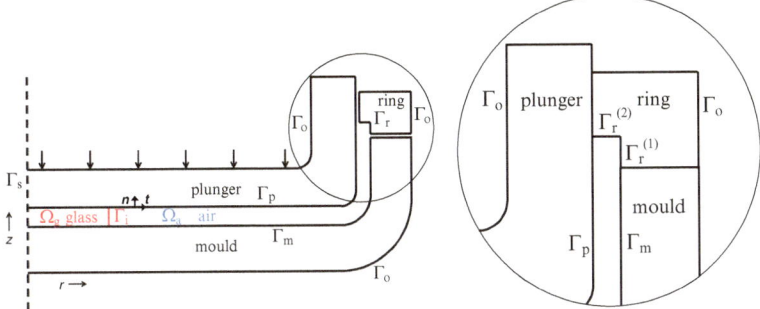

Fig. 27 Problem domains and boundaries of direct press model near the ring

$$
\begin{aligned}
\mathbf{u}^* \cdot \mathbf{n} &= 0, & \mathsf{T}^*\mathbf{n} \cdot \mathbf{t} &= 0, & &\text{on } \Gamma_s, \\
(\mathbf{u}^* - V_p\mathbf{e}_z) \cdot \mathbf{n} &= 0, & \left(\mathsf{T}^*\mathbf{n} + \beta_p^*(\mathbf{u}^* - V_p\mathbf{e}_z)\right) \cdot \mathbf{t} &= 0, & &\text{on } \Gamma_{g,p}, \\
\mathbf{u}^* \cdot \mathbf{n} &= 0, & \left(\mathsf{T}^*\mathbf{n} + \beta_m^*\mathbf{u}^*\right) \cdot \mathbf{t} &= 0, & &\text{on } \Gamma_{g,m}, \\
\mathbf{u}^* \cdot \mathbf{n} &= 0, & \left(\mathsf{T}^*\mathbf{n} + \beta_r^*\mathbf{u}^*\right) \cdot \mathbf{t} &= 0, & &\text{on } \Gamma_{g,r}, \\
\mathbf{u}^* \cdot \mathbf{n} &= 0, & \mathsf{T}^*\mathbf{n} \cdot \mathbf{t} &= 0, & &\text{on } \Gamma_{a,m} \cup \Gamma_{a,p} \cup \Gamma_{a,r}^{(1)}, \\
\mathsf{T}^*\mathbf{n} \cdot \mathbf{n} &= 0, & \mathsf{T}^*\mathbf{n} \cdot \mathbf{t} &= 0, & &\text{on } \Gamma_{a,r}^{(2)}.
\end{aligned}
\tag{141}
$$

Alternatively, one may prefer a small normal velocity of air on the impermeable equipment boundary to ensure that air does not get trapped during the simulation. The boundary conditions for the energy exchange problem are given by (14).

Essentially, a slender-geometry approximation such as in (3.2) is also possible for the direct press model as the glass becomes fairly long and thin during pressing (e.g. television screens, lenses). Unfortunately, in this case the influence of convection and radiation is too large to assume the temperature, and hence the viscosity, will remain constant for constant initial temperature is constant. As a result an analytical solution to the slender-geometry approximation cannot be found as easily as for the parison press. The slender-geometry approximation for the direct press is not considered in this paper.

5.2 Simulation Model

Obviously, the direct press model is much more complicated than the parison press model described in Sect. 3 and the blow model described in Sect. 4. It involves a coupled system of physical problems consisting of a Stokes flow problem, an energy exchange problem and a moving interface problem. In addition, the computational domain has a rather complicated geometry, which is changing in time, and is subdivided into several subdomains, each with different physical properties.

A simulation model for the two-dimensional direct press process is described in [24]. The model is able to compute and visualise the velocity field and the pressure in glass and air, as well as the stress and thermal deformation the glass and

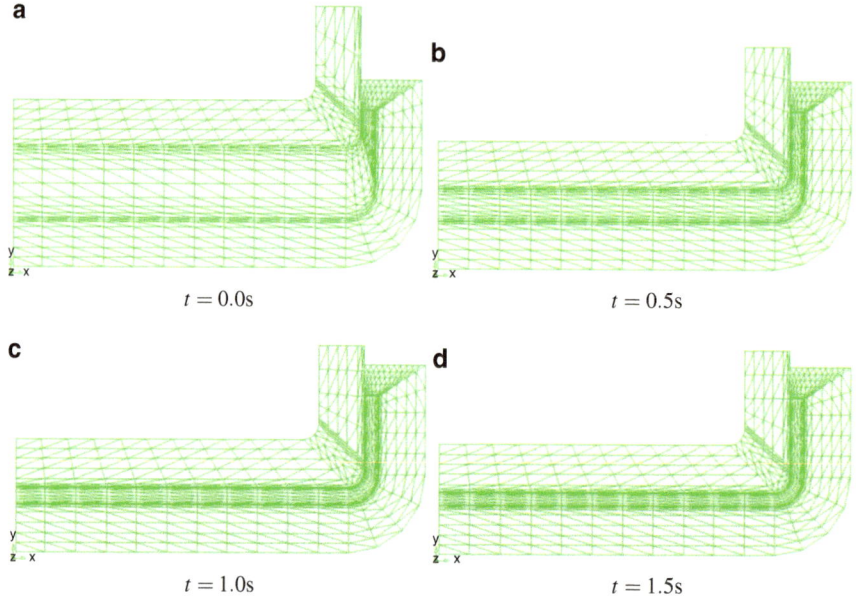

Fig. 28 Typical mesh for press simulation of a TV panel

mould undergo during the process. Furthermore, it can reproduce the temperature distribution in glass, air and equipment. A complete three-dimensional simulation tool with many additional features, such as several advanced radiation models and an annealing process simulation model, is described in [7]. Input parameters for the simulations are the gob volume, the equipment geometry, the initial temperature distribution and the physical properties of the glass and the equipment. Each time step the computational domain is discretised by means of FEM. Triangular Mini-elements are used for the discretisation of the flow problem. Figure 28 shows the deformation of a typical mesh in the press model from $t = 0$ s to 1.5 s, where $t = 1.5$ s is the end time of the process. The mesh for the lowest plunger position in Fig. 28d is provided as input for the press model. Then the elements in the flow domain are stretched in vertical direction by raising the plunger to its initial position to obtain the initial mesh (see Fig. 28a). In each subsequent time step a mesh is generated that fits the plunger position at that time following an arbitrary Euler–Lagrange approach, that is each mesh consists of the same numbers of elements along each dimension of the domain. Thus the mesh of the flow domain is compressed by the plunger, until the mesh for the lowest plunger position is regained. In this way time consuming remeshing is avoided, although remeshing is still necessary if the quality of the elements can become at issue as they are stretched or compressed [24].

Some modelling aspects of the direct press have already been discussed in Sects. 3 and 4.

- The motion of the plunger is modelled analogously to the parison press model (see Sects. 3.3, 3.4).

- The evolution of the glass-air interface can be modelled analogously to the blow model (see Sect. 4.2). The direct press model described in [7, 24] uses a Pseudo-Concentration Method [70] to track the interface.
- Air is replaced by a fictitious fluid (see Sect. 4.4).
- The energy equation is stabilised by means of a SUPG method (see Sect. 4.4).

5.3 Results

In this section two-dimensional simulations of the pressing of a TV panel are shown. This includes the visualisation of the flow velocity and temperature distribution, the tracking of the glass-air interfaces and the computation of the motion of the

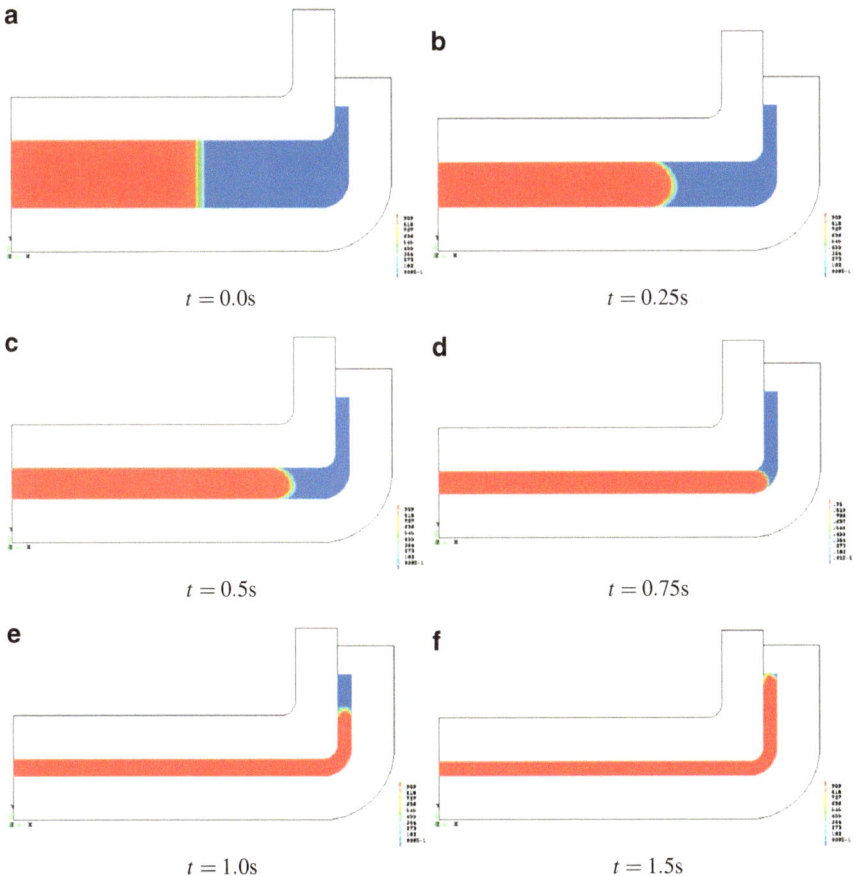

Fig. 29 Concentration of glass during the pressing of a TV panel

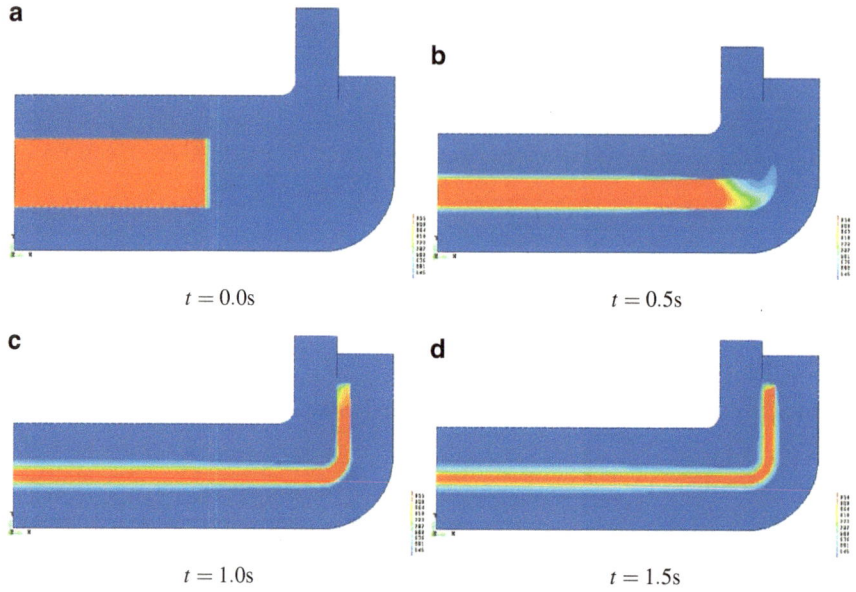

Fig. 30 Temperature distribution during the pressing of a TV panel

plunger. For more results of the simulation tool used the reader is referred to [7,24]. Moreover, see [16,35,76] for results of different direct press models.

Figure 29 visualises the (pseudo-)concentration of glass during pressing. In the figures glass is red and air is blue. The transition layer of the concentration between glass and air becomes thinner and smoother as the mesh is refined. A no-slip condition for glass at the equipment wall is used, which explains the strongly curved interface.

Figure 30 visualises the temperature distribution during pressing. The initial temperature of the glass is $1,000\,°C$ and the initial temperature of the air and the equipment is $500\,°C$. The glass gradually heats up the equipment at the contact surface. Apart from a thin boundary layer of roughly $0.2\,mm$, the temperature of the equipment does not become higher than $550\,°C$ within $1.5\,s$. By contrast, the glass rapidly heats up the air.

In order to verify the accuracy of numerical solutions, the volume conservation of the glass is examined. Figure 31 depicts the ratio of the glass volume to the initial glass volume as a function of time. It can be observed that the glass volume initially makes a relatively strong jump. Possibly, the solver has some difficulties in accurately meeting the initial conditions. At $t = 0.5\,s$ a loss volume of less than 1% is observed, which is a good result. The maximum change of volume during the simulation was 1.11%. The volume change can be further reduced by decreasing the time step, improving the mesh quality or by using higher order time discretisation schemes. Good volume conservation is also reported in [24].

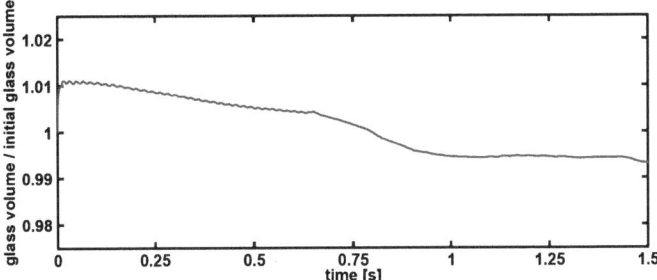

Fig. 31 Glass volume conservation

References

1. Adalsteinsson, D., Sethian, J.A.: A fast level set method for propagating interfaces. J. Comp. Phys. **118**, 269–277 (1995)
2. Allaart-Bruin, S.M.A., Linden, B.J.V.D., Mattheij, R.M.M.: Modelling the glass press-blow process. In: Di Bucchianico, A., Mattheij, R.M.M., Peletier, M.A. (eds.) Progress in Industrial Mathematics at ECMI 2004 (Proceedings 13th European Conference on Mathematics for Industry, Eindhoven, The Netherlands, June 21–25, 2004). Mathematics in Industry/The European Consortium for Mathematics in Industry, vol. 8, pp. 351–355. Springer, Berlin (2006)
3. Babcock, C.L.: Silicate Glass Technology Methods. Wiley, New York (1977)
4. Bernard, T., Ebrahimi Moghaddam, E.: Nonlinear model predictive control of a glass forming process based on a finite element model. In: Proceedings of the 2006 IEEE International Conference on Control Applications (2006)
5. Brooks, A.N., Hughes, T.J.R.: Stream-line upwind/petrov-galerkin formulation for convection dominated flows with particular emphasis on the incompressible navier-stokes equations. Comput. Methods Appl. Mech. Eng. **32**, 199–259 (1982)
6. Brown, M.: A review of research in numerical simulation for the glass-pressing process. Proc. IMechE Part B J. Eng. Manufact. **221**(9), 1377–1386 (2007)
7. Op den Camp, O., Hegen, D., Haagh, G., Limpens, M.: Tv panel production: Simulation of the forming process. Ceram. Eng. Sci. Proc. **24**(1), 1–19 (2003)
8. Chang, Y.C., Hou, T.Y., Merriman, B., Osher, S.: A level set formulation of eulerian interface capturing methods for incompressible fluid flows. J. Comp. Phys. **124**, 449–464 (1996)
9. Choi, J., Ha, D., Kim, J., Grandhi, R.V.: Inverse design of glass forming process simulation using an optimization technique and distributed computing. J. Mater. Process. Technol. **148**, 342–352 (2004)
10. Chopp, D.L.: Computing minimal surfaces via level set curvature flow. J. Comp. Phys. **106**(1), 77–91 (1993)
11. Chopp, D.L.: Some improvements on the fast marching method. SIAM J. Sci. Comput. **23**(1), 230–244 (2001)
12. Cormeau, A., Cormeau, I., Roose, J.: Numerical simulation of glass-blowing. In: Pittman, J.F.T., Zienkiewicz, O.C., Wood, R.D., Alexander, J.M. (eds.) Numerical Analysis of Forming Processes, pp. 219–237. Wiley, New York (1984)
13. Dijkstra, W.: Condition numbers in the boundary element method: Shape and solvability. Ph.D. thesis, Eindhoven University of Technology (2008)
14. Dijkstra, W., Mattheij, R.M.M.: Numerical modelling of the blowing phase in the production of glass containers. Electron. J. Bound. Elem. **6**, 1–23 (2008)
15. Dowling, W.C., Fairbanks, H.V., Koehler, W.A.: A study of the effect of lubricants on molten glass to heated metals. J. Am. Ceram. Soc. **33**, 269–273 (1950)

16. Dussere, G., Schmidt, F., Dour, G., Bernhart, G.: Thermo-mechanical stresses in cast steel dies during glass pressing process. J. Mater. Technol. **162–163**, 484–491 (2005)
17. Enright, D., Fedkiw, R., Ferziger, J., Mitchell, I.: A hybrid particel level set method for improved interface capturing. J. Comput. Phys. **183**, 83–116 (2002)
18. Falipou, M., Siclorof, F., Donnet, C.: New method for measuring the friction between hot viscous glass and metals. Glass Sci. Technol. **72**(3), 59–66 (1999)
19. Fowler, A.C.: Mathematical Models in the Applied Sciences. Cambridge University Press, Cambridge (1997)
20. Giannopapa, C.G.: Development of a computer simulation model for blowing glass containers. J. Manuf. Sci. Eng. **130** (2008)
21. Giannopapa, C.G., Groot, J.A.W.M.: A computer simulation model for the blow-blow forming process of glass containers. In: 2007 ASME Pressure Vessels and Piping Conf. and 8th Int. Conf. on CREEP and Fatigue at Elevated Temp. (2007)
22. Giegerich, W., Trier, W. (eds.): Glass Machines: Construction and Operation of Machines for the Forming of Hot Glass. Springer, Berlin (1969)
23. Grégoire, S., César de Sá, J.M.A., Moreau, P., Lochegnies, D.: Modelling of heat transfer at glass/mould interface in press and blow forming processes. Comput. Struct. **85**, 1194–1205 (2007)
24. Groot, J.A.W.M.: Analysis of the solver performance for stokes flow problems in glass forming process simulation models. Master's thesis, Eindhoven University of Technology (2007)
25. Groot, J.A.W.M., Giannopapa, C.G., Mattheij, R.M.M.: Development of a numerical optimisation method for blowing glass parison shapes. In: Proceedings of PVP 2008: ASME Pressure Vessels and Piping Division Conference (July 27–31, 2008)
26. Hirth, C.W., Nichols, B.D.: Volume of fluid (vof) method for the dynamics of free boundaries. J. Comp. Phys. **39**, 201–225 (1981)
27. Humpherys, C.E.: Mathematical modelling of glass flow during a pressing operation. Ph.D. thesis, University of Sheffield (1991)
28. Hyre, M.: Numerical simulation of glass forming and conditioning. J. Am. Ceram. Soc. **85**(5), 1047–1056 (2002)
29. John, V., Liakos, A.: Time-dependent flow across a step: The slip with friction boundary condition. Int. J. Numer. Meth. Fluids **50**, 713–731 (2006)
30. Johnson, C., Nävert, U., Pitkäranta, J.: Finite element methods for linear hyperbolic problems. Comput. Methods Appl. Mech. Eng. **45**, 285–312 (1984)
31. Kang, M., Fedkiw, R.P., Liu, X.D.: A boundary condition capturing method for multiphase incompressible flow. J. Sci. Comput. **15**(3), 323–360 (2000)
32. Kimmel, R., Sethian, J.A.: Computing geodesic paths on manifolds. Proc. Natl. Acad. Sci. USA **95**, 8431–8435 (1998)
33. Krause, D., Loch, H. (eds.): Mathematical Simulation in Glass Technology. Springer, Berlin (2002)
34. Laevsky, K.: Pressing of glass in bottle and jar manufacturing: Numerical analysis and computation. Ph.D. thesis, Eindhoven University of Technology (2003)
35. Li, G., Jinn, J.T., Wu, W.T., Oh, S.I.: Recent development and applications of three-dimensional finite element modeling in bulk forming processes. J. Mater. Process. Technol. **113**(2001), 40–45 (2001)
36. Li, L.X., Peng, D.S., Liu, J.A., Liu, Z.Q., Jiang, Y.: An experimental study on the lubrication behavior of a5 glass lubricant by means of the ring compression test. J. Mater. Process. Technol. **102**, 138–142 (2000)
37. Linden, B.J.V.D.: Radiative heat transfer in glass: The algebraic ray trace method. Ph.D. thesis, Eindhoven University of Technology (2002)
38. Linden, B.J.V.D., Mattheij, R.M.M.: A new method for solving radiative heat problems in glass. Int. J. Forming Process. **2**, 41–61 (1999)
39. Lochegnies, D., Moreau, P., Guilbaut, R.: A reverse engineering approach to the design of the blank mould for the glass blow and blow process. Glass Technol. **46**(2), 116–120 (2005)
40. Mattheij, R.M.M., Rienstra, S.W., Thije Boonkamp, J.H.M.T.: Partial Differential Equations: Modeling, Analysis, Computation. SIAM, Philadelphia (2005)

41. Moreau, P., Lochegnies, D., Oudin, J.: An inverse method for prediction of the prescribed temperature distribution in the creep forming process. Proc. Inst. Mech. Eng. J. Mech. Eng. Sci. **212**, 7–11 (1998)
42. Moreau, P., Lochegnies, D., Oudin, J.: Optimum tool geometry for flat glass pressing. Int. J. Forming Process. **2**, 81–94 (1999)
43. Nagtegaal, T.M.: Optical method for temperature profile measurements in glass melts. Ph.D. thesis, Eindhoven University of Technology (2002)
44. Noh, W.F., Woodward, P.: Simple line interface calculations. In: Vooren, A.I.V.D., Zandbergen, P.J. (eds.) Proceedings of the Fifth International Conference on Numerical Methods in Fluid Dynamics, Lecture Notes in Physics, vol. 59, pp. 330–340. Springer, New York (1976)
45. Oevermann, M., Klein, R., Berger, M., Goodman, J.: A projection method for two-phase incompressible flow with surface tension and sharp interface resolution. Tech. rep. ZIB-report, no. 00-17, Konrad-Zuse-Zentrum für Informationstechnik Berlin (2000)
46. Olsson, E., Kreis, G.: A conservative level set method for two phase flow. J. Comp. Phys. **210**, 225–246 (2005)
47. Peng, D., Kreis, G.: A pde-based fast local level set method. J. Comp. Phys. **155**, 410–438 (1999)
48. Pijl, S.P.V.D., Segal, A., Vuik, C., Wesseling, P.: A mass-conserving level-set method for modelling of multi-phase flows. Int. J. Numer. Meth. Fluids **47**, 339–361 (2005)
49. Pye, L.D., Montenero, A., Joseph, I.: Properties of Glass-Forming Melts. CRC, Boca Raton (2005)
50. Rienstra, S.W., Chandra, T.D.: Analytical approximations to the viscous glass-flow problem in the mould-plunger pressing process, including an investigation of boundary conditions. J. Eng. Math. **39**, 241–259 (2001)
51. Rosseland, S.: Note on the absorption of radiation within a star. Mon. Not. R. Astron. Soc. **84**, 525–528 (1924)
52. César de Sá, J.M.A.: Numerical modelling of glass forming processes. Eng. Comput. **3**, 266–275 (1986)
53. César de Sá, J.M.A.: Numerical modelling of incompressible problems in glass forming and rubber technology. Ph.D. thesis, University College of Swansea (1986)
54. César de Sá, J.M.A., Natal Jorge, R.M., Silva, C.M.C., Cardoso, R.P.R.: A computational model for glass container forming processes. In: Europe Conference on Computational Mechanics Solids, Structures and Coupled Problems in Engineering
55. Sadegh, N., Vachtsevanos, G.J., Barth, E.J., Pirovolou, D.K., Smith, M.H.: Modelling the glass forming process. Glass Technol. **38**, 216–218 (1997)
56. Scardovelli, R., Zaleski, S.: Direct numerical simulation of free-surface and interfacial flow. Annu. Rev. Fluid Mech. **31**, 567–603 (1999)
57. Sethian, J.A.: A fast marching level set method for montonically advancing fronts. Proc. Nat. Acad. Sci. **93**(4), 1591–1595 (1996)
58. Sethian, J.A.: Level Set Methods and Fast Marching Methods. Cambridge University Press, New York (1999)
59. Shakib, F.: Finite element analysis of the incompressible euler and navier-stokes equation. Ph.D. thesis, Stanford University (1989)
60. Sharp, D.E., Ginther, L.B.: Effect of composition and temperature on the specific heat of glass. J. Am. Ceram. Soc. **34**, 260–271 (1951)
61. Shelby, J.E.: Introduction to Glass Science and Technology, 2nd edn. The Royal Society of Chemistry, Cambridge (2005)
62. Shepel, S.V., Smith, B.L.: New finite-element/finite-volume level set formulation for modelling two-phase incompressible flows. J. Comp. Phys. **218**, 479–494 (2006)
63. Simmons, J.H., Simmons, C.J.: Nonlinear viscous flow in glass forming. Am. Ceram. Soc. Bull. **68**(11), 1949–1955 (1989)
64. Storck, K., Loyd, D., Augustsson, B.: Heat transfer modelling of the parison forming in glass manufacturing. Glass Technol. **39**(6) (1998)
65. Sussman, M., Almgren, A.S., Bell, J.B., Colella, P., Howell, L.H., Welcome, M.L.: An adaptive level set approach for incompressible two-phase flows. J. Comp. Phys. **148**, 81–124 (1999)

66. Sussman, M., Fatemi, E.: An efficient, interface-preserving level set re-distancing algorithm and its application to interfacial incompressible fluid flow. SIAM J. Sci. Comput. **20**(4), 1165–1191 (1995)
67. Sussman, M., Pucket, E.G.: A coupled level set and volume-of-fluid method for computing 3d and axisymmetric incompressible two-phase flows. J. Comp. Phys. **162**, 301–337 (2000)
68. Sussman, M., Smereka, P., Osher, S.: A level set approach for computing solutions to incompressible two-phase flow. J. Comp. Phys. **114**, 146–159 (1994)
69. Tasić, B.: Numerical methods for solving ode flow. Ph.D. thesis, Eindhoven University of Technology (2004)
70. Thompson, E.: Use of pseudo-concentration to follow creeping viscous flow during transient analysis. Int. J. Numer. Methods Fluids **7**, 749–761 (1986)
71. Tooley, F.V.: The Handbook of Glass Manufacture, Vol II. Aslee Publishing Co, New York (1984)
72. Tornberg, A.K., Engquist, B.: A finite element based level-set method for multiphase flow applications. Comput. Visual. Sci. **3**(1), 93–101 (2000)
73. Waal, H.D., Beerkens, R.G.C. (eds.): NCNG Handboek voor de Glasfabricage, 2edn. TNO-TPD-Glastechnologie (1997)
74. Williams, J.H., Owen, D.R.J., César de Sá, J.M.A.: The numerical modelling of glass forming processes. In: Bharwaj, H.C. (ed.) Collected Papers of the XIV International Congress on Glass, pp. 138–145. Indian Ceramic Society, Calcutta (1986)
75. Yigitler, K.: Numerical simulation of blowing of axisymmetric wide-mouth glass containers. In: Proceedings of the Colloquium on Modelling of Glass Forming
76. Zhou, H., Yan, B., Li, D.: Three-dimensional numerical simulation of the pressing process in tv panel production. Simulation **82**(3), 193–203 (2006)

Radiative Heat Transfer and Applications for Glass Production Processes

Martin Frank and Axel Klar

1 Introduction

In glass manufacturing, a hot melt of glass is cooled down to room temperature. The annealing has to be monitored carefully in order to avoid excessive temperature differences which may affect the quality of the product or even lead to cracks in the material. In order to control this process it is, therefore, of interest to have a mathematical model that accurately predicts the temperature evolution. The model will involve the direction-dependent thermal radiation field because a significant part of the energy is transported by photons. Unfortunately, this fact makes the numerical solution of the radiative transfer equations much more complex, especially in higher dimensions, since, besides position and time variables, the directional variables also have to be accounted for. Therefore, approximations of the full model that are computationally less time consuming but yet sufficiently accurate have to be sought. It is our purpose to present several recent approaches to this problem that have been co-developed by the authors.

This manuscript is organized as follows. In Sect. 2, we derive the underlying kinetic equation model for radiative transfer in glass. This model is supplemented by initial and boundary conditions. In addition, several versions of this model, that are later used, are introduced. For later reference and for the reader who wants to skip the derivation, the basic model is summarized in Sect. 2.6. Section 3 deals with direct numerical methods for the solution of the radiative transfer equations. These methods will later be used to compute benchmark results. Thus, we present convergence and robustness results. The rest of the discussion focuses on two approximation methods that have been co-developed by the authors, namely

M. Frank (✉)
University of Kaiserslautern, Erwin-Schrödinger-Strasse, 67663 Kaiserslautern, Germany
e-mail: frank@mathematik.uni-kl.de

A. Klar
University of Kaiserslautern, Erwin-Schrödinger-Strasse, 67663 Kaiserslautern, Germany and Fraunhofer ITWM, Fraunhofer Platz 1, 67663 Kaiserslautern, Germany
e-mail: klar@itwm.fhg.deá

A. Fasano (ed.), *Mathematical Models in the Manufacturing of Glass*,
Lecture Notes in Mathematics 2010, DOI 10.1007/978-3-642-15967-1_2,
© Springer-Verlag Berlin Heidelberg 2011

higher-order diffusion (Sect. 4) and moment methods (Sects. 5 and 6). These models are compared numerically in Sect. 7, where we also present results specifically related to glass cooling. Parts of this work have been taken from the articles [18, 23–25, 27, 47, 48, 76, 85].

2 Radiative Heat Transfer Equations for Glass

Radiative transfer has to compete with the two other modes of energy transfer, namely heat conduction and convection. In everyday life, these three effects can be seen at a cup of hot coffee. The cup itself gets warmer because of heat conduction between the coffee and the cup material. The warmth felt near the outside walls of the cup is due to radiation and the vapor emerging from the top of the cup carries energy by convection. The distinguishing features of the three modes are given in Table 1.

Radiation consists of electromagnetic waves, which have the same nature as visible light. The elementary particle of the radiation field is the photon. Heat is conducted in solids and fluids by free electrons and phonon–phonon interactions, whereas convection is energy transport by material transport.

While radiation can also be transported through the vacuum, conduction and convection need a medium. The conductive and convective heat flux is directly proportional to temperature differences. On the other hand, the celebrated Stefan–Boltzmann law states that the radiative heat flux is proportional to the difference of the fourth powers of the temperature. Because of this, radiation becomes the dominant effect at large temperatures.

Conduction and convection are local phenomena, which occur at the atomic length scale of approximately 10^{-9} m. Radiation on the other hand is a non-local phenomenon. The average distance a photon travels between two collisions can vary between the atomic length scale of 10^{-9} m up to 10^{10} m (distance earth-sun) and even more. As a consequence, the commonly used mathematical descriptions of radiative heat transfer and conduction/convection are different.

Table 1 Modes of energy transfer

	Radiation	Conduction	Convection
Energy transport by	photons	free electrons, Phonon interaction	material transport
Medium required	no	yes	yes
Temperature dependence	$q \sim T^4 - T_\infty^4$	$q \sim \nabla T$	$q \sim T - T_\infty$
Mean free path	$10^{-9} \sim 10^{10}\,m$	$\sim 10^{-9}\,m$	$\sim 10^{-9}\,m$
Depends on	x, t, Ω, ν	x, t	x, t

Furthermore, the physical quantities describing the radiation field depend on space, time, direction and frequency, while those used to describe conduction and convection depend only on space and time.

2.1 Fundamental Quantities

In this section we want to define the fundamental physical quantities describing the radiation field.

2.1.1 Intensity

The basic variable is the *spectral intensity*

$$\psi(t, x, \Omega, v), \tag{1}$$

the radiative energy flow per time, per area normal to the rays, per solid angle and per frequency. This means that $\psi \, dt \, dA \, d\Omega \, dv$ has the dimension of energy flux and is proportional to the number of photons. The spectral intensity depends on position $x \in \mathbf{R}^3$, time $t \in \mathbf{R}$, direction $\Omega \in S^2 = \{x \in \mathbf{R}^3 : \|x\| = 1\}$, and frequency $v \in \mathbf{R}^+$.

The *total intensity*

$$\psi(t, x, \Omega) = \int_0^\infty \psi(t, x, \Omega, v) dv \tag{2}$$

is the spectral intensity integrated over the whole spectrum.

2.1.2 Energy Flux

The *total energy* flux is defined as

$$E(t, x) = \varphi(t, x) = \int_{S^2} \psi(t, x, \Omega) d\Omega = \int_{S^2} \int_0^\infty \psi(t, x, \Omega, v) dv d\Omega. \tag{3}$$

In the context of moment models, this quantity is denoted by E, in the context of diffusion models it is traditionally denoted by φ or ϕ. The radiative energy is the zeroth order moment of the total intensity with respect to the direction Ω. Several other moments will play an important role in the following.

2.1.3 Heat Flux

Consider an infinitesimal surface element with outward normal n. The ingoing and outgoing spectral heat fluxes are

$$|F| = \left| -|F^{\text{in}}| + |F^b| \right|$$

$$= \int_{n \cdot \Omega < 0} (n \cdot \Omega) \psi d\Omega + \int_{n \cdot \Omega > 0} (n \cdot \Omega) \psi d\Omega$$

$$= \int_{S^2} (n \cdot \Omega) \psi d\Omega. \tag{4}$$

Thus the *spectral heat flux* is

$$F(t, x, v) = \int_{S^2} \Omega \psi(t, x, \Omega, v) d\Omega. \tag{5}$$

To obtain the *total heat flux*, we integrate over the spectrum,

$$F(t, x) = \int_{S^2} \int_0^\infty \Omega \psi(t, x, \Omega, v) dv d\Omega. \tag{6}$$

2.1.4   Radiation Pressure

The heat flux into a surface element dA is, as above,

$$(n \cdot \Omega) \psi dA d\Omega. \tag{7}$$

Thus the beam carries momentum at a rate

$$\frac{1}{c} (n \cdot \Omega) \psi n dA d\Omega. \tag{8}$$

The fraction of momentum falling onto dA is $|n \cdot \Omega|$. Therefore, the flow of momentum into dA in the normal direction is

$$\frac{1}{c} \psi |n \cdot \Omega|^2 dA d\Omega. \tag{9}$$

This must be counteracted by a pressure force pdA leading to the *spectral radiation pressure*

$$p = \frac{1}{c} \int_{S^2} \psi |n \cdot \Omega|^2 d\Omega. \tag{10}$$

The *spectral radiative pressure tensor* P_v is defined by

$$n^T P n = p, \tag{11}$$

thus

$$P(t, x, v) = \frac{1}{c} \int_{S^2} (\Omega \otimes \Omega) \psi(t, x, \Omega, v) d\Omega. \tag{12}$$

Here, $\Omega \otimes \Omega$ is the outer product (tensor product). The *total radiative pressure tensor* is

$$P(t,x) = \frac{1}{c} \int_0^\infty \int_{S^2} (\Omega \otimes \Omega) \psi(t,x,\Omega,v) d\Omega dv. \tag{13}$$

Not quite correctly, we will also call the second order moment of ψ, without the factor $\frac{1}{c}$, radiative pressure.

2.2 Blackbody Radiation

In this section we want to define a perfect absorber, also called blackbody, and derive the Planck equilibrium distribution. The Planckian plays a crucial role in the following.

Consider an electromagnetic wave that hits the surface of a medium. The wave can either be reflected at the surface or penetrate the medium. If the wave passes through the medium without attenuation, the medium is called *transparent*. If no radiation reemerges it is called *opaque*. Otherwise, in the case of partial attenuation, it is called *semitransparent*.

A *blackbody* or perfect absorber is defined to have an opaque surface that does not reflect any radiation. A blackbody is thus a maximal absorber. A simple thermo-dynamical argument [66] shows that it is also a perfect emitter at every frequency and into any direction.

The blackbody emissive power spectrum has first been derived by Max Planck in his famous work on Quantum Statistics [65]. In standard textbooks on Quantum Mechanics nowadays it is usually derived in the context of second quantization of the electromagnetic field. Here, we want to give a different derivation by entropy minimization/maximization which fits into the context of this work.

If $N(x,p)$ is the average number of photons with position x and momentum p in a phase space element of volume h^3, where h is Planck's constant, then

$$\int \int N(x,p) \frac{dxdp}{h^3} \tag{14}$$

is the number of photons in the phase space volume under consideration. Photons are integer-spin particles and obey Bose–Einstein statistics. According to a standard result [32] from statistical physics, the entropy of an ensemble of bosons is

$$S = -2k \int \int (N \log N - (N+1) \log(N+1)) \frac{dxdp}{h^3}. \tag{15}$$

Another standard result [32] relates the spectral intensity ψ and the number density N,

$$N = \frac{c^2}{2hv^3} \psi. \tag{16}$$

The momentum can be written in terms of frequency and direction ("spherical coordinates") as

$$p = \frac{h\nu}{c}\Omega,$$ (17)

thus

$$dp = \left(\frac{h}{c}\right)^3 \nu^2 d\nu d\Omega.$$ (18)

Consequently, the entropy density of the radiation field is

$$H = -\int\int \frac{2k\nu^2}{c^3}(N\log N - (N+1)\log(N+1))d\nu d\Omega.$$ (19)

We want to define the mathematical entropy as

$$H_R = -S.$$ (20)

According to the Second Law of Thermodynamics, the entropy is a non-decreasing function of time. Thus, the equilibrium distribution for a given temperature has to maximize S, or equivalently minimize H_R. This principle yields the Planck equilibrium distribution

$$B(\nu, T) = \frac{2h\nu^3}{c^2}\frac{1}{\exp(\frac{h\nu}{kT}) - 1},$$ (21)

which describes the emissive spectrum of a blackbody. Blackbody emissive power spectrum in nondimensional coordinates. In this derivation we made use of the Stefan–Boltzmann law,

$$B(T) = \int_0^\infty B(\nu, T)d\nu = \sigma_{SB}T^4,$$ (22)

with the Stefan–Boltzmann constant $\sigma_{SB} = 5.670 \cdot 10^{-8}\frac{W}{m^2 K^4}$, which gives the celebrated dependence of the total emissive power of a blackbody on the fourth power of its temperature.

2.3 The Transfer Equation

If the medium through which radiative energy travels is participating, then any incident beam will be affected by absorption and scattering while it travels through the medium. In the following, we want to consider a medium at rest (compared to the speed of light) and with constant refractive index. Furthermore, it is assumed that the medium is nonpolarizing and that it is in local thermodynamical equilibrium. For a very thorough discussion of these limitations see [88].

First we want to derive a discrete transfer equation and then, by passing to the limit, obtain the integro-differential equation describing radiative transfer.

Let us assume that there is only a finite set of directions (Ω_j) into which the photons can travel. Consider a beam into direction Ω_i which travels a distance Δs through the medium. Several effects can lead to the augmentation and reduction of the beam.

2.3.1 Absorption

When a photon hits an atom or molecule inside the medium with the right amount of energy it can be absorbed, thus leading to an excited state of the atom/molecule. The amount of absorbed photons is directly proportional to the distance traveled and to the number of photons itself. Thus the change in the spectral intensity due to absorption is

$$(\Delta\psi)_{\text{abs}} = -\kappa\psi\Delta s. \tag{23}$$

2.3.2 Scattering

The photons can hit atoms or molecules in the medium and change their direction. We assume that the energy (or frequency) of the photons does not change (elastic scattering). We denote the fraction of photons that change their direction from Ω_j to Ω_i by S_{ij}. Note that the normalization condition

$$\sum_i S_{ij} = 1 \tag{24}$$

has to hold. The win/loss balance reads

$$(\Delta\psi(\Omega_i))_{\text{scat}} = \sigma\left(-\psi(\Omega_i) + \sum_j S_{ij}\psi(\Omega_j)\right)\Delta s. \tag{25}$$

2.3.3 Emission

If the medium has a finite temperature then it also emits thermal radiation which is distributed as blackbody radiation. The emitted intensity along a path is again proportional to the length of the path. If the spectral intensity of the photons ψ were a Planckian itself there should be no net absorption/emission. Hence the proportionality constant must be κ. Thus the intensity change caused by emission is

$$(\Delta\psi)_{\text{em}} = \kappa B\Delta s. \tag{26}$$

2.3.4   Overall Balance

Drawing a balance of the different effects, we obtain the discrete transfer equation

$$\psi(s+\Delta s,\Omega_i) = \psi(s,\Omega_i)$$
$$+\Delta s\left(\kappa(B(T)-\psi(\Omega_i))+\sigma\left(\sum_j S_{ij}\psi(\Omega_j)-\psi(\Omega_i)\right)\right). \quad (27)$$

The sum on the right hand side can be interpreted as a numerical quadrature rule. The matrix S_{ij} can be interpreted as the evaluation of a function,

$$S_{ij} = s(\Omega_i,\Omega_j). \quad (28)$$

If we assume that the set of directions is continuous, then the summation over all directions becomes an integration over the unit sphere. We obtain for all $\Omega \in S^2$,

$$\psi(s+\Delta s,\Omega) = \psi(s,\Omega)$$
$$+\Delta s\left(\kappa(B(T)-\psi(\Omega))+\sigma\left(\int_{S^2} s(\Omega,\Omega')\psi(\Omega')d\Omega'-\psi(\Omega)\right)\right). \quad (29)$$

The normalization property (24) becomes

$$\int_{S^2} s(\Omega,\Omega')d\Omega' = 1. \quad (30)$$

A beam travels a distance Δx in a time $\frac{\Delta x}{c}$, where c is the speed of light. Thus we have

$$\psi(t+\Delta x/c,x+\Omega\Delta x,\Omega) = \psi(t,x,\Omega)$$
$$+\Delta x\left(\kappa(B(T)-\psi(t,x,\Omega))+\sigma\left(\int_{S^2} s(\Omega,\Omega')\psi(t,x\Omega')d\Omega'-\psi(t,x,\Omega)\right)\right). \quad (31)$$

Taking the limit $\Delta x \to 0$ we arrive at the *radiative transfer equation*. The frequency v can be incorporated as an additional parameter. At a position x and a time t, for all directions $\Omega \in S^2$, for all frequencies $v \in [0,\infty]$ it holds

$$\frac{1}{c}\partial_t\psi(t,x,\Omega,v)+\Omega\nabla\psi(t,x,\Omega,v)$$
$$= \kappa(B(v,T)-\psi(t,x,\Omega,v))+\sigma\left(\int_{S^2} s(\Omega,\Omega')\psi(t,x,\Omega',v)d\Omega'-\psi(t,x,\Omega,v)\right). \quad (32)$$

We will also consider the special case of a one-dimensional *slab geometry*. We consider a plate which is finite in one dimension and infinite in the other dimensions. Thus, the intensity depends only on one space variable and is axially symmetric. Hence the equation simplifies to

$$
\frac{1}{c}\partial_t \psi(t,x,\mu,v) + \mu\partial_x\psi(t,x,\mu,v)
$$
$$
= \kappa(2\pi B(v,T) - \psi(t,x,\mu,v)) + \sigma\left(\frac{1}{2}\int_{-1}^{1}\psi(t,x,\mu',v)d\mu' - \psi(t,x,\mu,v)\right).
$$
(33)

Here, μ is the cosine of the angle between direction and x-axis.

There are several other versions of the transfer equation, that we will consider in the following. First of all, in most applications, the time scale is much larger than the time the radiation needs to propagate into the medium. Thus we neglect the time-derivative and thus obtain the *steady* transfer equation. The absorption coefficient κ and the scattering coefficient σ can in general also depend on position and time. In the following, we want to assume *isotropic scattering*. This means that the scattering kernel is actually a constant, $s = \frac{1}{4\pi}$. For the purpose of glass, it often suffices to consider only absorption.

If frequency-dependence is not important, the so-called *grey* approximation can be used, meaning that all quantities are frequency-dependent. For glass manufacturing, however, frequency-dependence is important. Table 2 shows typical absorption coefficients for glass, depending on frequency. For smaller frequencies, absorption becomes very large. For all practical purposes, glass is perfectly opaque for frequencies smaller than a limit v_1.

2.4 Overall Energy Conservation

The radiation field, by emission, strongly depends on the temperature of the medium. On the other hand, by absorption, it also affects the temperature of the medium. For an overall energy balance we have to take into account this connection.

Table 2 Eight frequency bands for glass

Band ι	v_ι	$v_{\iota+1}$	κ_ι
1	∞	5	0.40
2	5	0.3333	0.50
3	0.3333	0.2857	7.70
4	0.2857	0.2500	15.45
5	0.2500	0.2222	27.98
6	0.2222	0.1818	267.98
7	0.1818	0.1666	567.32
8	0.1666	0.1428	7136.06
	0.1428	0	Opaque

Also, we have to consider the two other modes of energy transfer, heat conduction and convection.

The general energy conservation equation for a moving compressible fluid may be stated as [66]

$$\rho_m \frac{Du}{Dt} = \rho(\partial_t u + v\nabla u) = -\nabla q - p\nabla v + \mu\Phi + \dot{Q}''', \tag{34}$$

where u is the internal energy, v is the velocity vector, q is the total heat flux vector, Φ is the dissipation function, and $\dot{Q}'''$ is the heat generated within the medium.

If the medium interacts with the radiation field through emission, absorption and scattering, then the heat flux term q in (34) contains the radiative heat flux. The radiative contributions to the internal energy and the pressure tensor can be neglected [66].

If we assume that $du = c_m dT$ and furthermore that Fourier's law of heat conductivity holds,

$$q = q_{con} + F = -k\nabla T + F, \tag{35}$$

then (34) becomes

$$\rho_m c_m (\partial_t T + v\nabla T) = \nabla k\nabla T - p\nabla v + \mu\Phi + \dot{Q}''' - \nabla F. \tag{36}$$

In the following we want to restrict ourselves to a fluid at rest or a solid, i.e. $v = 0$. Furthermore we want to assume $\Phi = 0$ and $\dot{Q}''' = 0$. Thus we consider

$$\rho_m c_m \partial_t T = \nabla k\nabla T - \nabla F. \tag{37}$$

By integrating the transfer equation with respect to Ω and ν, we see that this can be written as

$$\rho_m c_m \partial_t T = \nabla k\nabla T - \int_0^\infty \int_{S^2} \kappa(\psi - B) d\Omega d\nu. \tag{38}$$

2.5 Boundary Conditions

Consider a beam of photons hitting a slab, as shown in Fig. 1. Some of the irradiation will be reflected at the surface. A fraction of the radiation which penetrates the medium will be absorbed, the remaining will be transmitted. Thus we define the quantities

$$\text{Reflectivity} \quad \rho = \frac{\text{reflected part of radiation}}{\text{incoming radiation}}$$

$$\text{Absorptivity} \quad \alpha = \frac{\text{absorbed part of radiation}}{\text{incoming radiation}}$$

$$\text{Transmittivity} \quad \tau = \frac{\text{transmitted part of radiation}}{\text{incoming radiation}}.$$

Fig. 1 Reflection, absorption, and transmission

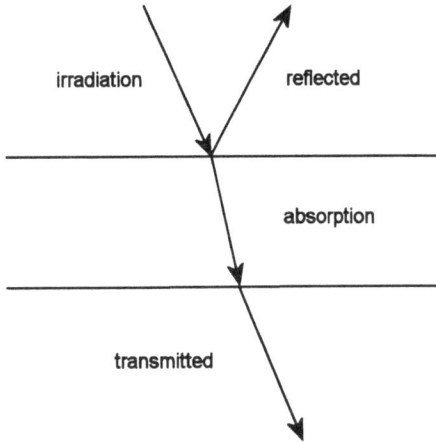

By definition, $\rho + \alpha + \tau = 1$. A medium is called opaque if $\tau = 0$, i.e. no radiation is transmitted. For a black body, $\alpha = 1$ and $\rho = \tau = 0$.

Radiative energy can also be emitted inside a medium and can be released through the surface. Since a blackbody is a perfect emitter, we define

$$\text{emissivity } \varepsilon = \frac{\text{energy emitted from surface}}{\text{energy emitted from a blackbody at the same temperature}}.$$

Radiative transfer is a long-range phenomenon. In principle, if we want to know the amount of radiation at one point x, we have to take into account radiation arriving from any direction and any point in space. Thus, an energy balance must be performed either over the whole space or over an enclosure bounded by opaque walls. When speaking of a wall or surface we actually mean a small layer (compared to the size of the enclosure) where radiation is reflected, absorbed and emitted.

Consider a domain bounded by an opaque surface and let n denote the outward normal vector. For a point on the boundary we have the following energy balance for all incoming directions, i.e. all Ω with $n \cdot \Omega < 0$,

$$\psi(x,t,\Omega) = \rho(\Omega')\psi(x,t,\Omega') + (1 - \rho(\Omega))\psi_b. \tag{39}$$

Here, $\Omega' = \Omega - 2(\Omega \cdot n)n$ is the outgoing direction that is reflected into Ω. Furthermore, $I_{\nu,b}$ is the amount of radiation emitted from the surface, cf. Fig. 2.

Mostly, we will assume that the incoming radiation is a Planckian at some temperature,

$$\psi_b = B(T_b). \tag{40}$$

The reflectivity ρ generally depends on the direction Ω. It can be computed using Snell's law. On the interface between two media with refraction indexes n_1 and n_2, the refraction angle (with respect to the normal) θ_2 of the transmitted ray and the incident angle θ_1 are related by

$$n_1 \sin \theta_1 = n_2 \sin \theta_2. \tag{41}$$

Fig. 2 Boundary condition

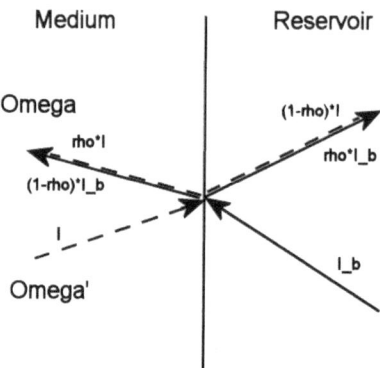

The reflectivity is then given by Fresnel's equation

$$\rho = \frac{1}{2}\left[\frac{\tan^2(\theta_1 - \theta_2)}{\tan^2(\theta_1 + \theta_2)} + \frac{\sin^2(\theta_1 - \theta_2)}{\sin^2(\theta_1 + \theta_2)}\right]. \tag{42}$$

In the case of total refection, $\rho = 1$.

The heat equation (38) can for example be supplemented with the following boundary conditions. Assuming that the opaque body surrounding the medium under consideration is a gas in a large reservoir, we can consider the heat flux through the boundary due to advection

$$k\frac{\partial T}{\partial n} = h(T_b - T). \tag{43}$$

Here, T_b is the outside temperature. We want to note that the value of the parameter h has to be determined by experiment. Also, we have to emphasize that the modeling of heat exchange by advection is actually quite sophisticated and still a subject of research. However, we will to assume in the following the simple boundary conditions stated above.

2.6 Summary

In order to model glass cooling we consider the following radiative transfer problem. For a point x in the domain $V \subset \mathbf{R}^3$

$$c_m \rho_m \frac{\partial T}{\partial t} = \nabla \cdot k\nabla T - \int_{\nu_1}^{\infty}\int_{S^2} \kappa(B - I)\, d\Omega dv, \tag{44a}$$

$$\forall v > v_1 : \ \mathbf{\Omega} \cdot \nabla \psi = \kappa(B - \psi) + \sigma(\phi - \psi), \tag{44b}$$

where

$$\phi(x,\Omega,\nu) = \frac{1}{4\pi}\int_{S^2} s(\Omega,\Omega')\psi(x,\Omega',\nu)d\Omega'$$

is the scattered intensity. Here, we have explicitly included the opaque band for frequencies in the interval $[0,\nu_1]$, where absorption is infinitely high and thus $\psi \equiv B$.

On the boundary, for $x \in \partial V$, the ingoing radiation is prescribed by semi-transparent boundary conditions

$$I(\Omega) = \rho(n\cdot\Omega)I(\Omega') + (1 - \rho(n\cdot\Omega))I_b(\Omega), \quad \forall n\cdot\Omega < 0, \tag{44c}$$

while the temperature is assumed to obey

$$kn\cdot\nabla T = h(T_b - T) + \alpha\pi\left(\frac{n_2}{n_1}\right)^2\int_0^{\nu_1} B(\nu,T_b) - B(\nu,T)\,d\nu. \tag{44d}$$

The additional term comes from the fact that in the opaque band $0 < \nu < \nu_1$ the medium behaves like a perfect black body.

At initial time $t = 0$, the temperature shall be given by

$$T(x,0) = T_0(x). \tag{44e}$$

In these equations, $\psi(t,x,\Omega,\nu)$ denotes the specific radiation intensity at point $x \in V$ traveling in direction $\Omega \in S^2$ with frequency $\nu > 0$ at time $t \geq 0$. The outside radiation I_b is assumed to be known for the ingoing directions (i.e. $n\cdot\Omega < 0$) on the boundary. We denote the outward normal on ∂V by n. Furthermore, $T(t,x)$ denotes the material temperature and T_b is the exterior temperature on the boundary. The equations contain as parameters the opacity κ, the scattering coefficient σ, the heat conductivity k and the convective heat transfer coefficient h. Moreover, B denotes Planck's function

$$B(\nu,T) = n_1^2\frac{2h_P\nu^3}{c^2}\left(e^{\frac{h_P\nu}{k_BT}} - 1\right)^{-1}$$

for black body radiation in glass which involves Planck's constant h_P, Boltzmann's constant k_B and the speed of light in vacuum c. The integration in the second term of the temperature boundary condition (44d) is done on the opaque interval of the spectrum $[0,\nu_1]$, where radiation is completely absorbed. At the interface between glass and surrounding air with refractive indices $n_1 > n_2$, respectively, light rays are reflected and refracted. This is modeled by the so-called semi-transparent boundary conditions (44c). The reflectivity $\rho \in [0,1]$ is the proportion of radiation that is reflected. It is equal to 1 if total reflection occurs i.e. if $\theta_1 > \theta_c$ where θ_c is the critical angle given by $\sin\theta_c = n_2/n_1$. Otherwise ρ is calculated according to Fresnel's equation

$$\rho(\mu) = \frac{1}{2}\left(\frac{\tan^2(\theta_1 - \theta_2)}{\tan^2(\theta_1 + \theta_2)} + \frac{\sin^2(\theta_1 - \theta_2)}{\sin^2(\theta_1 + \theta_2)}\right),$$

where the refraction angles θ_1 and θ_2 are given by $\cos\theta_1 = |n \cdot \Omega| = \mu$ and Snell's law of refraction

$$n_1 \sin\theta_1 = n_2 \sin\theta_2.$$

The solid angles of the reflected ray in (44c) is

$$\Omega' = \Omega - 2(n \cdot \Omega)n.$$

Finally, the hemispheric emissivity α of the boundary surface in (44d) is related to the reflectivity ρ by

$$\alpha = 2n_1 \int_0^1 1 - \rho(\mu)\, d\mu.$$

For these equations and other applications in glass manufacturing problems we refer, for example, to [39, 40, 46, 87], and the monographs [31] and [60].

Analytical results concerning the existence and uniqueness of solutions to the transfer equation itself and to the radiative heat transfer equations, where also energy conservation and additionally heat conduction are considered, have been obtained by many authors. A rather recent review on methods for transport equations can be found in [6], cf. also [5]. The transfer equation together with energy conservation is considered in [28, 58]. The issue of heat conduction is addressed in [37, 43, 44]. Convection, conduction and radiation is treated in [53, 69].

3 Direct Numerical Methods

The main difficulties in solving numerically the integro-differential equation (44) are the large set of unknowns and the coupling between the transport and the integral operators. For instance, ψ is a function of time variable t, space variable x, frequency variable v, and direction variable Ω. Solving the large linear system of algebraic equations induced by discretizing these variables is computationally very demanding.

Here, we focus on the solution of steady-state, mono-energetic, frequency decoupled, isotropic radiative transfer problems in three space dimensions. However, all the methods presented in this paper can be straightforwardly extended to the more general problem (44). Hence the radiative transfer equation we consider reads

$$\Omega \cdot \nabla\psi + (\sigma + \kappa)\psi = \sigma\phi + \kappa B \tag{45}$$

with boundary values $\psi = g$. The (45) models the changes of an intensity $\psi(x, \Omega)$ as particles are passing through the domain V at position point $x = (x, y, z)^T$ in the direction $\Omega = (\mu, \eta, \xi)^T$ and are subject to loses due to absorption κ and scattering σ, while their number grows due to the source B inside the domain V. We assume that σ and κ are nonnegative functions and we introduce the mean intensity ϕ as

$$\phi(x) = \frac{1}{4\pi} \int_{S^2} \psi(x, \Omega')\, d\Omega'. \tag{46}$$

We also define the scattering ratio γ and the opacity coefficient ϑ as

$$\gamma = \max_{x \in V} \left(\frac{\sigma(x)}{\sigma(x) + \kappa(x)} \right) \quad \text{and} \quad \vartheta = \min_{x \in V} \left(\sigma(x) + \kappa(x) \right) \text{diam}(V). \qquad (47)$$

Here $\text{diam}(V)$ is the diameter of the space domain V. In applications, γ and ϑ are used to characterize the convergence rates of the iterative methods and the diffusion limits in the optically thick medium.

Many numerical methods have been used to solve the (45). For a review on some of these methods see [52,75]. It is well known [1] that the standard Source Iteration (SI) becomes extremely costly when the scattering ratio $\gamma \approx 1$. The standard Diffusion Synthetic Acceleration (DSA) has been widely used to accelerate the source iteration [4, 10]. The SI and DSA methods can be seen respectively, as Richardson iteration and preconditioned Richardson iteration with the diffusion approach as preconditioner.

We implement SI, DSA and a Krylov subspace method to solve the (45). We also propose a fast multilevel algorithm [36] which uses the approximate inverse operator as a preconditioner and solves the linear system only in the coarse meshes. Numerical results show this algorithm to be faster than DSA in many regimes. The robustness, efficiency and convergence rates of these methods are illustrated by several numerical test examples in both one and two space dimensions. Comparison of the results obtained by different methods is also included in this section. The material in this section is taken from [76].

3.1 Ordinates and Space Discretizations

We start with a discrete ordinates discretization in angle. This corresponds to expanding the integrals on the unit sphere S^2 in terms of N weighted quadrature rules,

$$\int_{S^2} \psi(x, \Omega) d\Omega \simeq \sum_{l=1}^{N} w_l \psi(x, \Omega_l), \qquad (48)$$

where $\Omega_l = (\mu_l, \eta_l, \xi_l)^T$, for all $l = 1, \ldots, N$, with $N = n(n+2)$, and n is the number of direction cosines. Since $\Omega_l \in S^2$, we have

$$\mu_l^2 + \eta_l^2 + \xi_l^2 = 1, \qquad \text{for all} \quad l = 1, 2, \ldots, N.$$

We assume n an even number of quadrature points so that the points (μ_l, η_l, ξ_l) are nonzero, symmetric with respect to the x-, y- and z-axis and they are invariant under $90°$ rotations. Furthermore they satisfy the relation

$$\xi_i^2 = \xi_1^2 + 2 \frac{i-1}{n-2} (1 - 3\xi_1^2),$$

for $i = 1, 2, \ldots, n/2$ and $0 < \xi_1 < 1/3$.

In (48) w_l are the corresponding weights chosen to be positive and satisfy

$$\sum_{l=1}^{N} \Omega_l = 4\pi, \quad \sum_{l=1}^{N} \Omega_l \mu_l = 0, \quad \sum_{l=1}^{N} \Omega_l \eta_l = 0 \quad \text{and} \quad \sum_{l=1}^{N} \Omega_l \xi_l = 0.$$

In practice we choose $w_l = 4\pi/N$ and (μ_l, η_l, ξ_l) are set in such a way the above conditions are guaranteed. Let S_N be a given set of N discrete directions in S^2, then a semi-discrete formulation of (45) is

$$\mu_l \frac{\partial \psi_l}{\partial x} + \eta_l \frac{\partial \psi_l}{\partial y} + \xi_l \frac{\partial \psi_l}{\partial z} + (\sigma + \kappa)\psi_l = \sigma\phi(x) + \kappa q(x), \tag{49}$$

$\psi_l(x)$ denotes approximation to $\psi(x, \mu_l, \eta_l, \xi_l)$ and ϕ is given by

$$\phi(x) = \frac{1}{4\pi} \sum_{l=1}^{N} w_l \psi_l(x).$$

To discretize the (49) in space we suppose for simplicity, that the spatial domain is a box, $V = [a_x, b_x] \times [a_y, b_y] \times [a_z, b_z]$. Then we cover the domain V with a uniform numerical mesh defined by

$$V_h = \Big\{ x_{ijk} = (x_i, y_j, z_k)^T, \quad x_i = i\Delta x, \quad y_j = j\Delta y, \quad z_k = k\Delta z,$$

$$i = 0, 1\ldots, I, \quad j = 0, 1\ldots, J, \quad k = 0, 1\ldots, K \Big\},$$

where $x_0 = a_x$, $x_I = b_x$; $y_0 = a_y$, $y_J = b_y$; $z_0 = a_z$, $z_K = b_z$; and h denotes the maximum cell size. We define the averaged grid points

$$\Delta x = x_i - x_{i-1}, \quad \Delta y = y_j - y_{j-1}, \quad \Delta z = z_k - z_{k-1},$$

$$x_{i-\frac{1}{2}} = \frac{x_{i-1} + x_i}{2}, \quad y_{j-\frac{1}{2}} = \frac{y_{j-1} + y_j}{2}, \quad z_{k-\frac{1}{2}} = \frac{z_{k-1} + z_k}{2},$$

for $i = 1\ldots, I$, $j = 1\ldots, J$ and $k = 1\ldots, K$. By using the notation f_{ijk} to denote the approximation value of the function f at the grid point (x_i, y_j, z_k), the fully discrete approximation for the (45) can be written as

$$\mu_l \frac{\psi_{l,ijk} - \psi_{l,i-1jk}}{\Delta x} + \eta_l \frac{\psi_{l,ijk} - \psi_{l,ij-1k}}{\Delta y} + \xi_l \frac{\psi_{l,ijk} - \psi_{l,ijk-1}}{\Delta z}$$

$$+ \Big(\sigma_{i-\frac{1}{2}j-\frac{1}{2}k-\frac{1}{2}} + \kappa_{i-\frac{1}{2}j-\frac{1}{2}k-\frac{1}{2}} \Big) \psi_{l,i-\frac{1}{2}j-\frac{1}{2}k-\frac{1}{2}}$$

$$= \sigma_{i-\frac{1}{2}j-\frac{1}{2}k-\frac{1}{2}} \phi_{i-\frac{1}{2}j-\frac{1}{2}k-\frac{1}{2}} \kappa_{i-\frac{1}{2}j-\frac{1}{2}k-\frac{1}{2}} q_{i-\frac{1}{2}j-\frac{1}{2}k-\frac{1}{2}}, \tag{50}$$

where the cell averages values of ψ are given by

$$\psi_{l,i-1jk} = \frac{1}{\Delta x} \int_{y_{j-1}}^{y_j} \int_{z_{k-1}}^{z_k} \psi_l(x_i,y,z)dydz,$$

$$\psi_{l,ij-1k} = \frac{1}{\Delta y} \int_{x_{i-1}}^{x_i} \int_{z_{k-1}}^{z_k} \psi_l(x,y_j,z)dxdz,$$

$$\psi_{l,ijk-1} = \frac{1}{\Delta z} \int_{x_{i-1}}^{x_i} \int_{y_{j-1}}^{y_j} \psi_l(x,y,z_k)dxdy,$$

$$\psi_{l,ijk} = \frac{1}{\Delta x \Delta y \Delta z} \int_{x_{i-1}}^{x_i} \int_{y_{j-1}}^{y_j} \int_{z_{k-1}}^{z_k} \psi_l(x,y,z)dxdydz, \tag{51}$$

In this paper we use the Diamond difference method to approximate the fluxes in (51). The method consists on centred differences and approximating the function values at the cell centres $f_{l,i-\frac{1}{2}j-\frac{1}{2}k-\frac{1}{2}}$ by the average of their values at the eight neighbouring nodes as

$$f_{l,i-\frac{1}{2}j-\frac{1}{2}k-\frac{1}{2}} = \frac{1}{8} \Big[f_{l,i-1j-1k-1} + f_{l,i-1jk-1} + f_{l,i-1j-1k} + f_{l,i-1jk}$$
$$+ f_{l,ij-1k-1} + f_{l,ijk-1} + f_{l,ij-1k} + f_{l,ijk} \Big]. \tag{52}$$

Hence the discrete mean intensity $\phi_{i-\frac{1}{2}j-\frac{1}{2}k-\frac{1}{2}}$ in (50) is given by

$$\phi_{i-\frac{1}{2}j-\frac{1}{2}k-\frac{1}{2}} = \sum_{l=1}^{N} w_l \psi_{l,i-\frac{1}{2}j-\frac{1}{2}k-\frac{1}{2}}.$$

Other discretizations using Legendre polynomial collocation in ordinates and finite element or Petrov–Galerkin methods in space can be used in the same manner, we refer to [10, 52, 81, 82] for details. For the discretization of the boundary conditions in (49) we can proceed as follows:

when $\hat{x} = a_x$, the normal $n(\hat{x}_{0jk}) = (-1,0,0)^T$, then $n(\hat{x}_{0jk}) \cdot \Omega_l = -\mu_l$, and for $\mu_l > 0$ we have $\psi_{l,0jk} = g_{0jk}$
when $\hat{y} = a_y$, the normal $n(\hat{x}_{i0k}) = (0,-1,0)^T$, then $n(\hat{x}_{i0k}) \cdot \Omega_l = -\eta_l$, and for $\eta_l > 0$ we have $\psi_{l,i0k} = g_{i0k}$
when $\hat{z} = a_z$, the normal $n(\hat{x}_{ij0}) = (0,0,-1)^T$, then $n(\hat{x}_{ij0}) \cdot \Omega_l = -\xi_l$, and for $\xi_l > 0$ we have $\psi_{l,ij0} = g_{ij0}$

The other three cases can be discretized in a similar way. Needless to say that for a given $l = 1,2,\ldots,N$ no component of Ω_l is ever zero and only three of the above six cases can hold. Furthermore, in the discretization (50) there are $(I+1)(J+1)(K+1)$ unknowns $\psi_{l,ijk}$ and $IK + JI + JK + I + J + K + 1$ boundary equations.

3.2 Linear System Formulation

In order to simplify the notations and to get closer to a compact linear algebra formulation of (50), we first define the matrix entries

$$d_{l,i-\frac{1}{2}j-\frac{1}{2}k-\frac{1}{2}} = \frac{|\mu_l|}{\Delta x} + \frac{|\eta_l|}{\Delta y} + \frac{|\xi_l|}{\Delta z} + \frac{\sigma_{i-\frac{1}{2}j-\frac{1}{2}k-\frac{1}{2}} + \kappa_{i-\frac{1}{2}j-\frac{1}{2}k-\frac{1}{2}}}{8},$$

$$e_{l,i-\frac{1}{2}j-\frac{1}{2}k-\frac{1}{2}} = \frac{-|\mu_l|}{\Delta x} + \frac{-|\eta_l|}{\Delta y} + \frac{-|\xi_l|}{\Delta z} + \frac{\sigma_{i-\frac{1}{2}j-\frac{1}{2}k-\frac{1}{2}} + \kappa_{i-\frac{1}{2}j-\frac{1}{2}k-\frac{1}{2}}}{8}.$$

and

$$u_{l,i-\frac{1}{2}j-\frac{1}{2}k-\frac{1}{2}} = \frac{|\mu_l|}{\Delta x} + \frac{|\eta_l|}{\Delta y} + \frac{-|\xi_l|}{\Delta z} + \frac{\sigma_{i-\frac{1}{2}j-\frac{1}{2}k-\frac{1}{2}} + \kappa_{i-\frac{1}{2}j-\frac{1}{2}k-\frac{1}{2}}}{8},$$

$$\bar{u}_{l,i-\frac{1}{2}j-\frac{1}{2}k-\frac{1}{2}} = \frac{|\mu_l|}{\Delta x} + \frac{-|\eta_l|}{\Delta y} + \frac{|\xi_l|}{\Delta z} + \frac{\sigma_{i-\frac{1}{2}j-\frac{1}{2}k-\frac{1}{2}} + \kappa_{i-\frac{1}{2}j-\frac{1}{2}k-\frac{1}{2}}}{8},$$

$$\underline{u}_{l,i-\frac{1}{2}j-\frac{1}{2}k-\frac{1}{2}} = \frac{|\mu_l|}{\Delta x} + \frac{-|\eta_l|}{\Delta y} + \frac{-|\xi_l|}{\Delta z} + \frac{\sigma_{i-\frac{1}{2}j-\frac{1}{2}k-\frac{1}{2}} + \kappa_{i-\frac{1}{2}j-\frac{1}{2}k-\frac{1}{2}}}{8},$$

$$v_{l,i-\frac{1}{2}j-\frac{1}{2}k-\frac{1}{2}} = \frac{-|\mu_l|}{\Delta x} + \frac{|\eta_l|}{\Delta y} + \frac{|\xi_l|}{\Delta z} + \frac{\sigma_{i-\frac{1}{2}j-\frac{1}{2}k-\frac{1}{2}} + \kappa_{i-\frac{1}{2}j-\frac{1}{2}k-\frac{1}{2}}}{8},$$

$$\bar{v}_{l,i-\frac{1}{2}j-\frac{1}{2}k-\frac{1}{2}} = \frac{-|\mu_l|}{\Delta x} + \frac{|\eta_l|}{\Delta y} + \frac{-|\xi_l|}{\Delta z} + \frac{\sigma_{i-\frac{1}{2}j-\frac{1}{2}k-\frac{1}{2}} + \kappa_{i-\frac{1}{2}j-\frac{1}{2}k-\frac{1}{2}}}{8},$$

$$\underline{v}_{l,i-\frac{1}{2}j-\frac{1}{2}k-\frac{1}{2}} = \frac{-|\mu_l|}{\Delta x} + \frac{-|\eta_l|}{\Delta y} + \frac{|\xi_l|}{\Delta z} + \frac{\sigma_{i-\frac{1}{2}j-\frac{1}{2}k-\frac{1}{2}} + \kappa_{i-\frac{1}{2}j-\frac{1}{2}k-\frac{1}{2}}}{8},$$

Next, we define the vectors

$$\Psi_l \equiv \begin{pmatrix} \Psi_{l,0} \\ \vdots \\ \Psi_{l,K} \end{pmatrix} \in \mathbf{R}^{(I+1)(J+1)(K+1)}, \quad \text{with}$$

$$\Psi_{l,k} \equiv \begin{pmatrix} \Psi_{l,0k} \\ \vdots \\ \Psi_{l,Jk} \end{pmatrix} \in \mathbf{R}^{(I+1)(J+1)}, \qquad \Psi_{l,jk} \equiv \begin{pmatrix} \psi_{l,0jk} \\ \vdots \\ \psi_{l,Ijk} \end{pmatrix} \in \mathbf{R}^{(I+1)};$$

$$\Phi \equiv \begin{pmatrix} \Phi_{1-\frac{1}{2}} \\ \vdots \\ \Phi_{K-\frac{1}{2}} \end{pmatrix} \in \mathbf{R}^{IJK}, \quad \text{with}$$

$$\Phi_{k-\frac{1}{2}} \equiv \begin{pmatrix} \Phi_{1-\frac{1}{2}k-\frac{1}{2}} \\ \vdots \\ \Phi_{J-\frac{1}{2}k-\frac{1}{2}} \end{pmatrix} \in \mathbf{R}^{IJ}, \qquad \Phi_{j-\frac{1}{2}k-\frac{1}{2}} \equiv \begin{pmatrix} \phi_{1-\frac{1}{2}j-\frac{1}{2}k-\frac{1}{2}} \\ \vdots \\ \phi_{I-\frac{1}{2}j-\frac{1}{2}k-\frac{1}{2}} \end{pmatrix} \in \mathbf{R}^{I};$$

$$Q \equiv \begin{pmatrix} Q_{1-\frac{1}{2}} \\ \vdots \\ Q_{K-\frac{1}{2}} \end{pmatrix} \in \mathbf{R}^{IJK}, \quad \text{with}$$

$$Q_{k-\frac{1}{2}} \equiv \begin{pmatrix} Q_{1-\frac{1}{2}k-\frac{1}{2}} \\ \vdots \\ Q_{J-\frac{1}{2}k-\frac{1}{2}} \end{pmatrix} \in \mathbf{R}^{IJ}, \qquad Q_{j-\frac{1}{2}k-\frac{1}{2}} \equiv \begin{pmatrix} q_{1-\frac{1}{2}j-\frac{1}{2}k-\frac{1}{2}} \\ \vdots \\ q_{I-\frac{1}{2}j-\frac{1}{2}k-\frac{1}{2}} \end{pmatrix} \in \mathbf{R}^{I};$$

In what follows we define the matrix H_l (known as sweep matrix) for the first sweep case $\mu_l < 0$, $\eta_l < 0$, $\xi_l < 0$ and the other seven sweep cases can be derived similarly. In order to simplify the notation, we drop hereafter the space grids subscripts from the matrix entries unless otherwise stated. Thus,

$$H_l \equiv \begin{pmatrix} D_l & E_l & & & \\ & \ddots & \ddots & & \\ & & D_l & E_l & \\ & & & D_l & S \\ & & & & S \end{pmatrix} \in \mathbf{R}^{(I+1)(J+1)(K+1) \times (I+1)(J+1)(K+1)}, \qquad \text{with}$$

$$D_l \equiv \begin{pmatrix} D_l & U_l & & \\ & \ddots & \ddots & \\ & & D_l & U_l \\ & & & S \end{pmatrix} \in \mathbf{R}^{(I+1)(J+1) \times (I+1)(J+1)}, \qquad \text{with}$$

$$D_l \equiv \begin{pmatrix} d & u & & \\ & \ddots & \ddots & \\ & & d & u \\ & & & 1 \end{pmatrix} \in \mathbf{R}^{(I+1) \times (I+1)}, \quad U_l \equiv \begin{pmatrix} \underline{u} & \bar{u} & & \\ & \ddots & \ddots & \\ & & \underline{u} & \bar{u} \\ & & & 1 \end{pmatrix} \in \mathbf{R}^{(I+1) \times (I+1)};$$

$$E_l \equiv \begin{pmatrix} V_l & W_l & & \\ & \ddots & \ddots & \\ & & V_l & W_l \\ & & & S \end{pmatrix} \in \mathbf{R}^{(I+1)(J+1) \times (I+1)(J+1)}, \qquad \text{with}$$

$$V_l \equiv \begin{pmatrix} \underline{v} & \bar{v} & & \\ & \ddots & \ddots & \\ & & \underline{v} & \bar{v} \\ & & & 1 \end{pmatrix} \in \mathbf{R}^{(I+1) \times (I+1)}, \quad W_l \equiv \begin{pmatrix} v & \bar{e} & & \\ & \ddots & \ddots & \\ & & v & \bar{e} \\ & & & 1 \end{pmatrix} \in \mathbf{R}^{(I+1) \times (I+1)}.$$

$$S \equiv \begin{pmatrix} S & S & & \\ & \ddots & \ddots & \\ & & S & S \\ & & & S \end{pmatrix} \in \mathbf{R}^{(I+1)(J+1)\times(I+1)(J+1)}, \quad \text{with}$$

$$S \equiv \begin{pmatrix} 1 & 1 & & \\ & \ddots & \ddots & \\ & & 1 & 1 \\ & & & 1 \end{pmatrix} \in \mathbf{R}^{(I+1)\times(J+1)}.$$

$$\Sigma_l \equiv \begin{pmatrix} \Sigma_{l,1-\frac{1}{2}} & & & \\ & \ddots & & \\ & & \Sigma_{l,K-\frac{1}{2}} & \\ & & & 0 \end{pmatrix} \in \mathbf{R}^{(I+1)(J+1)(K+1)\times IJK}, \qquad \text{with}$$

$$\Sigma_{l,k-\frac{1}{2}} \equiv \begin{pmatrix} \Sigma_{l,1-\frac{1}{2}k-\frac{1}{2}} & & & \\ & \ddots & & \\ & & \Sigma_{l,J-\frac{1}{2}k-\frac{1}{2}} & \\ & & & 0 \end{pmatrix} \in \mathbf{R}^{(I+1)(J+1)\times IJ}, \quad \text{and}$$

$$\Sigma_{l,j-\frac{1}{2}k-\frac{1}{2}} \equiv$$

$$\begin{pmatrix} \frac{\sigma_{1-\frac{1}{2}j-\frac{1}{2}k-\frac{1}{2}}+\kappa_{i-\frac{1}{2}j-\frac{1}{2}k-\frac{1}{2}}}{8} & & & \\ & \ddots & & \\ & & \frac{\sigma_{i-\frac{1}{2}j-\frac{1}{2}k-\frac{1}{2}}+\kappa_{i-\frac{1}{2}j-\frac{1}{2}k-\frac{1}{2}}}{8} & \\ & & & 0 \end{pmatrix} \in \mathbf{R}^{(I+1)\times I}.$$

Using these definitions with Ψ and Φ being the unknowns, the fully discrete equation (50) can be written in matrix form as

$$\left(\begin{array}{ccc|c} H_1 & & & -\Sigma_1 \\ & \ddots & & \vdots \\ & & H_N & -\Sigma_N \\ \hline -\frac{w_1}{4\pi}S & \cdots & -\frac{w_N}{4\pi}S & I \end{array} \right) \begin{pmatrix} \Psi_1 \\ \vdots \\ \Psi_N \\ \Phi \end{pmatrix} = \begin{pmatrix} Q_1 \\ \vdots \\ Q_N \\ 0 \end{pmatrix}, \tag{53}$$

where I is the $IJK \times IJK$ identity matrix and 0 is the IJK null vector. The linear system (53) can be rewritten in common linear algebra notation as

$$Ax = b \tag{54}$$

with

$$
A \equiv \left(\begin{array}{ccc|c} H_1 & & & -\Sigma_1 \\ & \ddots & & \vdots \\ & & H_N & -\Sigma_N \\ \hline -\frac{w_1}{4\pi}S & \cdots & -\frac{w_N}{4\pi}S & I \end{array} \right), \quad x \equiv \left(\begin{array}{c} \Psi_1 \\ \vdots \\ \Psi_N \\ \Phi \end{array} \right), \quad \text{and} \quad b \equiv \left(\begin{array}{c} Q_1 \\ \vdots \\ Q_N \\ 0 \end{array} \right).
$$

3.3 Preconditioning Techniques

In computational radiative transfer the desired quantity is usually the mean intensity Φ which is a function only of position x. Therefore we use the Gaussian elimination to eliminate the intensity $\Psi_1, \ldots, \Psi_N$ from (53) and the reduced equation

$$
\left(I - \frac{1}{4\pi} \sum_{l=1}^{N} w_l S H_l^{-1} \Sigma_l \right) \Phi = \frac{1}{4\pi} \sum_{l=1}^{N} w_l S H_l^{-1} Q_l, \tag{55}
$$

is solved for Φ. We rewrite (55) in compact form as

$$
\left(I - \mathscr{A} \right) \Phi = f, \tag{56}
$$

where the Schur matrix $\mathscr{A}$ and the right hand side f are given by

$$
\mathscr{A} = \frac{1}{4\pi} \sum_{l=1}^{N} w_l S H_l^{-1} \Sigma_l \quad \text{and} \quad f = \frac{1}{4\pi} \sum_{l=1}^{N} w_l S H_l^{-1} Q_l.
$$

In this section we briefly discuss some numerical methods used in the literature to solve the linear system (56).

3.3.1 Source Iteration

The most popular iterative method to solve (55) is the Richardson iteration known in the radiative transfer community as Source Iteration (SI) method. Given an initial guess $\Phi^{(0)}$, the $(m+1)$-iterate solution is obtained by

$$
\Phi^{(m+1)} = \frac{1}{4\pi} \sum_{l=1}^{N} w_l S H_l^{-1} \left(Q_l + \Sigma_l \Phi^{(m)} \right), \quad m = 0, 1, \ldots. \tag{57}
$$

It is easy to see that iteration (57) is equivalent to preconditioned block Gauss–Seidel method applied to (54), where the preconditioner is the block lower triangle of the matrix A. Thus, if M is the block lower triangle of A,

$$M \equiv \begin{pmatrix} H_1 & & & \\ & \ddots & & \\ & & H_N & \\ -\frac{w_1}{4\pi}S & \cdots & -\frac{w_N}{4\pi}S & I \end{pmatrix},$$

x and b are as given in (54), then

$$Mx^{(m+1)} = (M - A)x^{(m)} + b,$$

and

$$x^{(m+1)} = (I - M^{-1}A)x^{(m)} + M^{-1}b. \tag{58}$$

Therefore the $(m+1)$-iterate mean intensity satisfy

$$\Phi^{(m+1)} = \frac{1}{4\pi}\sum_{l=1}^{N} w_l SH_l^{-1}\Psi_l^{(m+1)} = \frac{1}{4\pi}\sum_{l=1}^{N} w_l SH_l^{-1}\left(Q_l + \Sigma_l \Phi^{(m)}\right),$$

which is identical to (57).

Formal results from linear algebra [29, 35] demonstrate that the preconditioned Richardson iteration (58) converges rapidly as long as the norm of the matrix $(I - M^{-1}A)$ is small. This condition is ensured by taking the scattering ratio γ small, compare [1] for analysis. For $\gamma \ll 1$ the SI method converges rapidly, but for $\gamma \approx 1$ (large optical opacity) convergence becomes slow and may restrict the efficiency of the SI algorithm. The SI algorithm can be implemented as follows

Algorithm 1: SI algorithm

given the initial guess $\Phi^{(0)}$
do $m = 0, 1, \ldots, itmax$
do $l = 1, 2, \ldots, N$
a.   set $w_l \longleftarrow Q_l + \Sigma_l \Phi^{(m)}$
b.   solve for y_l: $H_l y_l = w_l$
c.   set $w_l \longleftarrow Sy_l$
end do
d.   compute $\Phi^{(m+1)} = \frac{1}{4\pi}\sum_{l=1}^{N} w_l w_l$
e.   compute $r^{(m)} = \Phi^{(m+1)} - \Phi^{(m)}$
if $\left(\frac{\|r^{(m)}\|_{L^2}}{\|r^{(0)}\|_{L^2}} \leq \tau\right)$ stop
end do

Here $itmax$ is the maximum number of the iterations m, τ is a given tolerance and $\|.\|_{L^2}$ is the discrete L^2-norm. The step (b) can be solved directly using Gaussian elimination known in computational radiative transfer as sweeping procedure. Additionally, for each direction Ω_l in S_N only one of the eight possible sweeps is needed.

3.3.2 Diffusion Synthetic Acceleration

Among the methods used to accelerate the SI algorithm are the synthetic acceleration procedures [1,4]. The procedures consist on splitting the SI in to two-step iterations. Thus, we denote $\psi^{(m+\frac{1}{2})}$ as first iteration for the SI in the continuous form of the problem (45),

$$\Omega \cdot \nabla \psi^{(m+\frac{1}{2})} + (\sigma + \kappa)\psi^{(m+\frac{1}{2})} = \frac{\sigma}{4\pi} \int_{S^2} \psi^{(m)}(x, \Omega')d\Omega' + \kappa q(x),$$
$$\psi^{(m+\frac{1}{2})}(\hat{x}, \Omega) = g(\hat{x}), \tag{59}$$

and an equation for $\psi^{(m+1)}$ is required in such a way to be more accurate approximation to ψ than $\psi^{(m+\frac{1}{2})}$. To perform this step with synthetic acceleration method, we subtract (59) from (45),

$$\Omega \cdot \nabla \left(\psi - \psi^{(m+\frac{1}{2})}\right) + (\sigma + \kappa)\left(\psi - \psi^{(m+\frac{1}{2})}\right) = \frac{\sigma}{4\pi} \int_{S^2} \left(\psi - \psi^{(m)}\right)(x, \Omega')d\Omega',$$
$$\left(\psi - \psi^{(m+\frac{1}{2})}\right)(\hat{x}, \Omega) = 0, \tag{60}$$

then (60) are replaced by an approximate problem. The Diffusion Synthetic Acceleration (DSA) method [4] approximates the (60) by the diffusion problem

$$-\nabla \cdot \left(\frac{1}{3(\sigma + \kappa)}\nabla \varphi\right) + \kappa \varphi = \frac{\sigma}{4\pi} \int_{S^2} \left(\psi - \psi^{(m)}\right)(x, \Omega')d\Omega', \quad x \in V,$$

Here $\varphi(x)$ is an approximation to the mean intensity

$$\varphi(x) \approx \frac{1}{4\pi} \int_{S^2} \left(\psi^{(m+1)} - \psi^{(m+\frac{1}{2})}\right)(x, \Omega')d\Omega'.$$

Thus the $(m+1)$-iterate mean intensity is given by

$$\phi^{(m+1)} = \phi^{(m+\frac{1}{2})} + \varphi.$$

Note that (61) does not depend on the angle variable Ω, is linear elliptic equation and simple to solve numerically with less computational cost and memory requirement.

In order to build a discretization for the diffusion problem (61) which is consistent to the one used for the radiative transfer equation (45), we consider the same grid structure and the same notations as those used in Sect. 2. Hence a space discretization for the (61) reads as

$$-\mathcal{D}_h^2 \left(\frac{1}{3(\sigma + \kappa)}\varphi\right)_{ijk} + \kappa_{i-\frac{1}{2}j-\frac{1}{2}k-\frac{1}{2}}\varphi_{i-\frac{1}{2}j-\frac{1}{2}k-\frac{1}{2}} = P_{i-\frac{1}{2}j-\frac{1}{2}k-\frac{1}{2}}, \tag{61}$$

where $p_{i-\frac{1}{2}j-\frac{1}{2}k-\frac{1}{2}} = \sigma_{i-\frac{1}{2}j-\frac{1}{2}k-\frac{1}{2}}\left(\phi_{i-\frac{1}{2}j-\frac{1}{2}k-\frac{1}{2}}^{(m+\frac{1}{2})} - \phi_{i-\frac{1}{2}j-\frac{1}{2}k-\frac{1}{2}}^{(m)}\right)$ and the difference

operator $\mathscr{D}_h^2$ is given by $\mathscr{D}_h^2 = \mathscr{D}_x^2 + \mathscr{D}_y^2 + \mathscr{D}_z^2$, with

$$\mathscr{D}_x^2(\beta\varphi)_{ijk} = \frac{\beta_{ijk}+\beta_{i+1jk}}{2}\frac{\varphi_{i+1jk}-\varphi_{ijk}}{(\Delta x)^2} - \frac{\beta_{i-1jk}+\beta_{ijk}}{2}\frac{\varphi_{ijk}-\varphi_{i-1jk}}{(\Delta x)^2},$$

$$\mathscr{D}_y^2(\beta\varphi)_{ijk} = \frac{\beta_{ijk}+\beta_{ij+1k}}{2}\frac{\varphi_{ij+1k}-\varphi_{ijk}}{(\Delta y)^2} - \frac{\beta_{ij-1k}+\beta_{ijk}}{2}\frac{\varphi_{ijk}-\varphi_{ij-1k}}{(\Delta y)^2},$$

$$\mathscr{D}_z^2(\beta\varphi)_{ijk} = \frac{\beta_{ijk}+\beta_{ijk+1}}{2}\frac{\varphi_{ijk+1}-\varphi_{ijk}}{(\Delta z)^2} - \frac{\beta_{ijk-1}+\beta_{ijk}}{2}\frac{\varphi_{ijk}-\varphi_{ijk-1}}{(\Delta z)^2}.$$

The functions $\kappa_{i-\frac{1}{2}j-\frac{1}{2}k-\frac{1}{2}}$, $\varphi_{i-\frac{1}{2}j-\frac{1}{2}k-\frac{1}{2}}$ and $p_{i-\frac{1}{2}j-\frac{1}{2}k-\frac{1}{2}}$ appeared in (61) are given as in formula (52). The gradient in the boundary conditions is approximated by upwinding without using ghost points. For example, on the boundary surface $x = a_x$ of the domain V, the boundary discretization is

$$\varphi_{\frac{1}{2}j+\frac{1}{2}k+\frac{1}{2}} - \frac{2}{3\left(\sigma_{\frac{1}{2}j-\frac{1}{2}k-\frac{1}{2}}+\kappa_{\frac{1}{2}j-\frac{1}{2}k-\frac{1}{2}}\right)}\frac{\varphi_{\frac{3}{2}j-\frac{1}{2}k-\frac{1}{2}}-\varphi_{\frac{1}{2}j-\frac{1}{2}k-\frac{1}{2}}}{\Delta x} = 0,$$

and similar work has to be done for the other boundaries. All together, the above discretization leads to a linear system of form

$$D\varphi = p, \tag{62}$$

where D is $IJK \times IJK$ nonsymmetric positive definite matrix obtained from the difference diffusion operator (61) with boundary conditions included, and p is IJK vector containing the right hand side term.

Once again the DSA method can be viewed as preconditioned Richardson iteration for the linear system (54) with the diffusion matrix D like preconditioner,

$$x^{(m+1)} = \left(I - D^{-1}A\right)x^{(m)} + D^{-1}b,$$

and D^{-1} is obtained by solving the diffusion linear system (62). In terms of Φ this is equivalent to

$$\Phi^{(k+1)} = \left(I - (I - D^{-1})\mathscr{A}\right)\Phi^{(k)} + (I - D^{-1})b.$$

The implementation of DSA method to approximate the solution of the radiative transfer equation (45) is carried out in the following algorithm

Algorithm 2: DSA algorithm

given the initial guess $\Phi^{(0)}$
do $m = 0, 1, \ldots, itmax$
(a) – (d) are similar to Algorithm 1 for the intermediate solution $\Phi^{(m+\frac{1}{2})}$

(e)   compute $p = \sigma\left(\Phi^{(m+\frac{1}{2})} - \Phi^{(m)}\right)$

(f)   solve for φ: $D\varphi = p$

(g)   set $\Phi^{(m+1)} = \Phi^{(m+\frac{1}{2})} + \varphi$

(h)   compute $r^{(m)} = \Phi^{(m+1)} - \Phi^{(m)}$

if $\left(\dfrac{\|r^{(m)}\|_{L^2}}{\|r^{(0)}\|_{L^2}} \le \tau\right)$ stop

end do

Note that the first lines in `Algorithm` 2 are similar to the `Algorithm` 1. However, the source iteration algorithm gives only the intermediate solution $\Phi^{(m+\frac{1}{2})}$ which has to be corrected by adding the solution φ obtained by the diffusion approach. Furthermore, if iterative methods are used for the diffusion approach, then an inner iteration loop has to be added to the iteration used by the SI algorithm and an outer SI iteration may require less accuracy from the inner iterations.

3.3.3   Krylov Subspace Methods

In general the matrices A and $\mathscr{A}$ in (54) and (56) respectively are nonsymmetric and not diagonally dominant. Furthermore, since σ and κ are nonnegative functions and S_N has nonzero directions, the matrix A has positive diagonal elements and nonpositive off-diagonal elements. In addition, if $e_{l,i-\frac{1}{2}j-\frac{1}{2}k-\frac{1}{2}} \le 0$, for all l, i, j, k, then the matrix A is weakly diagonally dominant. This condition is equivalent to

$$h = \max(\Delta x, \Delta y, \Delta z) \le \max_{ijk} \left(\frac{8|\mu_l|}{\sigma_{i-\frac{1}{2}j-\frac{1}{2}k-\frac{1}{2}} + \kappa_{i-\frac{1}{2}j-\frac{1}{2}k-\frac{1}{2}}}, \right.$$
$$\left. \frac{8|\eta_l|}{\sigma_{i-\frac{1}{2}j-\frac{1}{2}k-\frac{1}{2}} + \kappa_{i-\frac{1}{2}j-\frac{1}{2}k-\frac{1}{2}}}, \frac{8|\xi_l|}{\sigma_{i-\frac{1}{2}j-\frac{1}{2}k-\frac{1}{2}} + \kappa_{i-\frac{1}{2}j-\frac{1}{2}k-\frac{1}{2}}} \right), \quad \forall\, l, \qquad (63)$$

which means physically that the cell size is no more than eight mean free paths of the particles being simulated. Needless to say that the condition (63) gives the bound of the coarser mesh should be used in the computations.

In this paper we propose two Krylov subspace based methods, namely the BI-Conjugate Gradient Stabilized (Bicgstab) [86] and the Generalized Minimal Residual (Gmres(m)) [71], where m stands for the number of restarts for Krylov subspace used in the orthogonalization. Bicgstab method has been applied early in [82] to solve (56). The main idea behind these approaches is that the Krylov subspace methods can be interpreted as weighted Richardson iteration

$$x^{(m)} = \left(I - \alpha P^{-1}A\right)x^{(m-1)} + \alpha P^{-1}b, \quad 0 < \alpha < 2, \quad m = 1, \ldots, \qquad (64)$$

applied to the linear system (54), where the relaxation parameters α and the preconditioner P are variables within each iteration step. Note that when $\alpha = 1$ and $P = M$ the iteration (64) is reduced to the SI method.

The Bicgstab and Gmres(m) algorithms to solve the linear system (56) can be implemented in the conventional way as in [36,71,86], with the only difference that the dense matrix $\mathscr{A}$ can not be explicitly stored. All what is needed, however, is a subroutine that performs a matrix-vector multiplication as shown in the following algorithm

Algorithm 3 : Matrix-vector multiplication

given a vector u, to apply the matrix $\mathscr{A}$ to u we proceed as
do $l = 1, \ldots, N$
a.   set $v_l \longleftarrow \Sigma_l u$
b.   solve for w_l: $H_l w_l = v_l$
c.   set $v_l \longleftarrow S w_l$
end do
d.   set $u \longleftarrow u - \dfrac{1}{4\pi} \displaystyle\sum_{l=1}^{N} w_l v_l$

Note that only three vectors (u, v_l and w_l) are needed to perform the multiplication of the matrix $\mathscr{A}$ to the vector u. Moreover, only three calls for the algorithm 3 are required from the Bicgstab or Gmres(m) subroutines.

Preconditioned Bicgstab or Gmres(m) methods can be also used. For instance, in the case when the matrix $\mathscr{A}$ is diagonally dominant, the Bicgstab or Gmres(m) methods can be accelerated by using the diagonal as a preconditioner. This approach which requires additional computational work can be easily implemented. It is worth to say that incomplete Cholesky or ILU type preconditioners can not be used to solve (56) because the matrix $\mathscr{A}$ is never formed explicitly.

3.4 A Fast Multilevel Preconditioner

We describe in this section multilevel solvers for the linear system (56) using an approximate inverse operator as preconditioner on each level of the multigrid hierarchy. Multilevel methods were first applied to radiative transfer problems in [36]. The author proposes two different type of smoothings to approximate solutions for the one dimensional version of (45) in slab geometry.

To formulate multilevel solvers we first modify our notation slightly. Using the discretizations introduced in Sect. 2 we assume for simplicity, a given sequence of uniform, equidistant nested grids

$$V_1 \subset V_2 \subset \cdots \subset V_{L-1} \subset V_L = V_h,$$

on V with respective mesh sizes $\Delta x = \Delta y = \Delta z = 2^{-l}$, $l = 1, \ldots, L$. We use the subscripts l and L to refer to the coarse and fine level respectively. Therefore the problem statement (56) becomes

$$(I - \mathscr{A}_L)\Phi_L = f_L. \tag{65}$$

With M being the iteration matrix, multilevel can be written as follows

$$M\Phi_L^{(m+1)} + \left(I - \mathscr{A}_L - M\right)\Phi_L^{(m)} = f_L.$$

This formulation is equivalent to,

$$\Phi_L^{(m+1)} = \Phi_L^{(m)} + M^{-1}\left(f_L - (I - \mathscr{A}_L)\Phi_L^{(m)}\right) = \Phi_L^{(m)} + M^{-1}r^{(m)},$$

where r denotes the residual associated to (65) and is defined by

$$r = f_L - (I - \mathscr{A}_L)\Phi_L.$$

The preconditioner we consider in this section is the Atkinson–Brakhage approximate inverse [8] given as

$$M^{-1} = B_l^L = I + (I - \mathscr{A}_l)^{-1}\mathscr{A}_L. \tag{66}$$

Then the $(m+1)$-iterate solution for (65) is simply

$$\Phi_L^{(m+1)} = \Phi_L^{(m)} + B_l^L r^{(m)}.$$

Analysis of convergence for this kind of multilevel methods has been done in [36]. The central ideas in this analysis are the strong convergence and collective compactness of the operators generated by $\mathscr{A}_l$.

Let us first define the two-level (Grid2) algorithm. Applying the multilevel preconditioner (66) to the problem (65) we need the fine-to-coarse grid transfer operator $\mathscr{R}_L^l$ defined by

$$\mathscr{R}_L^l \Phi_L = \Phi_l,$$

and the coarse-to-fine grid transfer operator $\mathscr{P}_l^L$ defined by

$$\mathscr{P}_l^L \Phi_l = \Phi_L.$$

A natural way to choose these operators is, bilinear interpolation for $\mathscr{P}_l^L$ and simple injection for $\mathscr{R}_L^l$ as in the standard multigrid literature [30]. However, for radiative transfer equation with discontinuous variables these operators have to be changed to those given in [3] which are specially designed for problems with jumping coefficients. Note that to use these operators we require that any discontinuity of κ, σ or q in (45) is a spatial mesh point.

The Grid2 algorithm for solving (65) is detailed in the following steps

Algorithm 4: Grid2 algorithm

given the fine level $\{L, \mathscr{A}_L, f_L\}$, the coarse level $\{l, \mathscr{A}_l, f_l\}$, the initial guess $\Phi_L^{(0)}$ and the tolerance τ
do $m = 0, 1, \ldots, itmax$

a.   compute the residual $r^{(m)} = f_L - (I - \mathscr{A}_L)\Phi_L^{(m)}$

b.   set $u_L \longleftarrow \mathscr{A}_L r^{(m)}$

c.   restriction $u_l \longleftarrow \mathscr{R}_L^l u_L$

d.   solve on the coarse level for v_l: $(I - \mathscr{A}_l)v_l = u_l$
     Nyström interpolation

e.   compute $w_l \longleftarrow \mathscr{A}_l v_l$

f.   prolongation $w_L \longleftarrow \mathscr{P}_l^L w_l$

g.   set $z_L \longleftarrow u_L + w_L$

h.   compute the preconditioner $p \longleftarrow r + z_L$

i.   update the solution $\Phi_L^{(m+1)} = \Phi_L^{(m)} + p$

j.   compute the residual $r^{(m+1)} = f_L - (I - \mathscr{A}_L)\Phi_L^{(m+1)}$

$\quad$ if $\left(\dfrac{\|r^{(m+1)}\|_{L^2}}{\|r^{(0)}\|_{L^2}} \leq \tau \right)$ stop

end do

The step (d) usually solves the coarse problem exaclty using direct methods. However, since $\mathscr{A}_l$ is dense matrix which is never explicitly computed nor stored, iterative solvers are required to perfom the step (d). In our numerical examples we used Gmres(m) method which has been discussed in Sect. 4. Note that in our context the notation Grid2 for two-level algorithm does not necessarily mean that we consider two levels of mesh refinements. Thus, `Algorithm 4` is also applicable in cases where we have two different space discretizations on the same mesh $(L \neq l + 1)$.

The fully multilevel algorithm (Gridnest) or nested iteration as refered to in [30] can be implemented recursively as follows

`Algorithm 5`: Gridnest algorithm

$\quad$ given the finest level $\{l_{\max}, \mathscr{A}_{l_{\max}}, f_{l_{\max}}\}$, the coarsest level
$\quad$ $\{l_{\min}, \mathscr{A}_{l_{\min}}, f_{l_{\min}}\}$ and the tolerances $\{\tau_l\}$, $l = l_{\min}, \dots, l_{\max}$

a.   Solve on the coarsest level for $\Phi_{l_{\min}}$: $(I - \mathscr{A}_{l_{\min}})\Phi_{l_{\min}} = f_{l_{\min}}$
     do $k = l_{\min} + 1, \dots, l_{\max}$

b.   set $l \longleftarrow k - 1$

c.   set $L \longleftarrow k$

d.   compute the right hand side f_L

e.   set $\Phi_L^{(0)} \longleftarrow \mathscr{P}_l^L \Phi_l$

f.   set $\tau \longleftarrow \tau_l$

g.   call Grid2 to solve for Φ_L: $(I - \mathscr{A}_L)\Phi_L = f_L$
     end do

Some comments are in order. The steps (b)–(f) are needed only to set the inputs, fine level $\{L, \mathscr{A}_L, f_L\}$, coarse level $\{l, \mathscr{A}_l, f_l\}$, initial guess $\Phi_L^{(0)}$ and tolerance τ to the algorithm Grid2. Recall that Gridnest uses coarse levels to obtain improved initial guesses for fine level problems. The tolerance parameters $\{\tau_l\}$, which determine how many iterations of the multilevel algorithm to do on each level, can be considered either fixed or adaptively chosen during the course of computation.

3.5 Numerical Results

The methods discussed in the above sections are tested on a PC with AMD-K6 200 processors using Fortran compiler, see [76] for details. In all these methods the iterations are terminated when

$$\frac{\|r^{(m)}\|_{L^2}}{\|r^{(0)}\|_{L^2}} \leq 10^{-5}. \tag{67}$$

To solve the diffusion problem in DSA we used a preconditioned Bicgstab whereas, Gmres(10) is used in Grid2 and Gridnest to solve the coarse problems. These inner iterations are stopped as in (67) but with 10^{-2} instead of 10^{-5}. In all our computations for the two space dimension cases we used a discrete S_N-direction set with 60 directions from [22].

3.5.1 Radiative Transfer Equation in 1D Slab Geometry

Our first example is the (45) in 1D slab geometry

$$\mu\frac{\partial\psi}{\partial x} + (\sigma+\kappa)\psi = \frac{\sigma}{2}\int_{-1}^{1}\psi(x,\mu')d\mu' + \kappa q(x) \tag{68}$$
$$\psi(0,\mu) = 1, \quad \mu > 0, \quad \text{and} \quad \psi(1,\mu) = 0, \quad \mu < 0.$$

We set $q = 0$, we used 64 Gauss quadrature nodes in the discrete ordinates and a fine mesh of 512 gridpoints in the space dicrestization. The convergence results for two different values of σ and κ are shown in Fig. 3. The fast convergence of Grid2 is well demonstrated in both cases. Although the scattering ratio for the two cases is 0.99, Grid2 method shows strong reduction of number of iterations comparing to the other methods. Same observation is shown when the regime is optically thick ($\sigma = 99$, $\kappa = 1$).

Next we want to compare the efficiency of these methods in terms of CPU time and number of iterations when the scattering ratio γ runs in the range $(0,1)$. To this end we set $\sigma = 10$ and we vary κ keeping the fine gridpoints fixed to 512. In Fig. 4 we plot the scattering ratio versus the number of iterations in the left and versus the CPU time in the right. Grid2 and Gmres(40) preserve roughly the same amount of computational work (referring to number of iterations and CPU time) in the whole interval, while the SI and DSA become costly for values of γ near 1.

3.5.2 Radiative Transfer Equation with Thermal Source

The second example is the (45) in the unit square with a thermal source $q = B(T)$, with B is the frequency-integrated Planckian $B = a_R T^4$ with a_R is the radiation constant ($a_R = 1.8 \times 10^{-8}$). We fix the temperature to have the linear profile

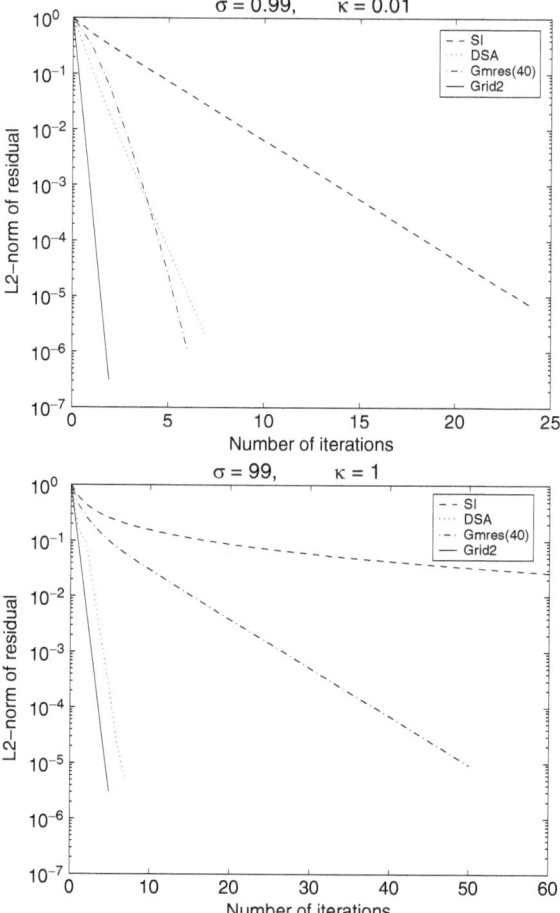

Fig. 3 Convergence plots for the radiative transfer equation in 1D slab geometry

$T(x,y) = 800x + 1,000$ and the boundary function $g(\hat{x},\hat{y}) = B(T(\hat{x},\hat{y}))$. In Fig. 5 we show the convergence plots for different values of σ and κ using a fine mesh of 257×257 gridpoints. In all cases Grid2 algorithm presents faster convergence behaviour than DSA even in the diffusion regime ($\sigma = 100$ and $\kappa = 1$).

We summarize in Table 3 the CPU time consumed for each method to perform the computations with the different values of σ and κ. In Table 4 we report, the number of gridpoints $I \times J$ at each level, the iteration counter m for that level, the number of iterations in Gmres(10) i_{Gmres}, the L^∞-norm of the residual, $\|r^{(m)}\|_{L^\infty}$, and the factor $\|r^{(m)}\|_{L^\infty} / \|r^{(m-1)}\|_{L^\infty}$.

The results presented in Table 4 do not include those obtained for $\sigma = 1$, $\kappa = 1$ or $\sigma = 100$, $\kappa = 1$ because the number of iterations m in Grid2 is very large. For instance, for $\sigma = 100$, $\kappa = 1$ this number surpasses 15 on the coarsest mesh.

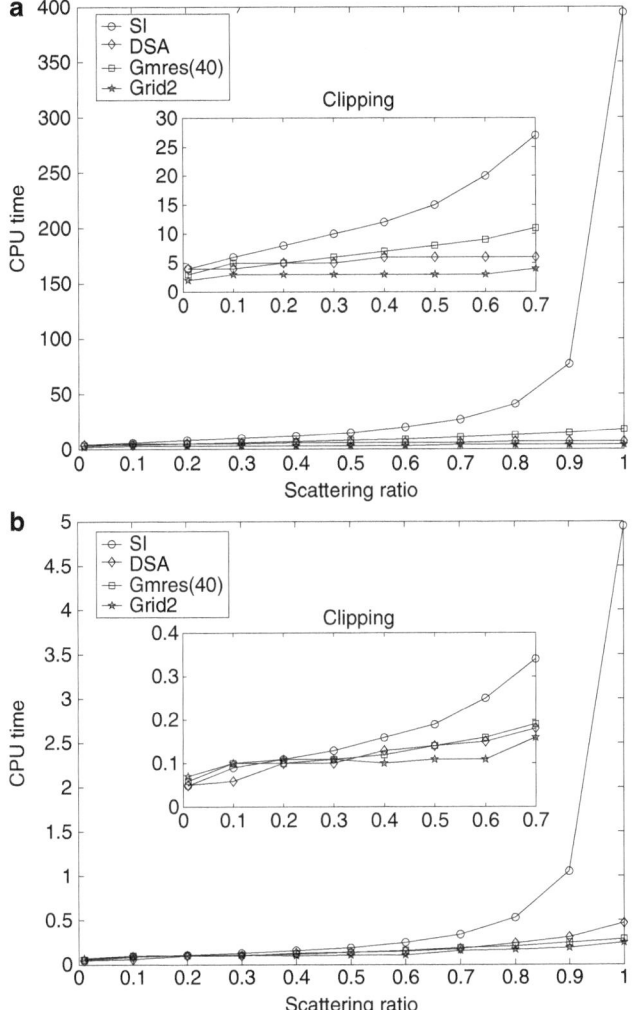

Fig. 4 Plots of scattering ratio γ versus the number of iterations in (**a**) and the CPU time (seconds) in (**b**). In both figures σ is fixed to 10

Nevertheless, in all these test cases we have observed that there is very little variation in the number i_{Gmres} and the reduction factor $\|r^{(m)}\|_{L^\infty}/\|r^{(m-1)}\|_{L^\infty}$ as the meshes are refined.

The results of these tables and Fig. 5 show various interesting features about the behaviours of the preconditioner used by each method. First, where the scattering ratio $\gamma = 0.99$ ($\sigma = 100$ and $\kappa = 1$), it is clear that the SI method is unacceptably slow to converge. The convergence rate is improved significantly by Gmres(40), and it is improved even more by DSA method but at the cost of extra work and storage.

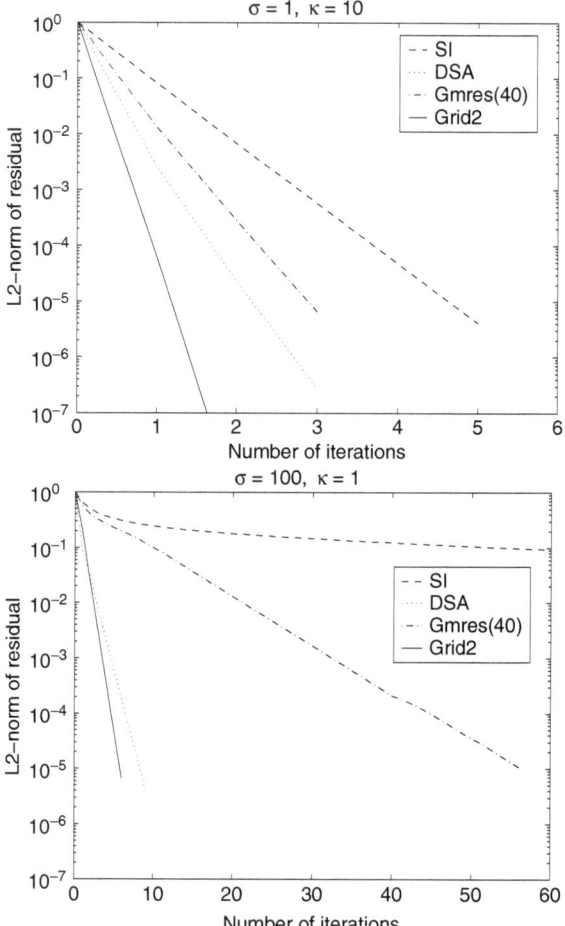

Fig. 5 Convergence plots for the radiative transfer equation with thermal source

Table 3 CPU time (in seconds)

	SI	DSA	Gmres(40)	Grid2
$\sigma = 1, \kappa = 10$	17.8	29.1	15.5	17.1
$\sigma = 1, \kappa = 1$	33.7	77.4	20.8	23.5
$\sigma = 100, \kappa = 1$	2389.8	82.3	174.4	121.7

The most effective method for solving this example, however, is the multilevel Grid2 and Gridnest methods. Second, the number of iterations i_{Gmres} in Gridnset remain nearly the same in all levels and is bounded by the number in the coarsest level.

Table 4 Results obtained by Gridnest for $\sigma = 1$ and $\kappa = 10$ at different levels

$I \times J$	m	i_{Gmres}	$\|r^{(m)}\|_{L^\infty}$	$\|r^{(m)}\|_{L^\infty} / \|r^{(m-1)}\|_{L^\infty}$
33×33	0		0.73E+00	
	1	2	0.32E-01	0.43E-01
	2	2	0.12E-02	0.37E-01
	3	3	0.56E-05	0.46E-04
65×65	0		0.11E-01	
	1	2	0.33E-03	0.30E-01
	2	2	0.12E-05	0.36E-04
129×129	0		0.83E-02	
	1	2	0.67E-03	0.81E-01
	2	2	0.53E-05	0.79E-04
257×257	0		0.22E-02	
	1	2	0.73E-04	0.33E-01
	2	2	0.15E-06	0.20E-04

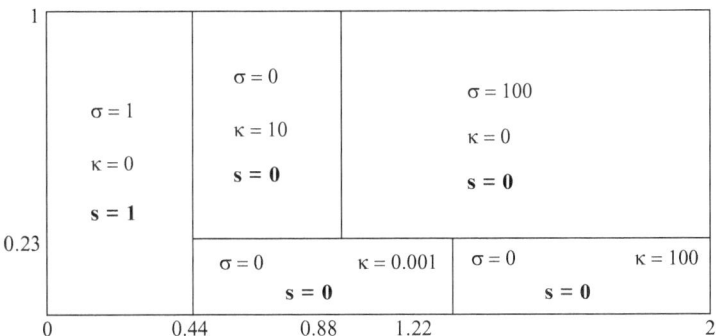

Fig. 6 Geometry and values of σ, κ, $s = \kappa q$ for the discontinuous equation

3.5.3 Radiative Transfer Equation with Discontinuous Variables

The aim of this example is to test the multilevel algorithm for radiative transfer problem with jumping coefficients. Thus, the problem statement is the (45) augmented by discontinuous scattering, absorption and source term [2]. The space domain geometry and the values of σ, κ and $s = \kappa B$ for each subdomain are given in Fig. 6. We take in the first run of this example vacuum boundary conditions whereas, in the second run we use the nonhomogeneous boundary condition g,

$$g(\hat{x}, \hat{y}) = \begin{cases} 1 & \text{when} \quad \hat{x} = 0 \text{ and } 0 \le \hat{y} \le 1, \\ 1 & \text{when} \quad \hat{y} = 0 \text{ and } 0 < \hat{y} \le 0.44, \\ 1 & \text{when} \quad \hat{y} = 1 \text{ and } 0 < \hat{y} \le 0.44, \\ 0 & \text{otherwise.} \end{cases}$$

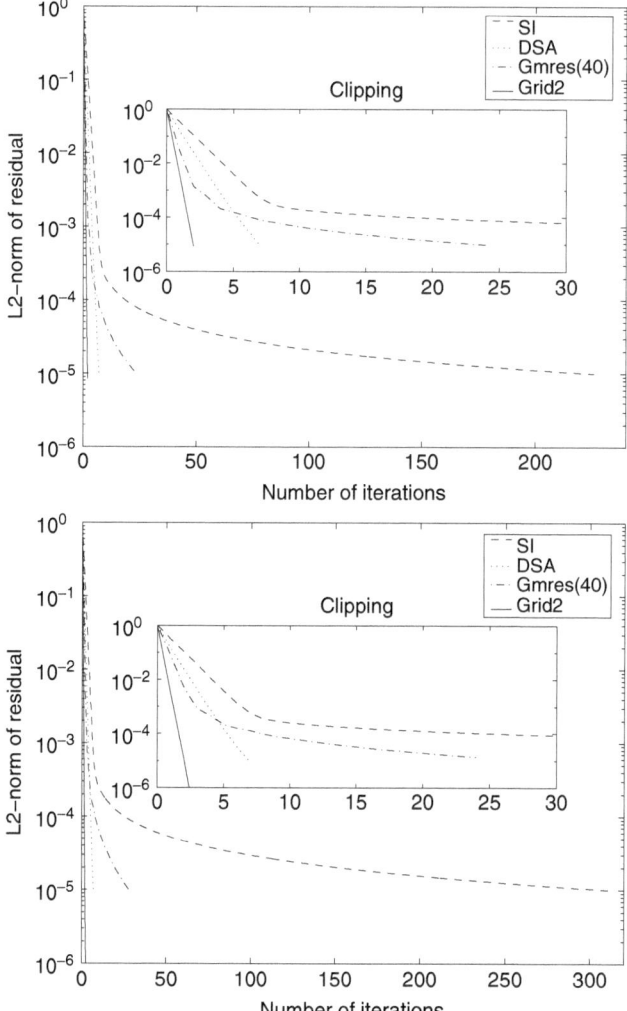

Fig. 7 Convergence rates for the discontinuous equation subject to vacuum boundary conditions (*top*) and nonhomogeneous boundary conditions (*bottom*)

Table 5 CPU time (in seconds)

	SI	DSA	Gmres(40)	Grid2
Vacuum boundary condition	794.04	304.42	192.25	205.22
Nonhomogeneous boundary condition	1100.18	397.53	110.07	201.91

The space domain is discretized uniformly into 400×200 gridpoints at the finest level. We display in Fig. 7 the convergence rates for the two runs. Table 5 provides the running time used by each method for these computations. A simple inspection

Table 6 Gridnest results for the discontinuous problem with vacuum boundary

$I \times J$	m	i_{Gmres}	$\|r^{(m)}\|_{L^\infty}$	$\|r^{(m)}\|_{L^\infty} / \|r^{(m-1)}\|_{L^\infty}$
100×50	0		0.13E+01	
	1	9	0.50E+00	0.39E+00
	2	9	0.13E+00	0.27E+00
	3	8	0.24E-01	0.19E+00
	4	7	0.14E-02	0.57E-01
	5	7	0.45E-05	0.32E-02
200×100	0		0.51E-01	
	1	8	0.10E-02	0.21E-01
	2	7	0.78E-05	0.73E-02
	3	7	0.32E-07	0.41E-02
400×200	0		0.79E-02	
	1	7	0.40E-04	0.51E-02
	2	7	0.13E-06	0.33E-02
800×400	0		0.38E-02	
	1	7	0.24E-04	0.64E-02
	2	7	0.94E-07	0.39E-02

of Fig. 7 shows that Grid2 algorithm solves this problem more effectively than SI or Gmres(40) methods and with less iterations than DSA method.

Furthermore, we note that the Gmres(40) method performs poorly after the seventh iteration in both runs. This may be partly due to the fact that the discontinuous σ and κ coefficients change the matrix structure very badly. While, it is not surprising that the SI algorithm performs very poorly in this case. An examination of the CPU time in Table 5 reveals that Gmres(40) consumes less computational work than the other methods. We have observed that the main part of the CPU time needed in DSA or Grid2 is used by Bicgstab or Gmres(10) to solve the diffusion problem in DSA or the coarse linear system in Grid2, respectively. However, by limiting the number of iterations in Bicgstab and Gmres(10) to 1, or increasing the tolerance from 10^{-2} to 10^{-1}, the results change favourably with significant advantage for Grid2.

Table 6 tabulates the results obtained by Gridnest using four levels. It can be clearly seen that there is very little variation in the number i_{Gmres} and the reduction factor $\|r^{(m)}\|_{L^\infty} / \|r^{(m-1)}\|_{L^\infty}$ as the meshes are refined.

3.5.4 Radiative Transfer Equation with Frequency Dependence

Our final example is the frequency-dependent problem

$$\Omega \cdot \nabla \psi(x, \Omega, v) + (\sigma + \kappa) \psi(x, \Omega, v) = \frac{\sigma}{4\pi} \int_{S^2} \psi(x, \Omega', v) d\Omega' + \kappa B(T, v), \quad (69)$$

with boundary condition $\psi = B(T(\hat{x}), v)$. Here $\psi = \psi(x, \Omega, v)$, $T = T(x)$, $\sigma = \sigma(x, v)$ and $\kappa = \kappa(x, v)$ denote respectively, the radiation intensity, the temperature, the scattering and the opacity within the frequency $v > 0$.

In order to discretize the (69) with respect to the frequency variable v, we assume N frequency bands $[v_\iota, v_{\iota+1}]$, $\iota = 1, \ldots, N$ with piecewise constant absorption,

$$\kappa = \kappa_\iota, \qquad \forall\, v \in [v_\iota, v_{\iota+1}] \quad \iota = 1, \ldots, N.$$

We define the frequency-averaged intensity in the band $[v_\iota, v_{\iota+1}]$ by

$$\psi_\iota = \int_{v_\iota}^{v_{\iota+1}} \psi(x, \Omega, v') dv'.$$

Then, the (69) are transformed to a system of N radiative transfer equations of the form

$$\Omega \cdot \nabla \psi_\iota(x, \Omega) + \left(\sigma_\iota + \kappa_\iota\right) \psi_\iota(x, \Omega) = \frac{\sigma_\iota}{4\pi} \int_{S^2} \psi_\iota(x, \Omega') d\Omega' + \kappa_\iota \int_{v_\iota}^{v_{\iota+1}} B(T, v') dv'.$$

$$(70)$$

To approximate the frequency integrals we used trapezoidal formula. In our numerical simulations we use eight frequency bands $[v_\iota, v_{\iota+1}]$, $\iota = 1, \ldots, 8$ from glass manufacturing [47]. These frequencies are given in Table 2 in the introduction.

We compute the solution of the (69) in a unit square, on the refined grid with 100×100 gridpoints, and 64 discrete ordinates. Hence the number of unknowns for each frequency band is 64×10^4, and these computations are done for the 8 frequencies such that the overall number of equations amounts 5.12×10^6. The scattering parameter σ is varying in the set $\{1, 10, 100\}$. For every frequency band we calculate it corresponding scattering ratio and, we store the number of iterations and the running time obtained by each method. The results are given in Table 7.

All the algorithms iterate the solution in a large iteration numbers for the first frequency bands (with large scattering ratio), then these numbers go decreasing as the frequency bands grow until their reach the minimum for the last frequency band. In this case when the size of scattering ratio is changing dramatically over the frequency bands, only the DSA and Grid2 algorithms lead to satisfying results for all frequencies and also in diffusive limit ($\sigma = 100$). The superiority of Grid2 is clearly demonstrated in Table 7.

4 Higher-Order Diffusion Approximations

Approximations that are widely used are the P_N approximations, cf. Sect. 5.1. A major drawback in higher dimensions and for complicated problems is the large number of equations which have to be solved. We propose the SP_N approximations as alternatives to the full glass equations. This class of approximations uses diffusion equations instead of the radiative transfer equations. The number of equations is

Table 7 Number of iterations and CPU time for the eight frequency-bands problem

	Band	Scattering ratio	SI	Gmres(40)	DSA	Grid2
	1	0.71428	13	6	6	3
	2	0.66666	13	6	6	3
	3	0.11494	7	5	4	3
	4	0.06077	6	4	4	3
$\sigma = 1$	5	0.03450	5	4	3	3
	6	0.00371	4	3	3	2
	7	0.00175	3	3	3	2
	8	0.00014	3	2	2	2
	CPU(sec) ——		30.45	26.36	39.38	29.29
	1	0.96153	117	17	8	5
	2	0.95238	114	17	8	5
	3	0.56497	21	8	5	4
	4	0.39292	14	7	5	3
$\sigma = 10$	5	0.26329	10	5	5	3
	6	0.03597	5	3	3	3
	7	0.01732	4	3	3	3
	8	0.00173	3	2	2	2
	CPU(sec) ——		108.30	36.61	35.13	38.40
	1	0.99601	1577	79	16	15
	2	0.99502	1637	79	17	14
	3	0.92850	151	22	15	10
	4	0.86617	80	16	14	9
$\sigma = 100$	5	0.78137	48	12	15	8
	6	0.27175	10	5	5	4
	7	0.14985	8	4	4	4
	8	0.01381	4	3	3	2
	CPU(sec) ——		1190.73	107.80	57.27	122.05

considerably reduced compared to the P_N equations. The method originates from neutron transport theory in nuclear physics where it was successfully introduced. Nevertheless, it suffered from a lack of theoretical foundation with the result that it was not completely accepted in the field. This has been remedied, however, during the last decade, such that the method has now been substantiated.

We want to study the optically thick regime where the opacity κ is large and the radiation is conveyed in a diffusion-like manner. Therefore, we rewrite the above equations in non-dimensional form introducing reference values which correspond to typical values of the physical quantities. In order to obtain a diffusion scaling we impose the relations

$$t_{ref} = c_m \rho_m \kappa_{ref} x_{ref}^2 \frac{T_{ref}}{I_{ref}}, \quad \text{and} \quad k_{ref} = \frac{I_{ref}}{\kappa_{ref} T_{ref}},$$

on these reference values and define the non-dimensional parameter

$$\varepsilon = \frac{1}{\kappa_{ref} x_{ref}} \tag{71}$$

which is small in the optically thick, diffusive regime. Then the rescaled equations read (without marking the scaled quantities):

$$\varepsilon^2 \frac{\partial T}{\partial t} = \varepsilon^2 \nabla \cdot k \nabla T - \int_{v_1}^{\infty} \int_{S^2} \kappa (B - \psi) \, d\Omega dv, \tag{72a}$$

$$\forall v > 0, \Omega \in S^2 : \quad \varepsilon \Omega \cdot \nabla \psi = \kappa (B - \psi). \tag{72b}$$

Here, we have neglected scattering. This can be incorporated in a straightforward way, however. The boundary condition for the temperature changes into

$$\varepsilon k \, n \cdot \nabla T = h(T_b - T) + \alpha \pi \left(\frac{n_2}{n_1}\right)^2 \int_0^{v_1} B(v, T_b) - B(v, T) \, dv. \tag{72c}$$

It is well known that an *outer* asymptotic expansion of (72a) and (72b) leads to equilibrium diffusion theory, which is, in the cases considered here, expected to be valid in the interior of V, see e.g. [46,67,68]. The diffusion or Rosseland approximation is

$$\frac{\partial T}{\partial t} = \nabla \cdot \left(k + k_r(T)\right) \nabla T, \quad \text{with} \quad k_r(T) = \frac{4\pi}{3} \int_{v_1}^{\infty} \frac{1}{\kappa} \frac{\partial B}{\partial T} \, dv.$$

However, this diffusion theory is not capable of describing boundary layers and the question arises whether more sophisticated approximations can suitably model the boundary layer effects. In the realm of neutron transport, such higher-order asymptotic corrections to diffusion theory exist, and are reasonably well understood; they are the so-called *simplified P_N (SP_N)* theories, see [9, 80]. These SP_N theories are, in fact, diffusion in nature i.e. diffusion equations or coupled systems of diffusion equations are employed. They contain boundary layer effects and can be remarkably accurate, much more accurate than the standard Rosseland approximation. In practice, these equations are viewed as an extended form of the classical diffusion theory. No separate boundary layer treatment is necessary because the boundary layers are included in the SP_N equations. For other approximate theories for the above equations and applications, see for example [17, 49]. The material of this section is taken from [47].

4.1 Asymptotic Analysis and Derivation of the SP$_N$ Approximations

To solve (72a) in the domain V, we write this equation as

$$\left(1 + \frac{\varepsilon}{\kappa} \Omega \cdot \nabla\right) \psi(x, t, \Omega, v) = B(v, T).$$

and apply Neumann's series to formally invert the operator

$$\psi = \left(1 + \frac{\varepsilon}{\kappa} \Omega \cdot \nabla\right)^{-1} B$$

$$\cong \left[1 - \frac{\varepsilon}{\kappa} \Omega \cdot \nabla + \frac{\varepsilon^2}{\kappa^2} (\Omega \cdot \nabla)^2 - \frac{\varepsilon^3}{\kappa^3} (\Omega \cdot \nabla)^3 + \frac{\varepsilon^4}{\kappa^4} (\Omega \cdot \nabla)^4 \cdots\right] B. \tag{73}$$

Integrating with respect to Ω and using the result

$$\int_{S^2} (\Omega \cdot \nabla)^n \, d\Omega = [1 + (-1)^n] : \frac{2\pi}{n+1} \nabla^n,$$

where $\nabla^2 = \nabla \cdot \nabla = \Delta$ is the Laplacian, we get

$$\varphi = \int_{S^2} \psi \, d\Omega = 4\pi \left[1 + \frac{\varepsilon^2}{3\kappa^2} \nabla^2 + \frac{\varepsilon^4}{5\kappa^4} \nabla^4 + \frac{\varepsilon^6}{7\kappa^6} \nabla^6 \cdots\right] B + \mathcal{O}(\varepsilon^8).$$

Hence,

$$4\pi B = \left[1 + \frac{\varepsilon^2}{3\kappa^2} \nabla^2 + \frac{\varepsilon^4}{5\kappa^4} \nabla^4 + \frac{\varepsilon^6}{7\kappa^6} \nabla^6\right]^{-1} \varphi + \mathcal{O}(\varepsilon^8)$$

$$= \left\{ 1 - \left[\frac{\varepsilon^2}{3\kappa^2} \nabla^2 + \frac{\varepsilon^4}{5\kappa^4} \nabla^4 + \frac{\varepsilon^6}{7\kappa^6} \nabla^6\right] \right.$$

$$+ \left[\frac{\varepsilon^2}{3\kappa^2} \nabla^2 + \frac{\varepsilon^4}{5\kappa^4} \nabla^4 + \frac{\varepsilon^6}{7\kappa^6} \nabla^6\right]^2$$

$$\left. - \left[\frac{\varepsilon^2}{3\kappa^2} \nabla^2 + \frac{\varepsilon^4}{5\kappa^4} \nabla^4 + \frac{\varepsilon^6}{7\kappa^6} \nabla^6\right]^3 \cdots \right\} \varphi + \mathcal{O}(\varepsilon^8),$$

so we have the asymptotic expansion

$$\forall \nu > 0: \quad 4\pi B = \left[1 - \frac{\varepsilon^2}{3\kappa^2} \nabla^2 - \frac{4\varepsilon^4}{45\kappa^4} \nabla^4 - \frac{44\varepsilon^6}{945\kappa^6} \nabla^6\right] \varphi + \mathcal{O}(\varepsilon^8). \tag{74}$$

If we discard terms of $\mathcal{O}(\varepsilon^4)$, $\mathcal{O}(\varepsilon^6)$ or $\mathcal{O}(\varepsilon^8)$ we obtain the SP_1, SP_2 and SP_3 approximations, respectively. All these equations contain the frequency ν as a parameter.

4.1.1 SP$_1$ and Diffusion Approximations

From (74), we obtain

$$4\pi B = \varphi - \frac{\varepsilon^2}{3\kappa^2} \nabla^2 \varphi + \mathcal{O}(\varepsilon^4)$$

and up to $\mathscr{O}(\varepsilon^4)$ we may write the equation in the form

$$\forall v > 0: \quad -\varepsilon^2 \nabla \cdot \frac{1}{3\kappa} \nabla \varphi + \kappa \varphi = \kappa(4\pi B). \tag{75a}$$

In this equation, v is simply a parameter. Thus, in practice, these equations would be solved independently for each frequency or frequency group and subsequently coupled via (72a). By (75a),

$$\int_{v_1}^\infty \int_{S^2} \kappa(B-I)\, d\Omega\, dv = \int_{v_1}^\infty \kappa(4\pi B - \varphi)\, dv = -\varepsilon^2 \int_{v_1}^\infty \nabla \cdot \frac{1}{3\kappa} \nabla \varphi\, dv + \mathscr{O}(\varepsilon^4).$$

Thus, the energy equation (72a) becomes up to $\mathscr{O}(\varepsilon^2)$:

$$\frac{\partial T}{\partial t} = \nabla \cdot k\nabla T + \int_{v_1}^\infty \nabla \cdot \frac{1}{3\kappa} \nabla \varphi\, dv. \tag{75b}$$

Equations (75b) and (75a) are the SP_1 *approximation* to (72a) and (72b). Since (75b) is only of order $\mathscr{O}(\varepsilon^2)$ the approximation is $\mathscr{O}(\varepsilon^2)$. Using $\varphi = 4\pi B + \mathscr{O}(\varepsilon^2)$ in (75b) one obtains up to $\mathscr{O}(\varepsilon^2)$

$$\begin{aligned}
\frac{\partial T}{\partial t} &= \nabla \cdot k\nabla T + \int_{v_1}^\infty \nabla \cdot \frac{1}{3\kappa} \nabla(4\pi B)\, dv \\
&= \nabla \cdot k\nabla T + \nabla \cdot \left(\frac{4\pi}{3} \int_{v_1}^\infty \frac{1}{\kappa} \frac{\partial B}{\partial T}\, dv \right) \nabla T,
\end{aligned} \tag{76}$$

i.e. we have obtained the conventional equilibrium *diffusion* or *Rosseland approximation* (73). However, (75a) permits a boundary layer behaviour near the boundary ∂V that is not present in (76).

4.1.2 SP₂ Approximation

From (74), we get for ε going to 0

$$4\pi B = \varphi - \frac{\varepsilon^2}{3\kappa^2} \nabla^2 \varphi - \frac{4\varepsilon^2}{15\kappa^2} \nabla^2 \left(\frac{\varepsilon^2}{3\kappa^2} \nabla^2 \varphi \right) + \mathscr{O}(\varepsilon^6).$$

This implies

$$\frac{\varepsilon^2}{3\kappa^2} \nabla^2 B = \varphi - 4\pi B + \mathscr{O}(\varepsilon^4).$$

Hence, with $\mathscr{O}(\varepsilon^6)$ error, the expansion above gives

$$4\pi B = \varphi - \frac{\varepsilon^2}{3\kappa^2} \nabla^2 \varphi - \frac{4\varepsilon^2}{15\kappa^2} \nabla^2 [\varphi - 4\pi B] = \varphi - \frac{\varepsilon^2}{3\kappa^2} \nabla^2 \left[\varphi + \frac{4}{5}(\varphi - 4\pi B) \right],$$

or equivalently,

$$-\varepsilon^2 \nabla \cdot \frac{1}{3\kappa} \nabla \left[\varphi + \frac{4}{5}(\varphi - 4\pi B) \right] + \kappa\varphi = \kappa(4\pi B). \tag{77}$$

Equation (77) implies

$$\int_{v_1}^{\infty} \int_{S^2} \kappa(B - I) \, d\Omega \, dv = -\varepsilon^2 \int_{v_1}^{\infty} \nabla \cdot \frac{1}{3\kappa} \nabla \left[\varphi + \frac{4}{5}(\varphi - 4\pi B) \right] dv + \mathcal{O}(\varepsilon^4).$$

Thus, the energy equation (72b) becomes up to $\mathcal{O}(\varepsilon^4)$

$$\frac{\partial T}{\partial t} = \nabla \cdot k \nabla T + \int_{v_1}^{\infty} \nabla \cdot \frac{1}{3\kappa} \nabla \left[\varphi + \frac{4}{5}(\varphi - 4\pi B) \right] dv. \tag{78}$$

These two equations can be written in a more advantageous way if we define

$$\xi = \varphi + \frac{4}{5}(\varphi - 4\pi B). \tag{79}$$

Introducing the new variable ξ into (77) and (78) we obtain the SP_2 *equations*:

$$\frac{\partial T}{\partial t} = \nabla \cdot k \nabla T + \int_{v_1}^{\infty} \nabla \cdot \frac{1}{3\kappa} \nabla \xi \, dv, \quad \text{and} \tag{80a}$$

$$-\varepsilon^2 \nabla \cdot \frac{3}{5\kappa} \nabla \xi + \kappa\xi = \kappa(4\pi B). \tag{80b}$$

There is a remarkable similarity between the SP_2 (80) and the SP_1 (75). This is because the SP_1 equations contain some, but not all, of the $\mathcal{O}(\varepsilon^4)$ correction terms. In the realm of neutron transport, the SP_2 approximation has not found favour because, in the presence of material inhomogenities, it yields discontinuous solutions. However, it is obvious that (80b) and (80a) can not produce a discontinuous solution.

Also, in the realm of neutron transport, the SP_1 and SP_2 solutions are not capable of exhibiting boundary layer behaviour, while the more complicated SP_3 solution described below does incorporate this in a remarkably accurate way. However, the radiative transfer SP_1 and SP_2 equations stated here can contain boundary layer behaviour. The SP_3 approximation derived in the following should capture significant radiative transfer boundary effects that are not captured by the SP_1 and SP_2 approximations.

4.1.3 SP₃ Approximation

Ignoring terms of $\mathcal{O}(\varepsilon^8)$ in (74), we get

$$4\pi B = \varphi - \frac{\varepsilon^2}{3\kappa^2}\nabla^2\left[\varphi + \frac{4\varepsilon^2}{15\kappa^2}\nabla^2\varphi + \frac{44\varepsilon^4}{315\kappa^4}\nabla^4\varphi\right] + \mathcal{O}(\varepsilon^8)$$

$$= \varphi - \frac{\varepsilon^2}{3\kappa^2}\nabla^2\left[\varphi + \left(1 + \frac{11\varepsilon^2}{21\kappa^2}\nabla^2\right)\left(\frac{4\varepsilon^2}{15\kappa^2}\varphi\right)\right] + \mathcal{O}(\varepsilon^8)$$

$$= \varphi - \frac{\varepsilon^2}{3\kappa^2}\nabla^2\left[\varphi + \left(1 - \frac{11\varepsilon^2}{21\kappa^2}\nabla^2\right)^{-1}\left(\frac{4\varepsilon^2}{15\kappa^2}\varphi\right)\right] + \mathcal{O}(\varepsilon^8) \qquad (81)$$

Hence, if we define

$$\varphi_2 \equiv \left(1 - \frac{11}{21}\frac{\varepsilon^2}{\kappa^2}\nabla^2\right)^{-1}\left(\frac{2\varepsilon^2}{15\kappa^2}\varphi\right), \qquad (82)$$

then (81) becomes up to $\mathcal{O}(\varepsilon^8)$:

$$4\pi B = \varphi - \frac{\varepsilon^2}{3\kappa^2}\nabla^2(\varphi + 2\varphi_2)$$

or

$$\forall v > 0: \quad -\varepsilon^2\nabla\cdot\frac{1}{3\kappa}\nabla(\varphi + 2\varphi_2) + \kappa\varphi = \kappa(4\pi B), \qquad (83a)$$

while (82) becomes

$$-\frac{11\varepsilon^2}{21\kappa^2}\nabla^2\varphi_2 + \varphi_2 = \frac{2}{15}\frac{\varepsilon^2}{\kappa^2}\nabla^2\varphi = \frac{2}{5}\left(\frac{\varepsilon^2}{3\kappa^2}\nabla^2\varphi\right)$$

$$= \frac{2}{5}\left[-4\pi B + \varphi - \frac{2\varepsilon^2}{3\kappa^2}\nabla^2\varphi_2\right]$$

or

$$\left(\frac{4}{15} - \frac{11}{21}\right)\frac{\varepsilon^2}{\kappa^2}\nabla^2\varphi_2 + \varphi_2 = \frac{2}{5}(\varphi - 4\pi B)$$

or eventually

$$\forall v > 0: \quad -\varepsilon^2\nabla\cdot\frac{9}{35\kappa}\nabla\varphi_2 + \kappa\varphi_2 - \frac{2}{5}\kappa\varphi = -\frac{2}{5}\kappa(4\pi B). \qquad (83b)$$

By (83a) we get up to $\mathcal{O}(\varepsilon^6)$

$$\int_{v_1}^{\infty}\int_{S^2}\kappa(B - I)\,d\Omega\,dv = -\varepsilon^2\int_{v_1}^{\infty}\nabla\cdot\frac{1}{3\kappa}\nabla(\varphi + 2\varphi_2)\,dv.$$

Thus, the energy equation (72b) becomes

$$\frac{\partial T}{\partial t} = \nabla\cdot k\nabla T + \int_{v_1}^{\infty}\nabla\cdot\frac{1}{3\kappa}\nabla(\varphi + 2\varphi_2)\,dv. \qquad (83c)$$

Equation (83c) together with the two approximate (83a, 83b) form the *SP₃ approximation* to the system (72a) and (72b).

These equations can be rewritten in a computationally more advantageous way. Let us calculate $\theta\{(83a)\} + \{(83b)\}$:

$$-\varepsilon^2 \nabla \cdot \frac{1}{\kappa} \nabla \left\{ \frac{\theta}{3}(\varphi + 2\varphi_2) + \frac{9}{35}\varphi_2 \right\} + \kappa \left\{ \theta\varphi + \varphi_2 - \frac{2}{5}\varphi \right\} = \kappa \left(\theta - \frac{2}{5} \right)(4\pi B).$$

We seek linear combinations of both equations such that the two functions in the brackets on the left are scalar multiples. More explicitly, we look for θ that fulfills the condition

$$\frac{\theta}{3}(\varphi + 2\varphi_2) + \frac{9}{35}\varphi_2 = \mu^2 \left(\theta\varphi + \varphi_2 - \frac{2}{5}\varphi \right), \tag{84}$$

where $\mu^2 > 0$ is a constant to be determined later. Equation (84) holds for arbitrary φ and φ_2 iff

$$\frac{\theta}{3} = \mu^2 \left(\theta - \frac{2}{5} \right) \quad \text{and} \quad \frac{2\theta}{3} + \frac{9}{35} = \mu^2.$$

The second of these equations may be solved for θ and we get a quadratic equation in μ^2:

$$\frac{1}{2}\mu^2 - \frac{9}{70} = \mu^2 \left(\frac{2}{3}\mu^2 - \frac{11}{14} \right).$$

Its discriminant is positive and thus we obtain two positive real solutions

$$\mu_1^2 = \frac{3}{7} - \frac{2}{7}\sqrt{\frac{6}{5}}, \quad \text{and} \quad \mu_2^2 = \frac{3}{7} + \frac{2}{7}\sqrt{\frac{6}{5}},$$

and the corresponding values of the scalar θ are

$$\theta_1 = \frac{9}{35} - \frac{3}{7}\sqrt{\frac{6}{5}}, \quad \text{and} \quad \theta_2 = \frac{9}{35} + \frac{3}{7}\sqrt{\frac{6}{5}}.$$

Now relation (84) implies, for $n = 1,2$,

$$\left(-\varepsilon^2 \nabla \cdot \frac{1}{\kappa}\nabla\mu_n^2 + \kappa \right) \left[\theta_n\varphi + \varphi_2 - \frac{2}{5}\varphi \right] = \left(\theta_n - \frac{2}{5} \right)\kappa(4\pi B). \tag{85}$$

This suggests that we define two new independent variables for $n = 1,2$

$$I_n = \frac{\theta_n\varphi + \varphi_2 - 2/5\varphi}{\theta_n - 2/5} = \varphi + \frac{1}{\theta_n - 2/5}\varphi_2 = \varphi + \gamma_n\varphi_2 \tag{86}$$

where

$$\gamma_n = \frac{1}{\theta_n - 2/5} = \frac{5}{7}\left[1 + (-1)^n 3\sqrt{\frac{6}{5}} \right].$$

The two equations in (85) are now

$$-\varepsilon^2 \mu_1^2 \nabla \cdot \frac{1}{\kappa} \nabla I_1 + \kappa I_1 = \kappa(4\pi B), \tag{87a}$$

$$-\varepsilon^2 \mu_2^2 \nabla \cdot \frac{1}{\kappa} \nabla I_2 + \kappa I_2 = \kappa(4\pi B). \tag{87b}$$

The advantage of this form of the SP_3 equations is that the diffusion equations are uncoupled. It will be seen below that there remains, however, a weak coupling in the boundary conditions.

The linear transformation of variables above is inverted according to the formulae

$$\varphi = \frac{\gamma_2 I_1 - \gamma_1 I_2}{\gamma_2 - \gamma_1}, \quad \text{and} \quad \varphi_2 = \frac{I_2 - I_1}{\gamma_2 - \gamma_1}. \tag{88}$$

Defining three constants

$$w_0 = \frac{1}{\gamma_2 - \gamma_1} = \frac{7}{30}\sqrt{\frac{5}{6}} = \frac{7}{36}\sqrt{\frac{6}{5}}, \quad \text{and} \tag{89a}$$

$$w_1 = \frac{\gamma_2}{\gamma_2 - \gamma_1} = \frac{1}{6}\left(3 + \sqrt{\frac{5}{6}}\right), \quad w_2 = \frac{-\gamma_1}{\gamma_2 - \gamma_1} = \frac{1}{6}\left(3 - \sqrt{\frac{5}{6}}\right) \tag{89b}$$

we can write $\varphi = w_1 I_1 + w_2 I_2$ and $\varphi_2 = w_0(I_2 - I1)$ and we have furthermore

$$\frac{1}{3}(\varphi + 2\varphi_2) = \frac{1}{3}(w_1 - 2w_0)I_1 + \frac{1}{3}(w_2 + 2w_0)I_2 = a_1 I_1 + a_2 I_2.$$

Here again we introduced constants

$$a_1 = \frac{w_1 - 2w_0}{3} = \frac{1}{30}\left(5 - 3\sqrt{\frac{5}{6}}\right), \quad \text{and} \quad a_2 = \frac{w_2 + 2w_0}{3} = \frac{1}{30}\left(5 + 3\sqrt{\frac{5}{6}}\right).$$

In this way, the SP_3 energy equation (83c) above becomes:

$$\frac{\partial T}{\partial t} = \nabla \cdot k\nabla T + \int_{V_1}^{\infty} \nabla \cdot \frac{1}{\kappa}\nabla(a_1 I_1 + a_2 I_2)\, dv. \tag{90}$$

4.2 Boundary Conditions for SP$_N$ Approximations

The boundary conditions for the SP_N equations in neutron transport come from a variational principle, see [9, 80]. Here, we use the boundary conditions developed for the transport case to state (and rewrite in a more suitable form) the boundary conditions for SP_1, SP_2 and SP_3 approximations to the transport problem (72b) with the boundary condition (44c).

We consider the transport equation (72b)

$$\forall v > 0: \quad \varepsilon \Omega \cdot \nabla \psi(x, \Omega) + \kappa I(x, \Omega) = \kappa B, \quad x \in V,$$

with semi-transparent boundary conditions on ∂V

$$\psi(x, \Omega) = \rho(n \cdot \Omega) \psi(x, \Omega') + (1 - \rho(n \cdot \Omega)) \psi_b(x, \Omega), \qquad n \cdot \Omega < 0.$$

Let us define the scalar flux as before

$$\varphi(x) = \int_{S^2} \psi(x, \Omega) \, d\Omega,$$

and define two integrals of the influx into the domain for $m = 1, 3$

$$\psi_m(x) = \int_{n \cdot \Omega < 0} (1 - \rho(n \cdot \Omega)) P_m(|\Omega \cdot n|) I_b(x, \Omega) \, d\Omega, \quad x \in \partial V. \tag{91}$$

Here, the Legendre polynomials of order 1 and 3 are used:

$$P_1(\mu) = \mu, \quad \text{and} \quad P_3(\mu) = \frac{5}{2}\mu^3 - \frac{3}{2}\mu.$$

Furthermore, it is convenient in the sequel to have the following integrals at hand:

$$r_1 = 2\pi \int_0^1 \mu \rho(\mu) \, d\mu, \qquad r_5 = 2\pi \int_0^1 P_3(\mu)\rho(\mu) \, d\mu,$$
$$r_2 = 2\pi \int_0^1 \mu^2 \rho(\mu) \, d\mu, \qquad r_6 = 2\pi \int_0^1 P_2(\mu)P_3(\mu)\rho(\mu) \, d\mu,$$
$$r_3 = 2\pi \int_0^1 \mu^2 \rho(\mu) \, d\mu, \qquad r_7 = 2\pi \int_0^1 P_3(\mu)P_3(\mu)\rho(\mu) \, d\mu,$$
$$r_4 = 2\pi \int_0^1 \mu P_3(\mu)\rho(\mu) \, d\mu,$$

The boundary conditions in [9, 80] were derived for the case $\rho = 0 = \text{const.}$ For semi-transparent boundary conditions the same arguments apply and the calculations can be analogously carried out, the only difference beeing modifications in the coefficients. We therefore content ourselves with stating the resulting equations. In the SP_1 approximation (75a), the boundary condition for φ is:

$$\forall v > 0: \quad (1 - 2r_1)\varphi(x) + (1 + 3r_2)\frac{2\varepsilon}{3\kappa}n \cdot \nabla \varphi(x) = 4\psi_1(x). \tag{92}$$

The boundary condition for φ in the SP_2 approximation (80b) is, see [80]:

$$(1 - 2r_1)\varphi + (1 + 3r_2)\frac{2\varepsilon}{3\kappa}n \cdot \nabla \left[\varphi + \frac{4}{5}(\varphi - 4\pi B)\right]$$
$$+ (1 - 4(3r_3 - r_1))\frac{1}{2}(\varphi - 4\pi B) = 4\psi_1. \tag{93}$$

These equations reduce to (92) if one deletes the $(\varphi - 4\pi B)$ terms in them. They can be written in a more advantageous way if we define ξ as in Sect. 2 which is equivalent to

$$\varphi = \frac{5}{9}\xi + \frac{4}{9}(4\pi B).$$

One obtains

$$(1 - 2r_1)\left(\frac{5}{9}\xi + \frac{4}{9}(4\pi B)\right) + (1 + 3r_2)\frac{2\varepsilon}{3\kappa}n \cdot \nabla\xi$$
$$+ (1 - 4(3r_3 - r_1))\frac{1}{2}\left(\frac{5}{9}\xi + \frac{4}{9}(4\pi B) - 4\pi B\right) = 4\psi_1,$$

or, using the abbreviations

$$\alpha_1 = \frac{5}{6}(1 - 4r_3) \quad \alpha_2 = \frac{1}{6}(-1 + 12r_1 - 20r_3)$$

the SP_2 boundary conditions for ξ in (80b) can be written:

$$\forall v > 0: \quad \alpha_1\xi(r) + (1 + 3r_2)\frac{2\varepsilon}{3\kappa}n \cdot \nabla\xi(r) = \alpha_2 4\pi B(T(x)) + 4\psi_1(x). \qquad (94)$$

The boundary conditions (92) and (93) were derived variationally, not from a boundary layer analysis. They should be accurate if $I_b(x, \Omega)$ is a reasonably smooth function of Ω, but they could be inaccurate if I_b is not smooth.

Finally, the SP_3 boundary conditions for φ and φ_2 in (83a) and (83b) are, see [9]: for all frequencies $v > 0$ and $x \in \partial V$ there must hold

$$(1 - 2r_1)\frac{1}{4}\varphi(x) + (1 - 8r_3)\frac{5}{16}\varphi_2(x) + (1 + 3r_2)\frac{\varepsilon}{6\kappa}n \cdot \nabla\varphi(x)$$
$$+ \left(\frac{1 + 3r_2}{3} + \frac{3r_4}{2}\right)\frac{2\varepsilon}{3\kappa}n \cdot \nabla\varphi_2(x) = \psi_1(x), \qquad (95a)$$

$$-(1 + 8r_5)\frac{1}{16}\varphi(x) + (1 - 8r_6)\frac{5}{16}\varphi_2(x) + 3r_4\frac{\varepsilon}{6\kappa}n \cdot \nabla\varphi(x)$$
$$+ \left(r_4 + \frac{3}{14}(1 + 7r_7)\right)\frac{\varepsilon}{\kappa}n \cdot \nabla\varphi_2(x) = \psi_3(x). \qquad (95b)$$

or formally

$$A_1\varphi(x) + A_2\varphi_2(x) + A_3\frac{\varepsilon}{\kappa}n \cdot \nabla\varphi(x) + A_4\frac{\varepsilon}{\kappa}n \cdot \nabla\varphi_2(x) = \psi_1(x)$$
$$B_1\varphi(x) + B_2\varphi_2(x) + B_3\frac{\varepsilon}{\kappa}n \cdot \nabla\varphi(x) + B_4\frac{\varepsilon}{\kappa}n \cdot \nabla\varphi_2(x) = \psi_3(x).$$

We still have to derive boundary conditions for I_1 and I_2. Using the formulae in (88), we can transform the boundary conditions for φ and φ_2 into boundary conditions for I_1 and I_2. The equations above then become

$$(A_1\gamma_2 w_0 - A_2 w_0)I_1 + (-A_1\gamma_1 w_0 + A_2 w_0)I_2$$
$$+ (A_3\gamma_2 w_0 - A_4 w_0)\frac{\varepsilon}{\kappa}n\cdot\nabla I_1$$
$$+ (-A_3\gamma_2 w_0 + A_2 w_0)\frac{\varepsilon}{\kappa}n\cdot\nabla I_2 = \psi_1$$
$$(B_1\gamma_2 w_0 - B_2 w_0)I_1 + (-B_1\gamma_1 w_0 + B_2 w_0)I_2$$
$$+ (B_3\gamma_2 w_0 - B_4 w_0)\frac{\varepsilon}{\kappa}n\cdot\nabla I_1$$
$$+ (-B_3\gamma_2 w_0 + B_2 w_0)\frac{\varepsilon}{\kappa}n\cdot\nabla I_2 = \psi_3$$

or, again formally rewritten for writing convenience,

$$C_1 I_1 + C_2 I_2 + C_3\frac{\varepsilon}{\kappa}n\cdot\nabla I_1 + C_4\frac{\varepsilon}{\kappa}n\cdot\nabla I_2 = \psi_1$$
$$D_1 I + D_2 I_2 + D_3\frac{\varepsilon}{\kappa}n\cdot\nabla I_1 + D_4\frac{\varepsilon}{\kappa}n\cdot\nabla I_2 = \psi_3.$$

We eliminate the gradient term $n\cdot\nabla I_2$ in the first equation and $n\cdot\nabla I_1$ in the second in order to get boundary conditions for the I_1 and I_2 equations, respectively. We find

$$(C_1 D_4 - D_1 C_4)I_1 + (C_3 D_4 - D_3 C_4)\frac{\varepsilon}{\kappa}n\cdot\nabla I_1$$
$$= -(C_2 D_4 - D_2 C_4)I_2 + (D_4\,\psi_1 - C_4\,\psi_3)$$
$$-(C_2 D_3 - D_2 C_3)I_2 + (C_3 D_4 - D_3 C_4)\frac{\varepsilon}{\kappa}n\cdot\nabla I_2$$
$$= (C_1 D_3 - D_1 C_3)I_1 - (D_3\,\psi_1 - C_3\,\psi_3)$$

so, if we set $D = C_3 D_4 - D_3 C_4$ and define constants

$$\alpha_1 = (C_1 D_4 - D_1 C_4)/D, \qquad \alpha_2 = -(C_2 D_3 - D_2 C_3)/D,$$
$$\beta_1 = -(C_1 D_3 - D_1 C_3)/D, \qquad \beta_2 = (C_2 D_4 - D_2 C_4)/D,$$

then we end up with SP_3 boundary conditions in the following form:

$$\alpha_1 I_1(x) + \frac{\varepsilon}{\kappa}n\cdot\nabla I_1(x) = -\beta_2 I_2(x) + (D_4\,\psi_1(x) - C_4\,\psi_3(x))/D, \qquad (96a)$$
$$\alpha_2 I_2(x) + \frac{\varepsilon}{\kappa}n\cdot\nabla I_2(x) = -\beta_1 I_1(x) - (D_3\,\psi_1(x) - C_3\,\psi_3(x))/D. \qquad (96b)$$

Equations (96a) and (96a) are the boundary conditions to go with the diffusion equations (87a) and (87b), respectively. The coupling of I_1 and I_2 in the boundary conditions is very weak.

Consider, for example, the simple case when there is no reflection $\rho = 0$ and $\psi_b(x, \Omega) = \psi_b(x)$ is isotropic. Then by (91),

$$\psi_m(x) = 2\pi \psi_b(x) \int_0^1 P_m(\mu) \, d\mu = \pi \psi_b(x) \begin{cases} 1, & \text{for } m = 1, \\ \frac{1}{4}, & \text{for } m = 3. \end{cases}$$

The constants $r_1, \ldots, r_7$ are all zero such that we find after some calculations $D = \frac{1}{144}\sqrt{6/5}$ and the constants in the boundary conditions are

$$\alpha_1 = \frac{5}{96}\left(34 + 11\sqrt{6/5}\right), \qquad \alpha_2 = \frac{5}{96}\left(34 + 11\sqrt{6/5}\right),$$
$$\beta_1 = \frac{5}{96}\left(2 - \sqrt{6/5}\right), \qquad \beta_2 = \frac{5}{96}\left(2 + \sqrt{6/5}\right).$$

Finally, the source terms in (96a) and (96b) become in this case

$$\left[6\psi_1(x) - 2\left(3 \pm 5\sqrt{\frac{6}{5}}\right)\psi_3(x)\right] = \pi\psi_b(x)\left[6 + 2\left(3 \pm 5\sqrt{\frac{6}{5}}\right)\frac{1}{4}\right]$$
$$= \frac{5}{2}\pi\psi_b(x)\left[3 \pm \sqrt{\frac{6}{5}}\right].$$

The approximate SP_N-theories stated above are simpler than transport theory because they do not contain the angular variable Ω. However, they do contain the frequency variable ν. It is formally possible to derive simpler theories in which the frequency is eliminated. We refer to [48].

5 Moment Models

First we briefly review the basics of the moment approach. Consider again the transport equation (32) for the radiation. We will assume isotropic scattering here. This equation is in fact a system of infinitely many coupled integro-differential equations that describes the distribution ψ of all photons in time, space and velocity space. On the one hand this system is computationally very expensive and on the other hand we are not interested in the photon distribution itself but in macroscopic quantities like the mean energy or mean flux of the radiation field. For instance, only the gradient of the radiative flux enters into the energy balance. The macroscopic quantities are moments of the distribution function. Let

$$\langle \cdot \rangle := \int_{S^2} \cdot \, d\Omega \tag{97}$$

denote the average over all directions. The energy, flux vector and pressure tensor of the radiation field are defined, respectively, as

$$E := \langle \psi \rangle, \quad F := \langle \Omega \psi \rangle, \quad P := \langle (\Omega \otimes \Omega) \psi \rangle. \tag{98}$$

To derive equations for the macroscopic quantities we multiply the transport equation by 1 and Ω and average over all directions. We obtain the conservation laws

$$\nabla F = \kappa(\langle B \rangle - E) \tag{99}$$

$$\nabla P = -(\kappa + \sigma)F. \tag{100}$$

These are four equations (the first is a scalar equation, the second has three components) for 10 unknowns (E scalar, F 3-component vector, P symmetric 3×3-matrix). Hence we have to pose an additional condition. Usually this condition is a constitutive equation for the highest moment P, expressed in terms of the lower moments E and F. This is referred to as the closure problem. The simplest approximation, the so-called P_1 approximation, is obtained if we assume that the underlying distribution is isotropic. Thus, we obtain $P = \frac{1}{3}E$ and therefore

$$\nabla F = \kappa(\langle B \rangle - E) \tag{101}$$

$$\nabla \frac{1}{3}E = -(\kappa + \sigma)F. \tag{102}$$

The general P_N closure is usually derived in a different way.

5.1 Spherical Harmonics

The Spherical Harmonics approach is one of the oldest approximate methods for radiative transfer [20, 33]. For the sake simplicity, we restrict our explanation to the case of slab geometry. The derivation for three-dimensional case can be found for example in [12] and also in standard textbooks [14,41,62]. The idea of the spherical harmonics approach is to express the angular dependence of the distribution function in terms of a Fourier series,

$$\psi(\mu) = \sum_{l=0}^{\infty} \psi_l^{SH} \frac{2l+1}{2} P_l(\mu), \tag{103}$$

where P_l are the Legendre polynomials. These form an orthogonal basis of the space of polynomials with respect to the standard scalar product on $[-1, 1]$,

$$\int_{-1}^{1} P_l(\mu)P_k(\mu)d\mu = \frac{2}{2l+1}\delta_{lk}. \tag{104}$$

In more space dimensions, one uses spherical harmonics, which are an orthogonal system on the unit sphere.

If we truncate the Fourier series at $l = N$ we have

$$\psi^{SH}(\mu) = \sum_{l=0}^{N} \psi_l^{SH} \frac{2l+1}{2} P_l(\mu). \tag{105}$$

One can obtain equations for the Fourier coefficients

$$\psi_l^{SH} = \int_{-1}^{1} \psi^{SH}(\mu) P_l(\mu) d\mu \tag{106}$$

by testing (32) with $P_l(\mu)$ and then integrating. Thus we get

$$\nabla \int_{-1}^{1} \mu P_l(\mu) \psi^{SH}(\mu) d\mu = \kappa(2\langle B \rangle \delta_{l0} - \psi_l^{SH}) + \sigma(\psi_0 \delta_{l0} - \psi_l^{SH}) \tag{107}$$

for the moments ψ_l^{SH} of the distribution function. Using the recursion relation

$$(l+1)P_{l+1}(\mu) + lP_{l-1}(\mu) = (2l+1)\mu P_l(\mu) \tag{108}$$

we obtain

$$\nabla \left(\frac{l+1}{2l+1} \psi_{l+1}^{SH} + \frac{l}{2l+1} \psi_{l-1}^{SH} \right) = \kappa(2\langle B \rangle \delta_{l0} - \psi_l^{SH}) + \sigma(\psi_0 \delta_{l0} - \psi_l^{SH}). \tag{109}$$

This is a linear system of first order partial differential equations. For a criterion on how many moments are sufficient for a given problem see [78].

The two most widely used boundary conditions are Mark [54, 55] and Marshak [56] boundary conditions. The idea of the Mark boundary conditions is to assign the values of the distribution at certain directions μ_i which are the zeros of the Legendre polynomial of order $N + 1$. That this is in fact a natural boundary condition becomes clear in the next section.

Marshak's boundary conditions, on the other hand, demand that the ingoing half moments of the distribution are prescribed, i.e. for the left boundary

$$\int_{0}^{1} P_l(\mu) \psi(\mu) d\mu. \tag{110}$$

This, in some sense, reflects the boundary conditions for the full equations.

5.2 Minimum Entropy Closure

The approximations based on the expansion of the distribution function into a polynomial or the equivalent diffusion approximations suffer from serious drawbacks. First, anisotropic situations are not correctly described. This becomes apparent most drastically for a ray of light, where $|P| = E$. Also, the distribution function can become negative and thus the moments computed from the distribution can become unphysical. Second, boundary conditions cannot be incorporated exactly. At a boundary we usually prescribe the ingoing flux only. Here we have to prescribe values for the full moments. These moments contain the unknown outgoing radiation. Moreover, a polynomial expansion cannot capture discontinuities in the angular photon distribution. Krook [42] remarks that at the boundary there is usually a discontinuity in the distribution between in- and outgoing particles.

In this section, we want to describe one idea which resolves the first problem. The idea is to use an Entropy Minimization Principle to obtain the constitutive equation for P. This principle has become the main concept of Rational Extended Thermodynamics [61].

We want to explain the Entropy Minimization Principle and its practical application by means of our simple moment system (99–100). To close the system we determine a distribution function ψ_{ME} that minimizes the radiative entropy

$$H_R^*(\psi) = \int_{\mathscr{S}^2} \int_0^\infty h_R^*(\psi) dv d\Omega \tag{111}$$

with

$$h_R^*(\psi) = \frac{2kv^2}{c^3}(n\log n - (n+1)\log(n+1)) \quad \text{where} \quad n = \frac{c^2}{2hv^3}\psi \tag{112}$$

under the constraint that it reproduces the lower order moments,

$$\langle \psi_{ME} \rangle = E \quad \text{and} \quad \langle \Omega \psi_{ME} \rangle = F. \tag{113}$$

The entropy is the the well-known entropy for bosons adapted to radiation fields [63, 70]. At first sight, it is not clear why the distribution should minimize the entropy when all that is known for non-equilibrium processes is that there exists an entropy inequality. But it can be shown [15] that the minimization of the entropy for given moments and the entropy inequality are equivalent.

The above minimization problem can be solved explicitly and the pressure can be written as [16]

$$P = D(f)E. \tag{114}$$

Here, $f = \frac{F}{E}$ is the relative flux,

$$D(f) = \frac{1 - \chi(f)}{2}I + \frac{3\chi(f) - 1}{2}\frac{f \otimes f}{|f|^2} \tag{115}$$

is the Eddington tensor and

$$\chi(f) = \frac{5 - 2\sqrt{4 - 3|f|^2}}{3} \tag{116}$$

is the Eddington factor. The Eddington tensor can always be written in the form (115) under the assumption that the intensity is symmetric about a preferred direction [50]. The minimum entropy Eddington factor satisfies the natural constraints

$$\text{tr}(D) = 1 \tag{117}$$
$$D(f) - f \otimes f \geq 0 \tag{118}$$
$$f^2 \leq \chi(f) \leq 1 \tag{119}$$

In the literature, the Eddington factor (116) has been derived based on many, apparently not connected, ideas. Levermore [50] assumed that there existed a reference frame in which the distribution was exactly isotropic and used the covariance of the radiation stress tensor. Anile et al. [7] derived it by collecting physical constraints on the Eddington factor and supposing the existence of an additional conservation law, where the conserved quantity behaves like the physical entropy near radiative equilibrium. The minimum entropy system was thoroughly investigated in [16,84]. Further variable Eddington factors have been proposed, cf. [50,59] and references therein.

The closed system has several desirable properties. The flux is limited in a natural way, i.e. $|f| < 1$. Physically, this corresponds to the fact that information cannot travel faster than the speed of light. Furthermore, the underlying distribution function is always positive. Also, the system can be transformed to a symmetric hyperbolic system [7], which makes it accessible to a general mathematical theory [21]. Again, Marshak type boundary conditions can be derived.

5.3 Flux-Limited Diffusion and Entropy Minimization

The classical diffusion approximation is a linear parabolic partial differential equation. In this equation, information is propagated at infinite speed. This can also be seen from the fact that the flux $|F|$ is not bounded by the energy E (relative flux $f < 1$). But this should hold, due to the definition of the moments. Thus the classical diffusion approximation contradicts fundamental physical concepts.

Therefore the concept of flux-limited diffusion has been introduced. A diffusion equation is called flux-limited if

$$|F| \leq E. \tag{120}$$

The following is a summary of [50]. We begin by writing the moment equations in the form

$$\nabla F = \kappa(\langle B \rangle - E) \tag{121}$$

$$\nabla(DE) = -(\sigma + \kappa)F, \tag{122}$$

with the Eddington tensor D. Two assumptions in the derivation of the classical diffusion equation will be modified. First, the Eddington tensor is only identically equal to $\frac{1}{3}$ for isotropic radiation. For a ray of light ("free-streaming"), on the other hand, we should have $|DE| = E$. Second, one should not neglect $\partial_t F$. Instead, we note that in the diffusive as well as in the free-streaming regime, the spatial and temporal derivatives of the relative flux $f = \frac{F}{E}$ and the Eddington tensor D can be neglected.

Rewriting the equations in terms of f and E we get

$$\nabla(fE) = \kappa(\langle B \rangle - E) \tag{123}$$

$$\nabla(DE) = -(\sigma + \kappa)fE. \tag{124}$$

The second equation becomes

$$\nabla(DE) = -(\sigma + \kappa)fE. \tag{125}$$

Inserting (123) into (125), we obtain

$$f\nabla f + \nabla((D - f \otimes f)E) + \bar{\sigma}fE = 0 \tag{126}$$

with $\bar{\sigma} = \frac{\kappa \langle B \rangle + \sigma E}{E}$. If we drop the derivatives of f and D, we arrive at

$$(D - f \otimes f)\nabla E + \bar{\sigma}fE = 0, \tag{127}$$

or

$$(D - f \otimes f)R = f \quad \text{with} \quad R = -\frac{1}{\bar{\sigma}}\frac{\nabla E}{E}. \tag{128}$$

The idea is now to

1. Choose D as a function of f
2. Solve $(D - f \otimes f)R = f$ for f
3. Insert $f(R)$ into the first moment equation to obtain a diffusion approximation

The first step shows how the concept of flux-limited diffusion is related to a (non-linear) moment closure. If

$$D = \frac{1 - \chi}{2}I + \frac{3\chi - 1}{2}\frac{f \otimes f}{|f|} \tag{129}$$

then f is an eigenvector of D and also of $(D - f \otimes f)$ with

$$(D - f \otimes f)f = (\chi - |f|^2)f. \tag{130}$$

Hence the equation $(D - f \otimes f)R = f$ has the solution

$$R = \frac{f}{\chi - f^2}. \tag{131}$$

Solving this equation for f and writing the result as

$$f = \lambda(R)R \tag{132}$$

we arrive at the closure

$$F = -\frac{1}{\sigma}\lambda\left(\frac{1}{\sigma}\frac{\nabla E}{E}\right)\nabla E. \tag{133}$$

If one chooses for D the minimum entropy Eddington factor then [50]

$$\lambda = \frac{3(1 - \beta^2)^2}{(3 + \beta^2)^2} \tag{134}$$

where β is implicitly given by

$$R = \frac{4\beta(3 + \beta^2)}{(1 - \beta^2)^2}. \tag{135}$$

The same boundary conditions as for the diffusion approximation can be used.

5.4 Partial Moments

In spite of its advantages the minimum entropy system still suffers from a major drawback. In Fig. 8 we show a numerical test case [11] with two colliding beams. The parameters are $\kappa = 2.5$, $\sigma = 0$. The temperature inside the medium is zero. At both sides, beams with a radiative temperature $T_R := \left(\frac{E}{\sigma_{SB}}\right)^{1/4}$, where σ_{SB} is Stefan–Boltzmann's constant, of 1000 and relative fluxes of $f = \pm 0.99$, respectively, enter. Figure 8 shows the radiative energy. The full moment model has a qualitatively wrong solution with two shocks. This is not surprising since this Eddington factor, as stated above, is related to radiation which is isotropic in a certain frame [50]. This assumption is violated in the test case above. The unphysical behavior can be remedied by combining Minimum Entropy with the partial moment idea described in the following.

The partial moment idea is somehow intermediate between the Discrete Ordinates approach and Moment Models. In Discrete Ordinates models the integral over all directions is discretized with a numerical quadrature rule. This yields a coupled system of finitely many transport equations, each describing transport into one direction.

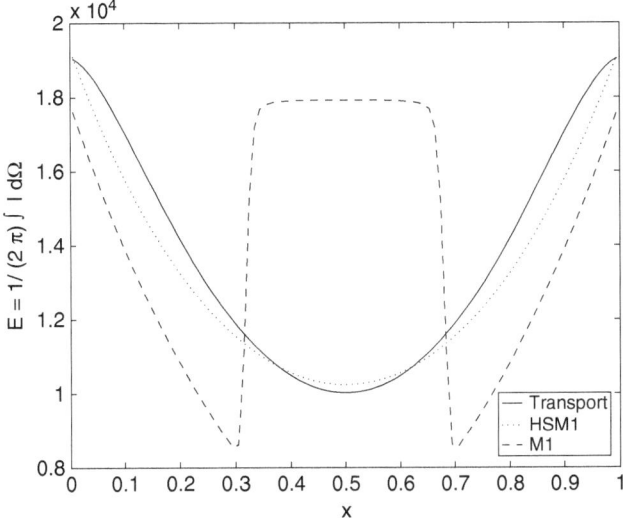

Fig. 8 Radiative energy. Artificial radiative shock wave in the full moment entropy (M_1) model

Let $\mathscr{A}$ be a partition of the unit sphere S^2, where $A \in \mathscr{A}$ denotes the set of the angular integration. Instead of integrating over all directions we average over each $A \in \mathscr{A}$ separately. Thus we define the average

$$\langle \cdot \rangle_A := \int_A \cdot \, d\Omega. \tag{136}$$

Again, we multiply the transport equation by 1 and Ω and average over each $A \in \mathscr{A}$ to obtain

$$\nabla F_A = \langle S \rangle_A \tag{137}$$

$$\nabla P_A = \langle \Omega S \rangle_A. \tag{138}$$

We define the corresponding partial moments by

$$E_A = \langle I \rangle_A \tag{139}$$

$$F_A = \langle \Omega I \rangle_A \tag{140}$$

$$P_A = \langle (\Omega \otimes \Omega) I \rangle_A. \tag{141}$$

To close this system we have to find an equation for the partial pressures P_A as functions of the partial energies E_A and partial fluxes F_A.

Examples for the choice of $\mathscr{A}$, which are used later, are

• For the full moment model we have $A = S^2$, i.e. the integral is over the full sphere.

- For the half moment model we have $A \in \{S^2_+, S^2_-\}$. Here, $S^2_+ = \{\Omega \in S^2 : \Omega_x > 0\}$ is the positive half sphere, where the x-component of Ω is positive, and $S^2_- = \{\Omega \in S^2 : \Omega_x < 0\}$ analogously is the negative half sphere.
- For the quarter moment model we have $A \in \{S^2_{++}, S^2_{+-}, S^2_{--}, S^2_{-+}\}$. Here, $S^2_{++} = \{\Omega \in S^2 : \Omega_x > 0, \Omega_y > 0\}$ is the quarter sphere in the first quadrant. Analogously, $S^2_{+-} = \{\Omega \in S^2 : \Omega_x > 0, \Omega_y < 0\}$ etc.

One could also choose other sets for the angular integration.

5.5 Partial Moment P_N Closure

The basic idea of the P_N closure is to expand the photon distribution into a polynomial. Here we use the same idea, but separately for both half ranges. This approach has been investigated in the literature in different forms and contexts and mostly in connection with boundary conditions, for example recently in [11]. Schuster and Schwarzschild [73, 74] introduce two constant distributions for left- and rightgoing photons (P_0 approximation). Krook [42], based on ideas of Sykes [79], considers half moment in one space dimension with a P_N closure. Sherman [77] compares full-P_N and half-P_N numerically in 1D. Özisik et al. [64] derive a half moment P_1 closure in spherical geometry. Further references can be found in [57], where also an octuple P_1 closure in cylindrical geometry is introduced. Similar ideas appear in related subjects, like gas dynamics, cf. [13] and references therein.

For the half moment P_1 system in one space-dimension, for instance, we assume that in each half range the distribution can be represented by a polynomial of degree one. The coefficients of the polynomial are determined by the constraint that the lower order half moments should be reproduced. The half moment P_1 system reads,

$$\partial_x F_+ = \kappa \left(\frac{1}{2} \langle B \rangle - E_+ \right) + \sigma \left(\frac{1}{2} (E_+ + E_+) - E_+ \right) \tag{142}$$

$$\partial_x (\chi_+ (f_+) E_+) = \kappa \left(\frac{1}{4} \langle B \rangle - F_+ \right) + \sigma \left(\frac{1}{4} (E_+ + E_+) - F_+ \right) \tag{143}$$

$$\partial_x F_- = \kappa \left(\frac{1}{2} \langle B \rangle - E_- \right) + \sigma \left(\frac{1}{2} (E_+ + E_+) - E_+ \right) \tag{144}$$

$$\partial_x (\chi_- (f_-) E_-) = \kappa \left(-\frac{1}{4} \langle B \rangle - F_- \right) + \sigma \left(-\frac{1}{4} (E_+ + E_+) - F_- \right). \tag{145}$$

The partial Eddington factors are

$$\chi_\pm (f_\pm) = -\frac{1}{6} \pm f_\pm \quad \text{with} \quad f_\pm = \frac{F_\pm}{E_\pm}. \tag{146}$$

We note that this is a hyperbolic system. The eigenvalues associated to the "+" moments are positive, while the eigenvalues associated to the "−" moments are negative, in accordance with physical intuition. This structure makes the formulation of accurate boundary conditions easy. We simply prescribe the ingoing half moments, in accordance with the conditions for the full equations. For more discussions, including existence and uniqueness results, and the explicit quarter moment P_1 closure in two space-dimensions we refer the reader to [72].

5.6 Partial Moment Entropy Closure

The partial moment entropy closure was introduced for radiative heat transfer in [19] and developed in [18, 25, 85]. For the sake of completeness we recall the procedure explained earlier. We have to find a distribution function ψ_{ME} that minimizes the radiative entropy H_R^* given by (111–112), under the constraint that it reproduces the lower order partial moments,

$$\langle \psi_{ME} \rangle_A = E_A \quad \text{and} \quad \langle \Omega \psi_{ME} \rangle_A = F_A \tag{147}$$

for all $A \in \mathscr{A}$. The minimizer is given by

$$\psi_{ME} = \sum_{A \in \mathscr{A}} \frac{1}{\alpha_A^4 (1 + \beta_A \cdot \Omega)^4} \mathbf{1}_A, \tag{148}$$

where α_A and β_A are Lagrange multipliers corresponding to the constraints. This formula differs from the one given in [19] since we consider frequency-averaged quantities here. It can be obtained from the minimizer in [19] by integration over ν.

In the case of $\mathscr{A} = \{S_+^2, S_-^2\}$, the half moments over this distribution can be computed explicitly.

Note that $E_\pm \geq 0, F_+ \geq 0, F_- \leq 0$. Multiplying the transfer equation with $m(\mu) = 1^+, 1^-, \mu^+, \mu^-)$ and integrating with respect to ν and μ we get

$$\varepsilon \partial_x F_+ = \kappa \left(\frac{1}{2} \langle B \rangle - E_+ \right) + \sigma \left(\frac{1}{2} (E_+ + E_+) - E_+ \right) \tag{149}$$

$$\varepsilon \partial_x \langle (\mu^+)^2 \psi \rangle = \kappa \left(\frac{1}{4} \langle B \rangle - F_+ \right) + \sigma \left(\frac{1}{4} (E_+ + E_+) - F_+ \right) \tag{150}$$

$$\varepsilon \partial_x F_- = \kappa \left(\frac{1}{2} \langle B \rangle - E_- \right) + \sigma \left(\frac{1}{2} (E_+ + E_+) - E_+ \right) \tag{151}$$

$$\varepsilon \partial_x \langle (\mu^-)^2 \psi \rangle = \kappa \left(-\frac{1}{4} \langle B \rangle - F_- \right) + \sigma \left(-\frac{1}{4} (E_+ + E_+) - F_- \right). \tag{152}$$

This system is closed by an entropy minimization principle. Then, the minimizer ψ_{ME} is determined by

$$H_R^*(\psi_{ME}) = \min_{\psi} \left\{ H_R^*(\psi) : \langle m(\mu) \psi \rangle = (E_+, E_-, F_+, F_-) \right\},$$

i.e. ψ_{ME} minimizes the entropy under the constraint of a reproduction of the half space moments with $E_+ \geq 0, E_- \geq 0, F_+ \geq 0, F_- \leq 0$ as above.

Solutions of similiar minimization problems are discussed in [16]. A straightforward computation shows that in the present case the unique solution $\psi_{ME} = \psi_{ME}(T,\mu) = \psi_{ME}(T,\mu,\nu)$ is given by

$$\psi_{ME}(T,\mu) = \frac{2h\nu^3}{c^2} \frac{1}{\exp(\frac{h\nu}{kT}(\alpha_-(1^- + \beta_-\mu^-) + \alpha_+(1^+ + \beta_+\mu^+))) - 1},$$

where $\alpha_+ > 0, \alpha_- > 0$ and $\beta_+ > -1, \beta_- < 1$ are determined by the constraints

$$\langle 1^\pm \psi_{ME} \rangle = \frac{\sigma_{SB}}{\pi} T^4 \frac{\beta_\pm^2 \pm 3\beta_\pm + 3}{3\alpha_\pm^4 (1 \pm \beta_\pm)^3} = E^\pm$$

$$\langle \mu^\pm \psi_{ME} \rangle = \frac{\sigma_{SB}}{\pi} T^4 \frac{\beta_\pm \pm 3}{6\alpha_\pm^4 (1 \pm \beta_\pm)^3} = F^\pm.$$

The temprature T is introduced here as a normalization parameter to measure the deviation from the usual Planckian. We mention in passing that in general maximization of entropy is a touchy business, see for example [34].

Having solved the minimization problem, one obtains

$$\frac{1}{c}\langle (\mu^\pm)^2 \psi_{ME} \rangle = \frac{\sigma}{c\pi} T^4 \frac{1}{3\alpha_\pm^4 (1 \pm \beta_\pm)^3} = \chi_\pm(f_\pm)E_\pm,$$

if we define the relative fluxes $f_\pm = \frac{F_\pm}{cE_\pm}$ and the Eddington factors

$$\chi_\pm(f_\pm) = \frac{8f_\pm^2}{1 \pm 6f_\pm + \sqrt{1 \pm 12f_\pm - 12f_\pm^2}}.$$

Approximating $\langle (\mu^\pm)^2 \psi \rangle \sim \langle (\mu^\pm)^2 \psi_{ME} \rangle$ in (149)–(152) and using the above computation we arrive at

$$\partial_x F_+ = \kappa \left(\frac{1}{2}\langle B \rangle - E_+ \right) + \sigma \left(\frac{1}{2}(E_+ + E_+) - E_+ \right) \tag{153}$$

$$\partial_x (\chi_+(f_+)E_+) = \kappa \left(\frac{1}{4}\langle B \rangle - F_+ \right) + \sigma \left(\frac{1}{4}(E_+ + E_+) - F_+ \right) \tag{154}$$

$$\partial_x F_- = \kappa \left(\frac{1}{2}\langle B \rangle - E_- \right) + \sigma \left(\frac{1}{2}(E_+ + E_+) - E_+ \right) \tag{155}$$

$$\partial_x (\chi_-(f_-)E_-) = \kappa \left(-\frac{1}{4}\langle B \rangle - F_- \right) + \sigma \left(-\frac{1}{4}(E_+ + E_+) - F_- \right). \tag{156}$$

The system (153–156) is a stationary hyperbolic equation with relaxation terms. An analysis of the equations has been performed in [26].

The eigenvalues associated to E_+ and F_+ are always positive while the eigenvalues associated to E_- and F_- are always negative. This result agrees with intuition since E_+ and F_+ describe transport to the right, while E_- and F_- describe transport to the left. Therefore we have to prescribe two boundary conditions on the left and right hand side, respectively. We use at $x = 0$:

$$F_+ = \langle \mu^+ \left[\rho(\mu) \psi_{ME}(T, -\mu) + (1 - \rho(\mu))B(T_{\text{out}}) \right] \rangle \tag{157}$$

$$\chi_+ E_+ = \langle (\mu^+)^2 \left[\rho(\mu) \psi_{ME}(T, -\mu) + (1 - \rho(\mu))B(T_{\text{out}}) \right] \rangle \tag{158}$$

and the analogous conditions at $x = 1$. Equations (153)–(158) are solved together with the temperature equation in the form

$$\varepsilon^2 \partial_t T = \varepsilon^2 k \partial_{xx} T - 2\pi \kappa \left(\frac{2\sigma_{SB}}{\pi} T^4 - (E_+ + E_-) \right) \tag{159}$$

and corresponding boundary conditions.

The Partial Moment Entropy approximation has a lot of desirable physical and mathematical properties. The underlying distribution function is always positive. Hence the relative flux and the speed of propagation are limited. The system is symmetriziable hyperbolic. This makes it accessible to a powerful mathematical theory guaranteeing well-posedness locally in time. Like the full moment entropy approximation [16], the system correctly approaches the diffusive limit and the free-streaming limit. The eigenvalues of the half moment and quarter moment entropy aproximation have a special structure. For the half moment case, the eigenvalues of the "+" direction are always positive, the eigenvalues of the "−" direction are always negative. Both are bounded in modulus by the speed of light c. This property makes very simple and accurate numerical schemes possible, for example kinetic schemes or upwind schemes. The formulation of accurate boundary conditions is again straight-forward.

6 Frequency-Averaged Moment Equations

Moment models are obtained by testing (32) with functions depending on direction, in our case $(1, \mu)^T$, then integrating the result over all the directions and frequencies. Then, the system does only depend on time and space variables, and is hence far cheaper to solve. However, this has a cost since we are not always able to reproduce neither frequency dependent problems nor very stiff directional configurations such as the collision of two opposite beams [11, 19].

In order to solve this difficulty, we do not average over all directions and all frequencies but distinguish photons going to the left and to the right and different frequency bands.

Let

$$\langle g \rangle_m^+ = \int_{v_{m-\frac{1}{2}}}^{v_{m+\frac{1}{2}}} \int_0^1 g \, d\mu \, dv \quad \text{and} \quad \langle g \rangle_m^- = \int_{v_{m-\frac{1}{2}}}^{v_{m+\frac{1}{2}}} \int_{-1}^0 g \, d\mu \, dv \tag{160}$$

denote the average over all right/left-going photons in the mth frequency band $[v_{m-\frac{1}{2}}, v_{m+\frac{1}{2}}[$. We denote the bands as half-open intervals to have mathematically disjoint sets. However, since only integrals over the bands matter, one could also use closed intervals. The moments $E_{R,m}^+ = \langle \psi \rangle_m^+$, $F_{R,m}^+ = \langle \Omega \psi \rangle_m^+$ and $P_{R,m}^+ = \langle (\Omega \otimes \Omega) \psi \rangle_m^+$ are respectively the radiative energy, the radiative flux and the radiative pressure inside the mth group and the positive half-space. The quantities for the negative half space are defined in analogy.

Testing (32) with $(1, \mu)^T$ and averaging with the above defined averages we get

$$\partial_x F_{R,m}^+ = \hat{\kappa}_m^+ a \theta_{m,+}^4 - \tilde{\kappa}_m^+ E_{R,m}^+ + \tilde{\sigma}_m^+ \left(\frac{E_{R,m}^+ + E_{R,m}^-}{2} - E_{R,m}^+ \right) \tag{161}$$

$$\partial_x P_{R,m}^+ = \hat{\kappa}_m^+ \frac{a}{2} \theta_{m,+}^4 - \check{\kappa}_m^+ F_{R,m}^+ - \tilde{\sigma}_m^+ \left(\frac{E_{R,m}^+ + E_{R,m}^-}{4} - F_{R,m}^+ \right) \tag{162}$$

and

$$\partial_x F_{R,m}^- = \hat{\kappa}_m^- a \theta_{m,-}^4 - \tilde{\kappa}_m^- E_{R,m}^- + \tilde{\sigma}_m^- \left(\frac{E_{R,m}^+ + E_{R,m}^-}{2} - E_{R,m}^- \right) \tag{163}$$

$$\partial_x P_{R,m}^- = -\hat{\kappa}_m^- \frac{a}{2} \theta_{m,-}^4 - \check{\kappa}_m^- F_{R,m}^- - \tilde{\sigma}_m^- \left(-\frac{E_{R,m}^+ + E_{R,m}^-}{4} - F_{R,m}^- \right), \tag{164}$$

where we have used the following frequency averages of the frequency dependent quantites κ and σ:

$$\hat{\kappa}_m^+ = \frac{\langle \kappa B \rangle_m^+}{\langle B \rangle_m^+}, \quad \tilde{\kappa}_m^+ = \frac{\langle \kappa \psi \rangle_m^+}{\langle \psi \rangle_m^+}, \quad \check{\kappa}_m^+ = \frac{\langle \kappa \mu \psi \rangle_m^+}{\langle \mu \psi \rangle_m^+} \quad \text{and} \quad \tilde{\sigma}_m^+ = \frac{\langle \sigma \psi \rangle_m^+}{\langle \psi \rangle_m^+}. \tag{165}$$

6.1 Entropy Minimization

For each m (161)–(164) is a system of 4 equations for 6 unknown moments. To obtain a well-posed system one usually expresses the highest moment, here $P_{R,m}^\pm$, as a function of the lower order moments, here $E_{R,m}^\pm$ and $F_{R,m}^\pm$. This is referred to as "closure" of the system.

To close the system here, we use entropy minimization, see [7, 16, 51, 61]. Compare [19] for the grey half space model and [83] for the multigroup full space model.

Let us first recall the definition of the radiative entropy,

$$h_R(I) = \frac{2kv^2}{c^3} \left[n_I \ln n_I - (n_I + 1) \ln(n_I + 1) \right] \tag{166}$$

where

$$n_I = \frac{c^2}{2hv^3} \psi. \tag{167}$$

According to the entropy minimization principle, we determine a distribution function that minimizes the radiative entropy under the constraint that it reproduces the lower order moments,

$$H_R(\psi_{ME}) = \min_{\psi} \left\{ H(\psi) = \sum_m (\langle h_R(\psi) \rangle_m^+ + \langle h_R(\psi) \rangle_m^-) :: \right| $$
$$\forall m: \langle \psi \rangle_m^{\pm} = E_m^{\pm} \text{ and } c \langle \mu \psi \rangle_m^{\pm} = F_m^{\pm} \right\}. \tag{168}$$

This gives the closure function,

$$\psi_{ME}(\Omega, v) = \sum_m 1_{[v_{m-\frac{1}{2}}; v_{m+\frac{1}{2}}]} \frac{2hv^3}{c^2} \left[\exp(\frac{hv}{k}(\alpha_m^+(1+\beta_m^+\mu^+)+\alpha_m^-(1+\beta_m^-\mu^-)) - 1 \right]^{-1} \tag{169}$$

where $\alpha_m^{\pm}, \beta_m^{\pm}$ are Lagrange multipliers, that are defined to reproduce the moments.

6.2 Inversion of the System

The next step is to express the Lagrange multipliers $\alpha_m^{\pm}, \beta_m^{\pm}$ as functions of $E_{R,m}^{\pm}$, $F_{R,m}^{\pm}$ and to substitute

$$P_{R,m}^{\pm} \approx \langle \psi_{ME}(\alpha_m^{\pm}, \beta_m^{\pm}) \rangle_m^{\pm} = \langle \psi_{ME}(E_{R,m}^{\pm}, F_{R,m}^{\pm}) \rangle_m^{\pm}. \tag{170}$$

Hence we obtain a system for $E_{R,m}^{\pm}$ and $F_{R,m}^{\pm}$.

For the grey half space model [19], the Lagrange multipliers as functions of the moments can be computed explicitly. However, with the introduction of multigroup variables this is not the case anymore. Integrations require the knowledge of the following function,

$$\Xi(\eta) = \int_0^{\eta} \xi^3 [\exp(\xi) - 1]^{-1} d\xi. \tag{171}$$

For example,

$$E_m^+ = \frac{1}{c} \int_0^1 \int_{v_{m-\frac{1}{2}}}^{v_{m+\frac{1}{2}}} \psi_{ME} dv d\mu$$
$$= \int_0^1 \frac{2k^4}{h^3 c^3} (\alpha_m^+(1+\beta_m^+\mu^+))^{-1} (\Xi(v_{m+\frac{1}{2}}') - \Xi(v_{m-\frac{1}{2}}')) d\mu \tag{172}$$

with $v' = \frac{hv}{k}\alpha^+(1+\beta^+\mu^+)$. Unfortunately, except for $\eta = 0$ and $\eta = +\infty$ there is no analytic expression of Ξ. A numerical calculation would be too expensive since we have to be very accurate. Therefore, in [83] an approximation was introduced, which we can also be used here,

$$\Xi(\eta) \simeq C_\infty + \exp(-C_*\eta) \sum_{i=0}^{imax} C_i \eta^i \qquad (173)$$

The constants C_i are chosen so that the approximation has a very good behaviour in the vicinity of $\eta = 0$. For our applications, taking $imax = 5$ is sufficient.

Once this approximation is made, it is possible to integrate and hence to compute the Lagrange multipliers of the minimization problem as functions of the moments. Then, we are able to compute the radiative pressures as functions of the radiative energies and fluxes. Moreover, we can show that we can write the pressures in Eddington form, $P_R^\pm = D_R^\pm E_R^\pm$, where

$$D_m^\pm = \frac{(1-\chi_m^\pm)}{2} : I_d + \frac{(3\chi_m^\pm - 1)}{2} : \frac{F_m^\pm \otimes F_m^\pm}{\left\|F_m^\pm\right\|^2}. \qquad (174)$$

The scalars $\chi_m^\pm$ are called Eddington factors.

6.3 Properties

The multigroup half space model keeps the interesting properties of the other moment models closed by entropy minimization, that is to say

- The main physical properties remain: conservation of the total energy and dissipation of the total entropy. Moreover, the addition of multigroup allows to have a better balanced-energy in the case of strongly frequency-dependant problems.
- The model naturally limits the flux. This property can be expressed as follows:

$$\forall m, : \frac{F_m^\pm}{E_m^\pm} < 1 \qquad (175)$$

This means that the photons cannot travel faster than the speed of the light. We note that this important property is often not satisfied by macroscopic models.
- For 1D problems, it is very easy to make a simple numerical scheme that can efficiently solve every possible angular configuration. This is done only by using upwind schemes (see [19]). We chose to develop only a four-moments model to obtain a simple and very competitive model. However, in some situations one might need more moments to capture the physical solution [78].
- The cost of the method is low and can be lowered to be less than the number of groups times the cost of the half space model by doing a pressure precalculation.

These properties are the most important ones but it is to note that the multigroup half space model keeps all the properties (and limitations) of both the half space [19] and multigroup full space [83] models.

7   Numerical Comparisons

7.1   Numerical Results

The approximations presented above have different mathematical structures. The Discrete Ordinates, Spherical Harmonics and partial P_N equations are linear first order partial differential equations. The minimum entropy and the partial moment entropy system are nonlinear hyperbolic first order partial differential equations. On the other hand the diffusion and flux-limited diffusion equations are parabolic equations, whereas the SP_N equations are elliptic/parabolic. We remark that, although they are closely related, the minimum entropy moment model and flux-limited diffusion with the same Eddington factor are not completely equivalent, but can in fact have very different solutions. For example, the solutions for the minimum entropy system can have shocks whereas this is impossible for flux-limited diffusion.

In the following Figures we show some numerical comparisons of the different models. The abbreviations in the legends mean

- *S40/Transport:* Discrete Ordinates Solution with 40 directions
- *P1:* P_1 approximation with Marshak boundary conditions
- *SP1:* SP_1/Diffusion approximation with Marshak boundary condition
- *FLD:* flux-limited diffusion with minimum entropy Eddington factor and Marshak boundary conditions
- *HSP1:* half P_1 approximation
- *HSM1:* half moment entropy approximation
- *Quarter Space:* quarter moment entropy approximation

The *transport* solution has been obtained with a direct discretization as described above. The parabolic equations *SP1* and *FLD* have been discretized with a standard finite difference scheme. For the balance laws *P1*, *HSP1*, *HSM1* and *Quarter Space* we used kinetic schemes based on the distribution function from the moment closure. All of the latter systems have eigenvalues in modulus less than the speed of light. Thus, similar CFL conditions hold. To be valid in the diffusive limit, the kinetic schemes can be modified to become asymptotic preserving, cf. [19] for a simple analysis in 1D.

7.2   Grey Transport

First, we investigate the transport equation (with fixed temperature) without frequency dependence. In Fig. 9 we consider a given temperature profile in the unit

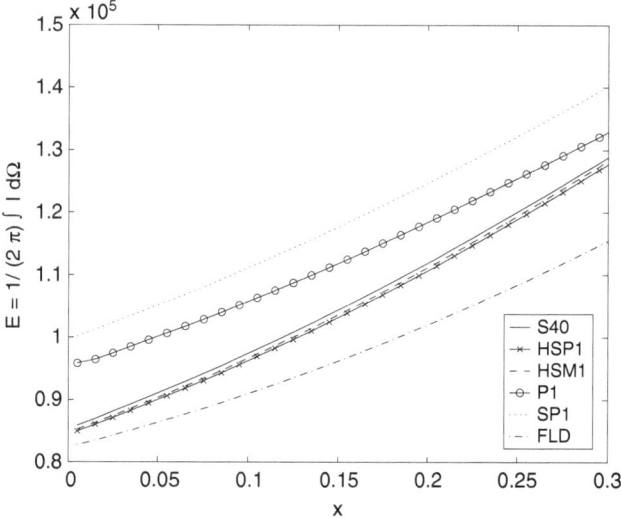

Fig. 9 Steady radiative energy for a fixed temperature profile $T(x) = 1000 + 800x$ in the interval $[0, 1]$, $\kappa = 1$, $\sigma = 0.1$

interval $[0, 1]$, $T(x) = 1000 + 800x$. This temperature enters into the Planck source term $\langle B \rangle$ via Stefan–Boltzmann's law

$$\langle B \rangle (T) = \sigma_{SB} T^4. \tag{176}$$

At the boundary we prescribe black body radiation at the corresponding temperature as ingoing radiation. In Fig. 9 we see that the high order Discrete Ordinates solution (considered as benchmark result) and the half moment approximations agree very well, whereas P_1, SP_1 and flux-limited diffusion differ significantly.

This becomes more striking in the 2D example in Fig. 10. The P_1 and SP_1 approximations are unable to capture the simple anisotropy in this test case, whereas the quarter moment model and the solution of the full equations agree very well.

7.3 Grey Cooling

Here we apply the above methods to a cooling problem. We use an initial temperature of $1,000$ K and an outside temperature of 300 K. The parameters a, ε, κ are set equal to 1. α and the reflectivity are chosen equal to 0. k is chosen equal to 1 and 0.1, i.e. we consider two situations where heat conduction and radiation are dominating, respectively. u is chosen equal to zero. The gridsize is $\Delta x = 0.01$, $\Delta t = 10^{-4}$. For the radiative transfer solution a Gaussian quadrature with 64 points is used for the angular discretization. We use the above first order finite difference discretization

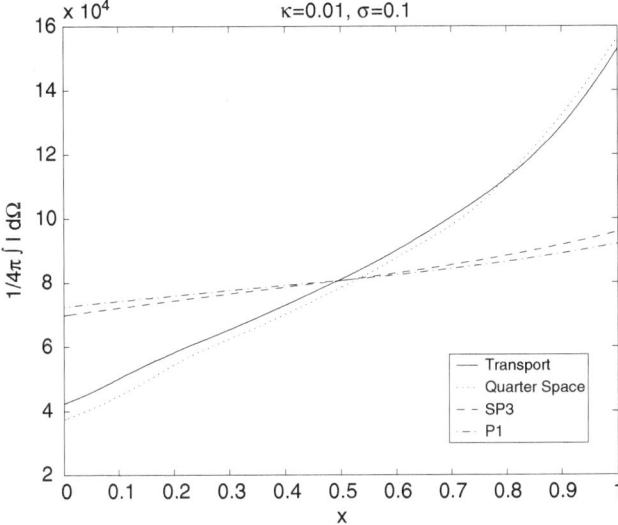

Fig. 10 Steady radiative energy for a fixed temperature profile $T(x) = 1000 + 400(x+y)$ in $[0,1]^2$, $\kappa = 0.01$, $\sigma = 0.1$. Cut along the diagonal

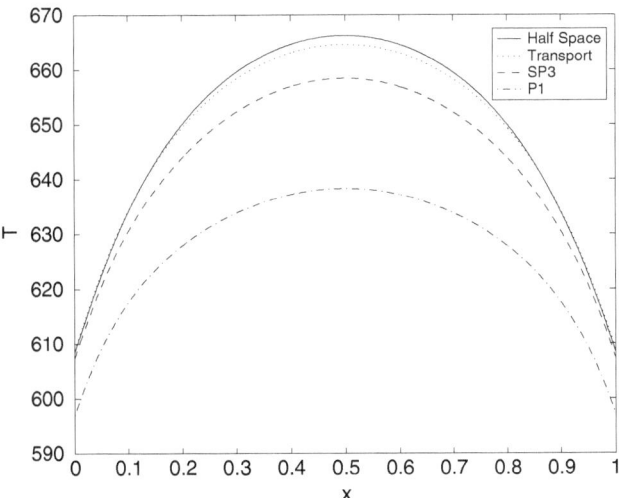

Fig. 11 Temperatures at time $t = 0.01$ with $k = 1$

in space and a Newton iteration to obtain an approximate solution of the nonlinear equations (153–156). The calculated temperatures are shown in Figs. 11 and 13. The mean intensities, i.e. $E_+ + E_-$ in the half moment case and $< \mu I >$ for the radiative transfer solutions, are shown in Figs. 12 and 14. The results obtained with the half space moment method, the P_1 approximation, the SP_3 approximation [47]

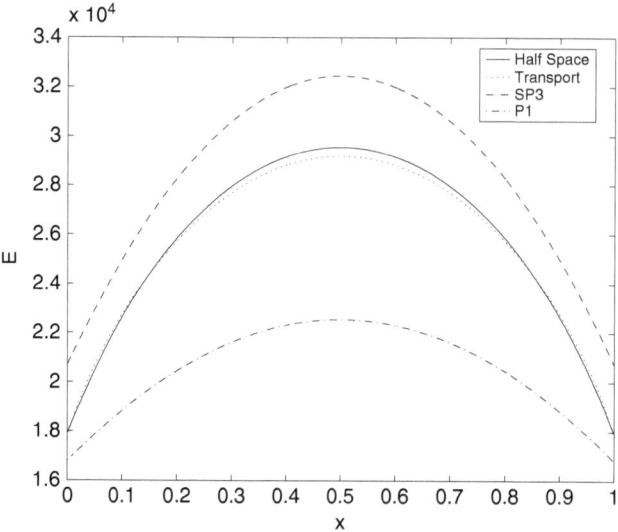

Fig. 12 Mean intensity at time $t = 0.01$ with $k = 1$

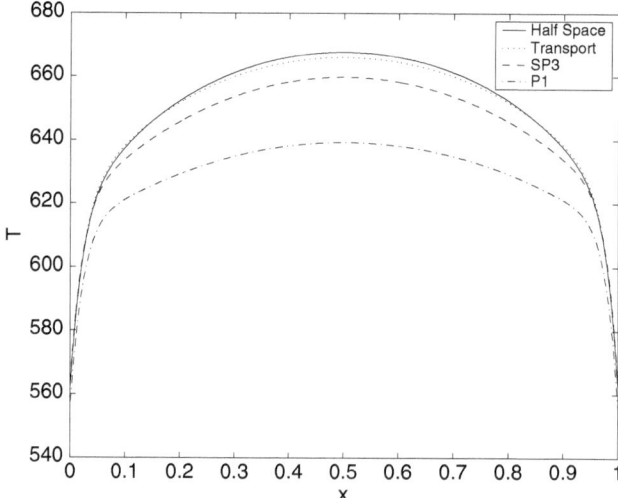

Fig. 13 Temperatures at time $t = 0.01$ with $k = 0.1$

and the solution of the full transport equation are compared. We note that the usual Rosseland or diffusion approximation [45] gives in all cases results which are far less accurate than the solutions considered here. As can be seen in the figures the half moment method outperforms the other methods in both cases.

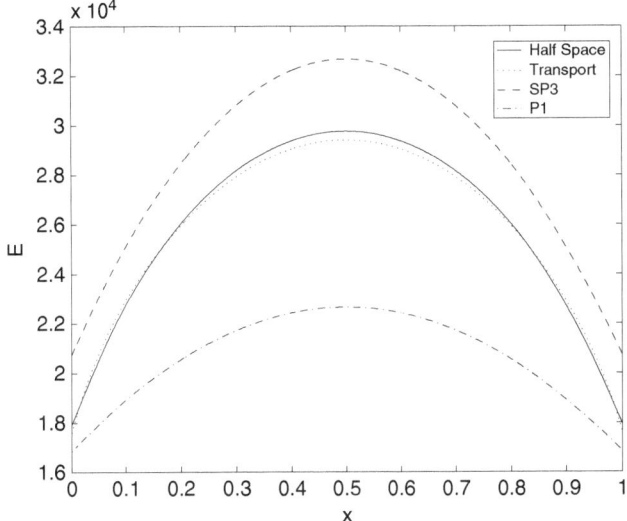

Fig. 14 Mean intensity at time $t = 0.01$ with $k = 0.1$

7.4 Multigroup Transport

First we consider only the equation for the radiative intensity with a fixed matter temperature profile. We divide the spectrum into four bands $[\lambda_{i-\frac{1}{2}}, \lambda_{i+\frac{1}{2}}[\ (]\nu_{i+\frac{1}{2}}, \nu_{i-\frac{1}{2}}]$ respectively) with piecewise constant κ_i on $[\lambda_{i-\frac{1}{2}}, \lambda_{i+\frac{1}{2}}[$. We used $\lambda_{\frac{1}{2}} = 0\ \mu m$, $\lambda_{\frac{3}{2}} = 1.035\ \mu m$, $\lambda_{\frac{5}{2}} = 2.07\ \mu m$, $\lambda_{\frac{7}{2}} = 7\ \mu m$ and $\lambda_{\frac{9}{2}} = \infty$ and $\sigma = 0$.

In Figs. 15–17 we compare the results obtained with the half space moment model to the solution of the full RHT equations using a source iteration as well as diffusive P_1 and SP_3 approximations. For details on these equations we refer the reader to [47]. The classical Rosseland approximation gives in all cases considered here far less accurate results.

For the radiative energy

$$E_R = \int_0^\infty \int_{-1}^1 \psi d\mu d\nu = \sum_m (E_{R,m}^+ + E_{R,m}^-) \tag{177}$$

we define, in analogy to Stefan's law, the radiative temperature

$$T_R := \left(\frac{2\pi E_R}{a} \right)^{1/4}. \tag{178}$$

The parameters corresponding to Fig. 15 are $\kappa_1 = 100\ m^{-1}$, $\kappa_2 = 1\ m^{-1}$, $\kappa_3 = 10\ m^{-1}$, $\kappa_4 = \infty$ and represent a rather diffusive, optically thick physical regime.

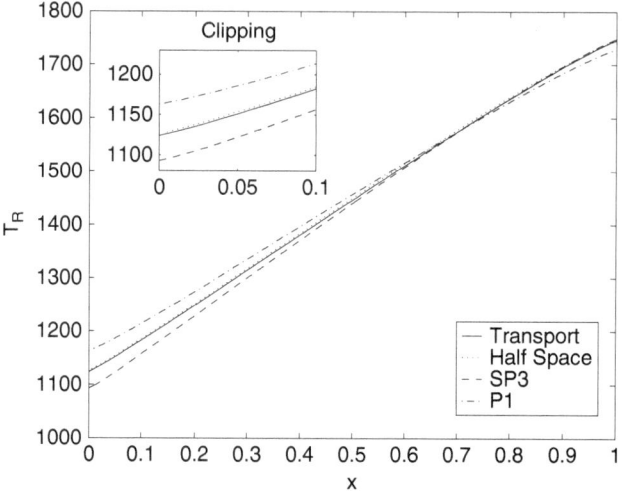

Fig. 15 Steady radiative temperature for a fixed matter temperature profile, $T(x) = 1000 + 800x$, $T_b(0) = 1000$, $T_b(1) = 1800$. Diffusive regime. Four frequency bands

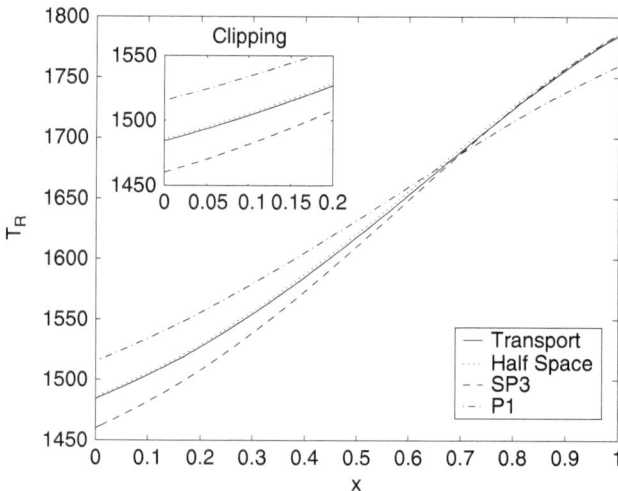

Fig. 16 Steady radiative temperature for a fixed matter temperature profile, $T(x) = 500 + 1500x$, $T_b(0) = 500$, $T_b(1) = 2000$. Transport regime. Four frequency bands

The half space model performs better than the diffusive approximations which are designed for this physical situation. The differences become more striking in Fig. 16 where we chose a rather opposite physical regime with large photon mean free path, $\kappa_1 = 0.1 \ m^{-1}$, $\kappa_2 = 0.01 \ m^{-1}$, $\kappa_3 = 1 \ m^{-1}$, $\kappa_4 = \infty$. We chose the same absorption coefficients in Fig. 17. However, while in the first two cases the boundary

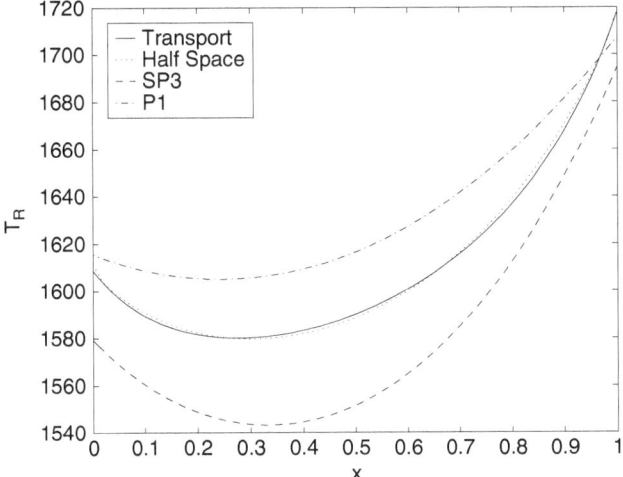

Fig. 17 Steady radiative temperature for a fixed matter temperature profile, $T(x) = 500$, $T_b(0) = 1500$, $T_b(1) = 2000$. Transport regime. Four frequency bands

temperature agreed with the interior matter temperature we chose here a much higher boundary temperature which corresponds to heat flux entering the medium. The half space model is far more accurate than the diffusive approximations.

7.5 Multigroup Cooling

In our next test case we consider the transport equation coupled to the heat equation. We use $k = h = 1$, $\alpha = 0$ and $\rho = 0$. The outside temperature is $T_b = 1,000$ at the left and $T_b = 1,800$ at the right boundary. The scattering and absorption coefficients are chosen as in our second and third uncoupled test cases. In Fig. 18 we show the steady radiative temperature. Again, the new half space model agrees best with the full transport solution.

7.6 Adaptive methods for the Simulation of 2-d and 3-d Cooling Processes

The application that we study here is the cooling of a glass cube representing a typical fabrication step in glass manufacturing. We consider clean glass, which means that the treatment of scattering can be omitted. The frequencies are approximated by an eight-band model. The values used are given in Table 2. Furthermore, we set

$$k = 1, \quad h = 0.001, \quad T_b = 300$$

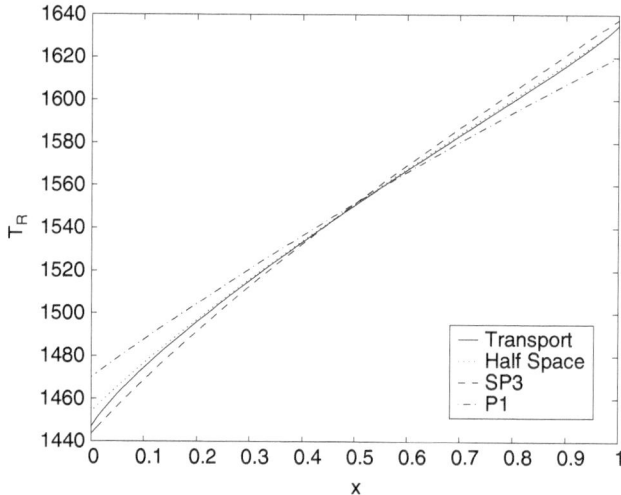

Fig. 18 Steady radiative temperature for the coupled equations, $T_b(0) = 1,000$, $T_b(1) = 1,800$. Transport regime. Four frequency bands

and start with a uniform temperature distribution $T_0(x) = 1,000$. The time integration is stopped at $t = 0.001$.

We use a space-time adaptive method described in detail in [38]. We validate the SP_N-solutions with numerical solutions to the full RHTE. The full RHTE is solved by a diamond differencing discretization coupled with a discrete ordinate method which uses 60 directions [10, 76]. This is for the present situation sufficient to obtain an accurate solution for the transport problem provided the spatial grid is chosen fine enough.

7.6.1 Two-Dimensional Glass Cooling

We consider an infinitely long square glass block which allows us to use a two-dimensional approximation on the scaled square domain $\Omega = [0, 1]^2$.

In Fig. 19, we show temperature distributions at the final time $t_e = 0.001$ obtained for the SP_3-approximation. As expected, the strongest cooling takes place in the corners of the computational domain. The meshes automatically chosen by our adaptive approach are highly refined at the boundary caused by the steep temperature gradients there. In this case, a stable uniform discretization of the two-dimensional RHTE requires the solution of a linear system with more than 4.8 million unknowns in each time step, whereas the dimension of the linear algebraic systems for the adaptive SP_3-approximation is not greater than 272,000.

In Fig. 20 the SP_N-solutions are compared to the full RHTE- and Rosseland approximation. In particular, they reconstruct the temperature near the boundary much more accurately than the Rosseland approximation which is often used in engineering practice.

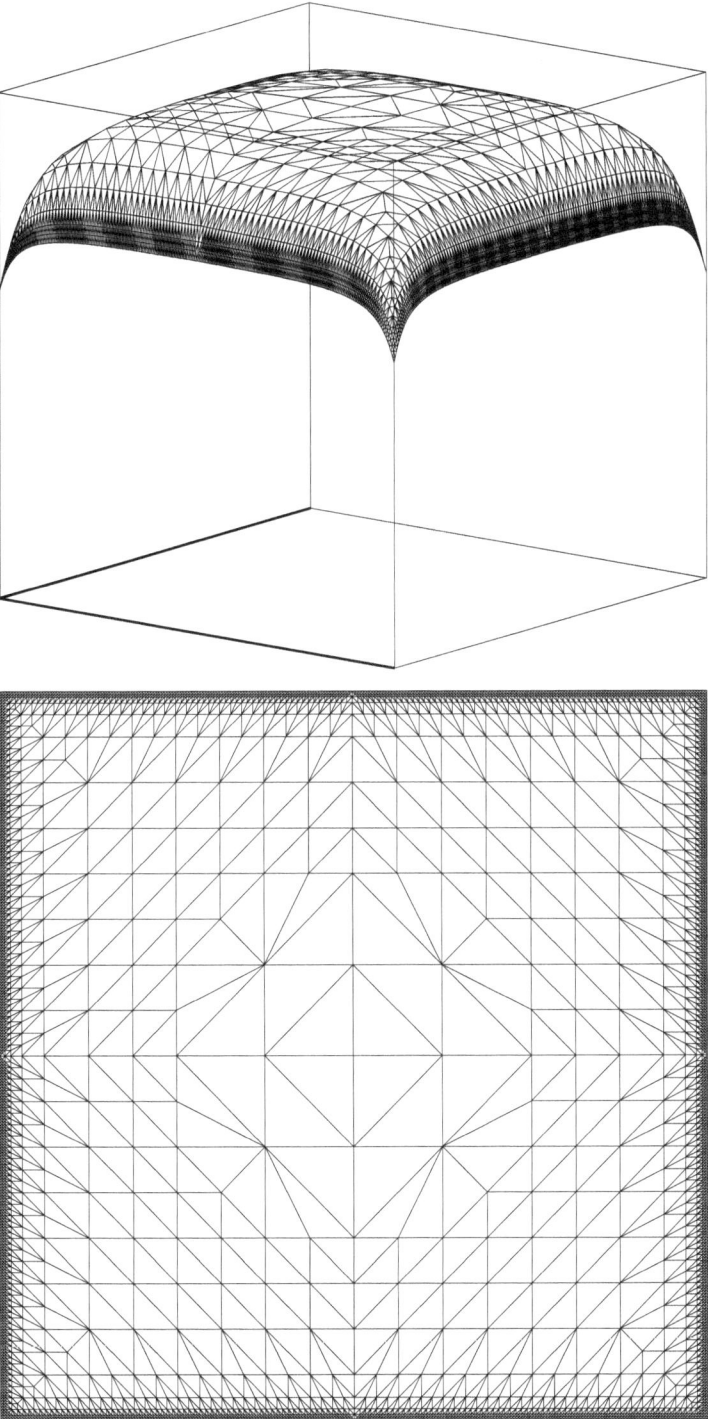

Fig. 19 Two-dimensional temperature distributions and spatial meshes on $\Omega = [0,1]^2$ resulting from SP_3-approximations at $t_e = 0.001$. The temperature axis ranges from 300 to 1,000. Strong refinement takes place in the boundary layer due to the large temperature gradients there

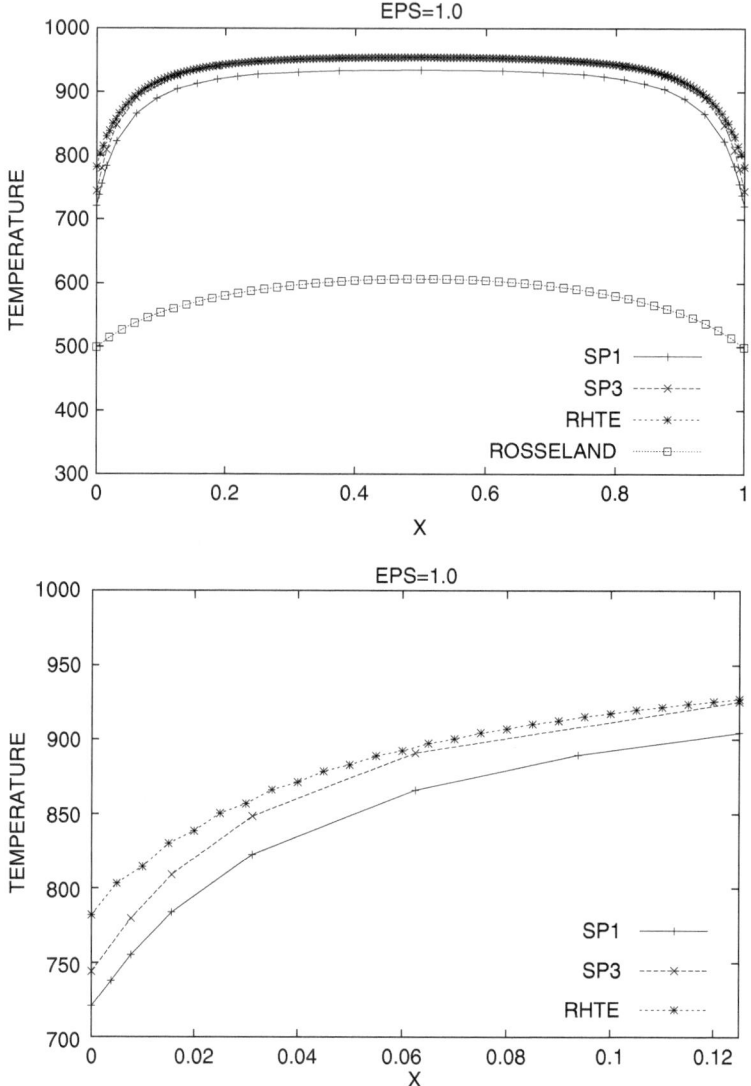

Fig. 20 Comparison of two-dimensional temperature distributions at $t_e = 0.001$ along the line $y = 0.5$ obtained from different radiation models. The SP_3-solution matches very well with the RHTE solution inside the glass cube. Some differences are visible in the boundary region. Both SP_N-approximations give much more accurate results than the Rosseland approximation

The time steps in the adaptive procedure increase rapidly by two orders of magnitude reflecting the ongoing diffusive smoothing in the boundary layer. Altogether 9 and 24 time steps are needed. In contrast, a uniform time discretization yielding the same accuracy, requires 100 steps.

Concerning the computation times the parameters discussed above lead to the following results: Using the method described above without adaptivity in space and time for Rosseland, SP_1 and SP_3 the computational effort is approximately doubled

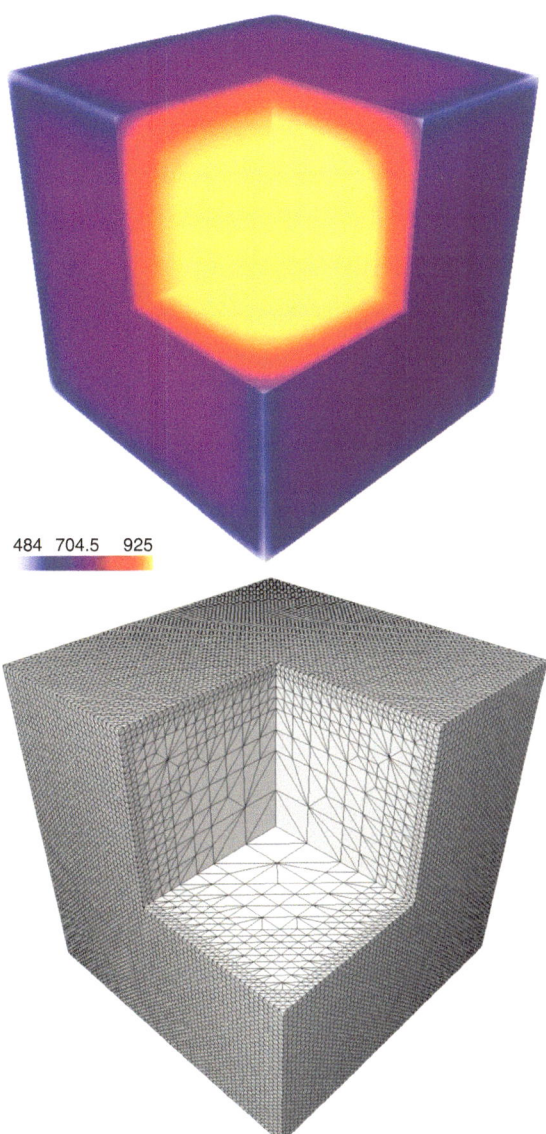

Fig. 21 Three-dimensional temperature distribution and adaptive spatial mesh on $\Omega = [0,1]^3$ resulting from the SP_3-approximation at $t_e = 0.001$. We removed one small cube to present details from inside the glass cube. Refinement takes place in the boundary layer due to the large temperature gradients there. The adaptive mesh consists of $82,705$ grid points

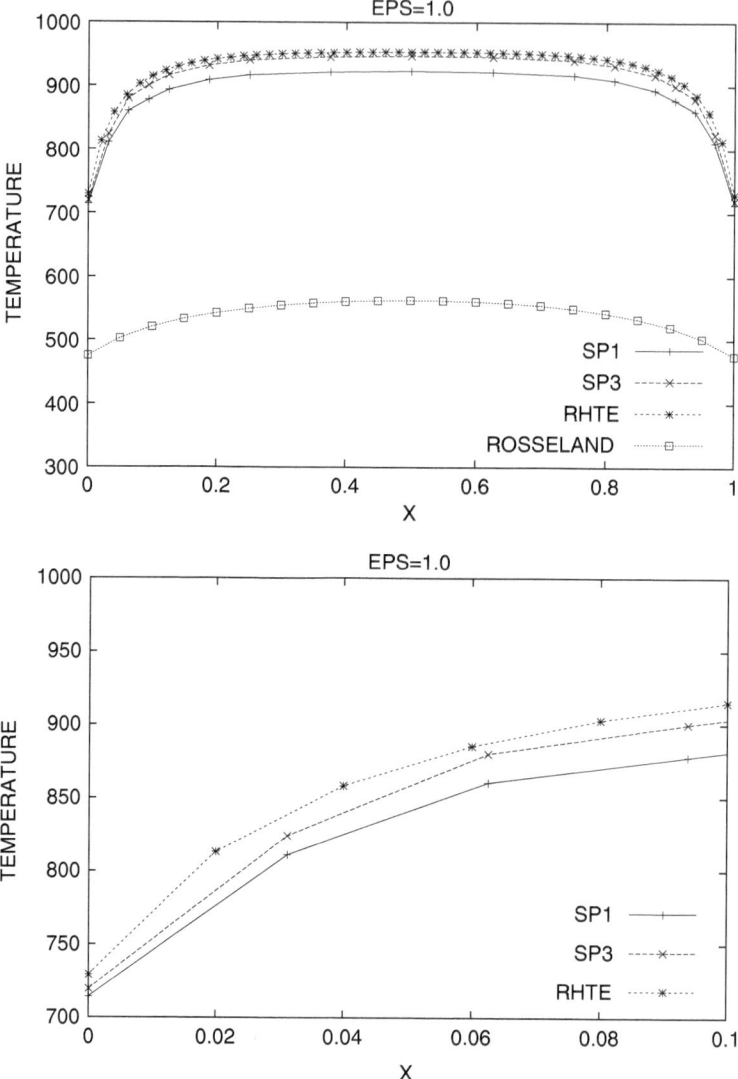

Fig. 22 Comparison of three-dimensional temperature distributions at $t_e = 0.001$ along the line $y = z = 0.5$ obtained from different radiation models. The SP_3-solution matches very well with the RHTE solution, whereas the Rosseland approximation gives quite poor results

from Rosseland to SP_1 and from SP_1 to SP_3. The solution of the RHT problem using the multigrid method described in [76] takes again approximately twice as much time as the SP_3 solution for the same accuracy. Adaptivity in space yields a factor of 3–5 in computation time for the present situation and adaptivity in time yields a factor of 10–50.

7.6.2 Three-Dimensional Glass Cooling

We consider a glass block which is represented by a scaled cube $\Omega = [0, 1]^3$.

Figure 21 displays the SP_3-solution and the adaptive three-dimensional grid chosen by our method for $TOL_x = 0.01$. As already observed in the two-dimensional case, the SP_N-solutions approximate the temperature computed from the full RHTE very well. In contrast to the Rosseland approximation, they exhibit physically correct boundary layers as can be seen from Fig. 22. To accurately capture these boundary layers, the use of local refinement is essential.

The mesh shown in Fig. 21 consists of 82,705 nodes, leading to a linear system of order 1,405,985. A uniform method requires approximately 250,000 grid points to reach a comparable solution quality. The solution of the full RHTE is done on a $100 \times 100 \times 100$-grid, yielding a linear system with 480 million unknowns which has to be solved in each time step. The comparison of the computation times yields similiar results as in 2-D.

To conclude, these investigations show that the SP_N-equations and moment methods described above are a relatively inexpensive way to improve the accuracy of classical diffusion models. Compared to the solution of full radiative heat transfer equations, the complexity and computer time are considerably reduced. Further reduction can be achieved by fully adaptive discretization methods steered by robust a posteriori error estimators.

Acknowledgements We wish to thank all our collaborators and co-authors, in particular B. Dubroca, T. Götz, J. Lang, E.W. Larsen, M. Seaïd, G. Thömmes, R. Turpault and R. Pinnau. Parts of this work have been taken from the articles [18, 23–25, 27, 38, 47, 48, 76, 85]. This work was supported by German Research Foundation DFG under grants KL 1105/7 and 1105/14.

References

1. Adams, M.L., Larsen, E.W.: Fast iterative methods for deterministic particle transport computations. Prog. Nucl. Energy **40**, 3–159 (2002)
2. Adams, M.L.: Subcell balance formulations for radiative transfer on arbitrary grids. Transp. Theory and Stat. Phys. **26**, 385–431 (1997)
3. Alcouffe, R., Brandt, A., Dendy, J., Painter, J.: The multigrid method for diffusion equations with strongly discontinuous coefficients. SIAM J. Sci. Stat. Comp. **2**, 430–454 (1981)
4. Alcouffe, R.: Diffusion synthetic acceleration methods for the diamond-differenced discrete-ordinates equations. Nucl. Sci. Eng. **64**, 344–355 (1977)
5. Agoshkov, V.: On the existence of traces of functions in spaces used in transport theory problems. Sov. Math. Dokl. **33**, 628–632 (1986)
6. Agoshkov, V.: Boundary value problems for transport equations. Birkhäuser, Boston (1998)
7. Anile, A.M., Pennisi, S., Sammartino, M.: A thermodynamical approach to Eddington factors. J. Math. Phys. **32**, 544–550 (1991)
8. Atkinson, K.E.: Iterative variants of the Nyström method for the numerical solution of integral equations. Numer. Math. **22**, 17–31 (1973)
9. Brantley, P.S., Larsen, E.W.: The simplified P_3 approximation. Nucl. Sci. Eng. **134**, 1–21 (2000)

10. Brown, P.N.: A linear algebraic development of diffusion synthetic acceleration for three-dimensional transport equations. SIAM. J. Numer. Anal. **32**, 179–214 (1995)
11. Brunner, T.A., Holloway, J.P.: One-dimensional Riemann solvers and the maximum entropy closure. J. Quant. Spectrosc. Radiat. Transfer **69**, 543–566 (2001)
12. Cheng, P.: Dynamics of a radiating gas with applications to flow over a wavy wall. AIAA J. **4**, 238–245 (1966)
13. Clause, P.-J., Mareschal, M.: Heat transfer in a gas between parallel plates: Moment method and molecular dynamics. Phys. Rev. A **38**, 4241–4252 (1988)
14. Davison, B.: Neutron transport theory. Oxford University Press, Oxford (1958)
15. Dreyer, W.: Maximisation of the entropy in non-equilibrium. J. Phys. A **20**, 6505–6517 (1987)
16. Dubroca, B., Feugeas, J.L.: Entropic moment closure hierarchy for the radiative transfer equation. C. R. Acad. Sci. Paris Ser. I **329**, 915–920 (1999)
17. Dubroca, B.: Thèse d'Etat, Dept. of Mathematics, University of Bordeaux (2000)
18. Dubroca, B., Frank, M., Klar, A., Thömmes, G.: Half space moment approximation to the radiative heat transfer equations. Z. Angew. Math. Mech. **83**, 853–858 (2003)
19. Dubroca, B., Klar, A.: Half moment closure for radiative transfer equations. J. Comput. Phys. **180**, 584–596 (2002)
20. Eddington, A.: The Internal Constitution of the Stars. Dover, New York (1926)
21. Fischer, A.E., Marsden, J.E.: The Einstein evolution equations as a first-order quasi-linear hyperbolic system I. Commun. Math. Phys. **26**, 1–38 (1972)
22. Fiveland, W.A.: The selection of discrete ordinate quadrature sets for anisotropic scattering. ASME HTD. Fundam. Radiat. Heat Transf. **160**, 89–96 (1991)
23. Frank, M.: Partial Moment Models for Radiative Transfer. PhD thesis, TU Kaiserslautern (2005)
24. Frank, M.: Approximate models for radiative transfer. Bull. Inst. Math. Acad. Sinica (New Series) **2**, 409–432 (2007)
25. Frank, M., Dubroca, B., Klar, A.: Partial moment entropy approximation to radiative transfer. J. Comput. Phys. **218**, 1–18 (2006)
26. Frank, M., Pinnau, R.: Analysis of a half moment model for radiative heat transfer equations. Appl. Math. Lett. **20**, 189–193 (2007)
27. Frank, M., Seaïd, M., Janicka, J., Klar, A., Pinnau, R.: A comparison of approximate models for radiation in gas turbines. Prog. Comput. Fluid Dyn. **4**, 191–197 (2004)
28. Golse, F., Perthame, B.: Generalized solution of the radiative transfer equations in a singular case. Commun. Math. Phys. **106**, 211–239 (1986)
29. Greenbaum, A.: Iterative Methods for Solving Linear Systems. SIAM, Philadelphia (1997)
30. Hackbusch, W.: Multi-Grid Methods and Applications. Springer Series in Computational Mathematics, vol. 4. Springer, New York (1985)
31. Howell, R., Siegel, J.R.: Thermal Radiation Heat Transfer, 3rd edn. Taylor & Francis, New York (1992)
32. Huang, K.: Introduction to Statistical Physics. Taylor and Francis, New York (2001)
33. Jeans, J.H.: The equations of radiative transfer of energy. Mon. Not. R. Astron. Soc. **78**, 28–36 (1917)
34. Junk, M.: Domain of definition of levermore's five-moment system. J. Stat. Phys. **93**, 1143–1167 (1998)
35. Kelley, C.T.: Iterative Methods for Linear and Nonlinear Equations. SIAM, Philadelphia (1995)
36. Kelley, C.T.: Multilevel Source Iteration Accelerators for the Linear Transport Equation in Slab Geometry. Transp. Theory Stat. Phys. **24**, 679–707 (1995)
37. Kelley, C.T.: Existence and uniqueness of solutions of nonlinear systems of conductive radiative heat transfer equations. Transp. Theory Stat. Phys. **25**, 249–260 (1996)
38. Klar, A., Lang, J., Seaid, M.: Adaptive solutions of SPN-Approximations to radiative heat transfer in glass. Int. J. Therm. Sci. **44**, 1013–1023 (2005)
39. Klar, A., Schmeiser, C.: Numerical passage from radiative heat transfer to nonlinear diffusion models. Math. Mod. Meth. Appl. Sci. **11**, 749–767 (2001)

40. Klar, A., Siedow, N.: Boundary layers and domain decomposition for radiative heat transfer and diffusion equations: Applications to glass manufacturing processes. Eur. J. Appl. Math. **9–4**, 351–372 (1998)
41. Korganoff, V.: Basic Methods in Transfer Problems. Dover, New York (1963)
42. Krook, M.: On the solution of equations of transfer. Astrophys. J. **122**, 488 (1955)
43. Laitinen, M.T., Tiihonen, T.: Integro-differential equation modelling heat transfer in conducting, radiating and semitransparent materials. Math. Meth. Appl. Sci. **21**, 375–392 (1998)
44. Laitinen, M.T., Tiihonen, T.: Conductive-radiative heat transfer in grey materials. Quart. Appl. Math. **59** 737–768 (2001)
45. Larsen, E.W., Keller, J.B.: Asymptotic solution of neutron transport problems for small mean free path. J. Math. Phys. **15**, 75 (1974)
46. Larsen, E.W., Pomraning, G., Badham, V.C.: Asymptotic analysis of radiative transfer problems. J. Quant. Spectr. Radiati. Transf. **29**, 285–310 (1983)
47. Larsen, E.W., Thömmes, G., Klar, A., Seaïd, M., Götz, T.: Simplified P_N approximations to the equations of radiative heat transfer in glass. J. Comput. Phys. **183**, 652–675 (2002)
48. Larsen, E.W., Thoemmes, G., Klar, A. : New frequency-averaged approximations to the equations of radiative heat transfer. SIAM Appl. Math. **64** 565–582 (2003)
49. Lentes, F.T., Siedow, N.: Three-dimensional radiative heat transfer in glass cooling processes. Glastech. Ber. Glass Sci. Technol. **72**, 188–196 (1999)
50. Levermore, C.D.: Relating Eddington factors to flux limiters. J. Quant. Spectroscop. Radiat. Transf. **31**, 149–160 (1984)
51. Levermore, C.D.: Moment closure hierarchies for kinetic theories. J. Stat. Phys. **83** (1996)
52. Lewis, E.E., Miller, W.F. Jr., Computational Methods of Neutron Transport. Wiley, New York (1984)
53. Lopez-Pouso, O.: Trace theorem and existence in radiation. Adv. Math. Sci. Appl. **10**, 757–773 (2000)
54. Mark, J.C.: The spherical harmonics method, part I. Tech. Report MT 92, National Research Council of Canada (1944)
55. Mark, J.C.: The spherical harmonics method, part II. Tech. Report MT 97, National Research Council of Canada (1945)
56. Marshak, R.E.: Note on the spherical harmonic method as applied to the milne problem for a sphere. Phys. Rev. **71**, 443–446 (1947)
57. Mengüc, M.P., Iyer, R.K.: Modeling of radiative transfer using multiple spherical harmonics approximations. J. Quant. Spectrosc. Radiat. Transf. **39** (1988), 445–461.
58. Mercier, B.: Application of accretive operators theory to the radiative transfer equations. SIAM J. Math. Anal. **18**, 393–408 (1987)
59. Minerbo, G.N.: Maximum entropy Eddington factors. J. Quant. Spectrosc. Radiat. Transf. **20**, 541–545 (1978)
60. Modest, M.F.: Radiative Heat Transfer, 2nd edn. Academic, San Diego (1993)
61. Müller, I., Ruggeri, T.: Rational extended thermodynamics. Springer, New York (1998)
62. Murray, R.L.: Nuclear reactor physics. Prentice Hall, New Jersey (1957)
63. Ore, A.: Entropy of radiation. Phys. Rev. **98**, 887 (1955)
64. Özisik, M.N., Menning, J., Hälg, W.: Half-range moment method for solution of the transport equation in a spherical symmetric geometry. J. Quant. Spectrosc. Radiat. Transf. **15**, 1101–1106 (1975)
65. Planck, M.: Distribution of energy in the spectrum. Ann. Phys. **4**, 553–563 (1901)
66. Pomraning, G.C.: The equations of radiation hydrodynamics. Pergamon, New York (1973)
67. Pomraning, G.C.: Initial and boundary conditions for equilibrium diffusion theory. J. Quant. Spectrosc. Radiat. Transf. **36**, 69 (1986)
68. Pomraning, G.C.: Asymptotic and variational derivations of the simplified P_N equations. Ann. Nucl. Energy **20**, 623 (1993)
69. Porzio, M.M., Lopez-Pouso, O.: Application of accretive operators theory to evolutive combined conduction, convection and radiation. Rev. Mat. Iberoam. **20**, 257–275 (2004)

70. Rosen, P.: Entropy of radiation. Phys. Rev. **96**, 555 (1954)
71. Saad, Y., Schultz, M.H.: GMRES: A generalized minimal residual algorithm for solving non-symmetric linear systems. SIAM. J. Sci. Statist. Comput. **7**, 856–869 (1986)
72. Schäfer, M., Frank, M., Pinnau, R.: A hierarchy of approximations to the radiative heat trans-fer equations: Modelling, analysis and simulation. Math. Meth. Mod. Appl. Sci. **15**, 643–665 (2005)
73. Schuster, A.: Radiation through a foggy atmosphere. Astrophys. J. **21**, 1–22 (1905)
74. Schwarzschild, K.: Über das Gleichgewicht von Sonnenatmosphären, Akad. Wiss. Göttingen. Math. Phys. Kl. Nachr. **195**, 41–53 (1906)
75. Seaïd, M.: Notes on Numerical Methods for Two-Dimensional Neutron Transport Equation, Technical Report Nr. **2232**, TU Darmstadt (2002)
76. Seaid, M., Klar, A.: Efficient Preconditioning of Linear Systems Arising from the Discretiza-tion of Radiative Transfer Equation, Challenges in Scientific Computing. Springer, Berlin (2003)
77. Sherman, M.P.: Moment methods in radiative transfer problems. J. Quant. Spectrosc. Radiat. Transf. **7**, 89–109 (1967)
78. Struchtrup, H.: On the number of moments in radiative transfer problems. Ann. Phys. **266**, 1–26 (1998)
79. Sykes, J.B.: Approximate integration of the equation of transfer. Mon. Not. R. Astron. Soc. **111**, 377 (1951)
80. Tomašević, D.I., Larsen, E.W.: The Simplified P_2 Approximation. Nucl. Sci. Eng. **122**, 309–325 (1996)
81. Turek, S.: An efficient solution technique for the radiative transfer equation. IMPACT, Com-put. Sci. Eng. **5**, 201–214 (1993)
82. Turek, S.: A generalized mean intensity approach for the numerical solution of the radiative transfer equation. Computing **54**, 27–38 (1995)
83. Turpault, R.: Construction d'une modèle M1-multigroupe pour les équations du transfert ra-diatif. C. R. Acad. Sci. Paris Ser. I **334**, 1–6 (2002)
84. Turpault, R.: A consistent multigroup model for radiative transfer and its underlying mean opacities. J. Quant. Spectrosc. Radiat. Transf. **94**, 357–371 (2005)
85. Turpault, R., Frank, M., Dubroca, B., Klar, A.: Multigroup half space moment approximations to the radiative heat transfer equations. J. Comput. Phys. **198**, 363–371 (2004)
86. Van der Vorst, H.A.: BI-CGSTAB: A fast and smoothly converging variant of BI-CG for the solution of nonsymmetric linear systems. SIAM. J. Sci. Statist. Comput. **13**, 631–644 (1992)
87. Viskanta, R., Anderson, E.E.: Heat transfer in semitransparent solids. Adv. Heat Transf. **11**, 318 (1975)
88. Viskanta, R., Mengüc, M.P.: Radiation heat transfer in combustion systems. Prog. Energy Combust. Sci. **13**, 97–160 (1987)

Radiative Heat Transfer and Applications for Glass Production Processes II

Norbert Siedow

1 Introduction

Glass is a man-made material which is used for many thousands of years. It plays an important role in our everydays life, in modern architecture, in science and many other fields.

N. Siedow (✉)

Fraunhofer-Institut für Techno- und Wirtschaftsmathematik Kaiserslautern, Germany
e-mail: norbert.siedow@itwm.fraunhofer.de

A. Fasano (ed.), *Mathematical Models in the Manufacturing of Glass*,
Lecture Notes in Mathematics 2010, DOI 10.1007/978-3-642-15967-1_3,
© Springer-Verlag Berlin Heidelberg 2011

Glass is one of the oldest materials in the world. The oldest finds date back to the stone age in around 700 B.C.. In Egypt, the organized production of glass into jewelry items and small vessels commenced in around 3000 B.C.. From 1500 B.C. onwards, hollow glass was manufactured into ointment jars and oil containers in Egypt. Window glass dates back to the Gothic era of the twelfth century and the first glass rolling process to 1688. More details about the history of glass and glass manufacturing can be found at the homepage of the Federal Association of the German Glass Industry (www.bvglas.de).

Today the glass industry is discovering new applications for glass based on state of the art technology and recent scientific findings. One of the most recent applications for glass is its use as a building material. Glass is also used as an insulating material in the form of glass fiber, it is used to make optical fibres for telephone calls or TV in communications technology. Glass has become a key component in displays and semi-conductors.

Glass products and their manufacturing require continuous adjustments to customer needs and constant process optimization is necessary to ensure and improve the quality of the glass products. Beside expensive and time consuming laboratory experiments numerical simulation has become a key technology. "Strict numerical treatment, i.e., the mathematical simulation of product behavior and all aspects of production processes, is a must for every material-producing company" [5]. At a workshop on "Modeling Needs of the Glass Industry" [8] Choudhary and Huff from Owens–Corning, OH (USA), wrote: "In a complex process such as glassmaking, several variables are monitored and controlled ... . From the perspective of modeling, the greatest interest is in the control of various temperatures."

The knowledge of the temperature is important in almost all stages of the glass production and glass processing. During the melting process in the glass tank the temperature influences the homogeneity of the glass melt. Undesired cracks could be the result of a wrong cooling of the glass. Thus, the knowledge of the right temperature is important to guarantee the quality of the final products. Glass is a semitransparent material. Besides heat conduction and heat convection radiation plays an important role. Especially for high temperatures heat transfer by radiation is the dominant process. Conduction and convection are local phenomena. The mean free path for molecular collisions is generally very small. Thermal radiation, on the other side, is generally a global phenomenon. The average distance a photon travels before interacting with a molecule may be very short (e.g. absorption in metal), but can also be very long (e.g. sun rays).

The simulation of radiative heat transfer is a challenging problem for engineers, physicists and mathematicians since many years. The coupling of heat conduction and thermal radiation results in a high-dimensional non-linear system of partial differential equations. For the numerical solution of the partial differential equations different methods can be used. Besides classical approaches like ray tracing techniques or full discretization methods like the method of discrete ordinates asymptotic approaches offer a good opportunity to derive alternative numerical methods (Rosseland approximation, PN- and SPN-approximation, improved diffusion methods). From the practical point of view it is necessary that these methods can be used together with commercial software packages. The methods should give sufficiently accurate results in acceptable time. In the first part of the lecture we will discuss numerical models for fast radiative heat transfer simulation. The most used method in industrial application is the Rosseland approximation, which treats radiation as a correction of heat conduction. This fast method gives sufficiently accurate results only for optically thick materials. Therefore, we present two diffusion approximation methods, which are nearly as fast as the Rosseland approximation but valid also for semitransparent materials like glass. These methods conserve all geometrical information about the glass domain under consideration.

Fast numerical methods are the fundamental basis for optimization and inverse problems like the control of glass cooling to avoid high permanent thermal stress inside the glass or indirect measurements which are very often the only way to measure the temperature inside hot glasses or glass melts. Spectral remote sensing is an example for an indirect measurement method. Using a pyrometer or thermocamera the spectral radiative intensity is measured for each wavelength. For the temperature reconstruction a Fredholm's integral equation of the first kind has to be solved. From the mathematical point of view one has to solve an ill-posed inverse problem. In the second part of the lecture we will discuss ill-posed inverse problems – examples, the reason of ill-posedness and methods to overcome this difficulty. As an first example coming from glass industry we discuss spectral remote sensing. The second industrial application deals with the reconstruction of the initial temperature condition. Measuring temperature and heat flux at the whole boundary of the glass domain or parts of it we present an other possibility to reconstruct the temperature inside the glass body.

2   Models for Fast Radiative Heat Transfer Simulation

2.1   Introduction

For many industrial applications the modeling of heat transfer processes is of utmost importance. This applies especially to radiative heat transfer in semitransparent materials like glasses. The temperature is the most important parameter in almost all stages of glass making and glass processing. During the glass melting the temperature influences the homogeneity of the glass melt, the drop temperature influences the following forming process and finally the cooling of the hot glass decides about the frozen thermal stresses inside the glass product. Undesired effects like breakage could be the result of a wrong cooling. Hence, the knowledge of the exact temperature distribution is necessary to control the production processes and the final quality of the products. To determine the temperature one can use measurements or simulate the heat transfer process using computers. In contrast to opaque materials, heat can be transported in semi-transparent glass by phononic vibrational heat conduction as well as via thermal radiation which is important particularly for high

Fig. 1 Internal optical transmittance of a 1-mm-thick glass pane

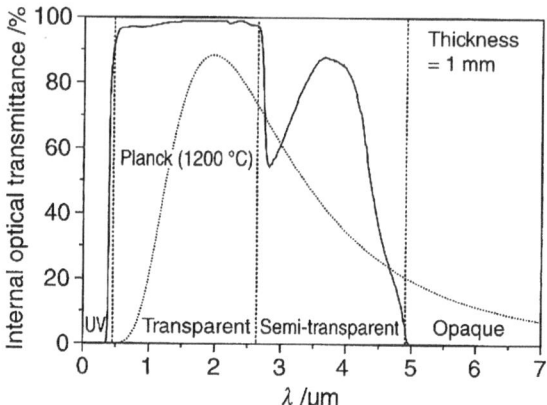

temperatures. Whereas heat conduction acts on a microscale in a $nm-$range thermal radiation acts on a macroscale from the $nm-$ to the $cm-$range. Figure 1 taken from [5] shows the internal optical transmittance of a typical optical glass.

The radiative field is divided into an opaque, semitransparent, and a transparent wavelength range. In the wavelength range $\lambda > 5\,\mu m$ the glass is opaque, that is no heat can be transported by radiation within the glass. The mean free path length of photons is very small. Only the surface of the glass can exchange heat via radiation with the surroundings. In the semi-transparent wavelength range, (in Fig. 1: $2.5 < \lambda \leq 5\mu m$), heat can be transported by radiation within the glass. In the transparent wavelength range, (in Fig. 1: $0.5 < \lambda \leq 2.5\,\mu m$), the radiative field does not interact with the glass.

Very good surveys about radiative heat transfer one can find in [1] and [7]. Furthermore we refer to [5], where the application to glass industry is in the focus of consideration, and to [13] and [14] for the numerical treatment of the radiative transfer equation.

From the mathematical point of view for the simulation of the temperature distribution of hot glasses one has to deal with a coupled system of partial differential equations. Let D be a three-dimensional domain of absorbing and emitting glass. The heat transfer in D is described by the energy equation

$$c_m \rho_m \frac{\partial T}{\partial t}(\mathbf{r},t) = \nabla \cdot (k_h \nabla T(\mathbf{r},t)) - \nabla \cdot \mathbf{q}_r(T), \quad (\mathbf{r},t) \in D \times (0,t^*], \qquad (1)$$

where c_m is the specific heat, ρ_m the glass density, k_h the heat conductivity, T the temperature depending on the space position $\mathbf{r}$ and time t. $\mathbf{q}_r(T)$ denotes the radiative flux vector, whereas $\nabla \cdot \mathbf{q}_r \neq 0$ only for the semitransparent wavelength region. The glass has a given initial temperature distribution

$$T(\mathbf{r},0) = T_0(\mathbf{r}). \qquad (2)$$

The net heat flux through the surface at the boundary ∂D is defined by heat convection and diffuse surface radiation for those wavelength for which the glass is opaque.

$$k_h \frac{\partial T}{\partial n}(\mathbf{r}_b, t) = \gamma[T_a - T(\mathbf{r}_b, t)] + \varepsilon \pi \int\limits_{opaque} [B_a(T_a, \lambda) - B_a(T(\mathbf{r}_b, t), \lambda)] d\lambda,$$

$$\mathbf{r}_b \in \partial D, \ 0 < t \le t^*, \tag{3}$$

where n is the outer normal vector to the boundary point $\mathbf{r}_b$, γ denotes the heat transfer coefficient, $B_a(T_a, \lambda)$ Planck's function (mentioned below by formula (7)) depending on T_a the temperature of the surroundings and the wavelength λ, and ε denotes the emissivity of the glass. For a detailed derivation and discussion of the boundary condition we refer to [6].

The radiative flux vector $\mathbf{q}_r$ is defined as the first moment with respect to the direction Ω of the radiative intensity $I(\mathbf{r}, \Omega, \lambda)$, depending on space position $\mathbf{r}$, direction Ω, and wavelength λ,

$$\mathbf{q}_r(\mathbf{r}) = \int\limits_0^\infty \int\limits_{S^2} I(\mathbf{r}, \Omega, \lambda) \Omega d\Omega d\lambda. \tag{4}$$

S^2 denotes the unit sphere. The radiative intensity is described by the radiative transfer equation

$$\Omega \cdot \nabla I(\mathbf{r}, \Omega, \lambda) + \kappa(\lambda) I(\mathbf{r}, \Omega, \lambda) = \kappa(\lambda) B_g(T(\mathbf{r}, t), \lambda), \tag{5}$$

with specular reflecting boundary conditions

$$I(\mathbf{r}_b, \Omega, \lambda) = \rho(\Omega) I(\mathbf{r}_b, \Omega', \lambda) + (1 - \rho(\Omega)) n_g^2 B(T_a, \lambda), \tag{6}$$

$$\Omega' = \Omega - 2(\mathbf{n} \cdot \Omega)\mathbf{n}.$$

Here $\kappa(\lambda)$ denotes the absorption coefficient, n_g the refractive index of glass and $B_g(T, \lambda)$ Planck's function for glass

$$B_g(T, \lambda) = \frac{2hc_0^2}{\lambda^5 n_g^2 \left(e^{\frac{hc_0}{n_g \lambda k_B T}} - 1\right)} \tag{7}$$

($c_0 = 2.998 \times 10^8\,\text{m/s}$ speed of light, $h = 6.626 \times 10^{-34}\,Js$ Planck constant, $k_B = 1.381 \times 10^{-23}\,J/K$ Boltzmann constant). $B_a(T, \lambda)$ denotes Planck's function for air with refractive index $n_a = 1$.

The system (1)–(6) is non-linear and highdimensional. To solve the system numerical methods are required which offer a sufficient accurate solution in appropriate computational time.

A detailed survey about different numerical methods for solving the radiative transfer equation can be found in [1]. The majority of methods can be divided into four classes:

(a) Diffusion approximation methods
    Typical examples are the Rosseland approximation or the below described methods developed at ITWM.
(b) Methods, where the directional dependence is expressed by a series of special basis functions
    Representants are the P_N- and SP_N-approximation.
(c) Methods using a discretization of the angular dependence
    Here we mention the discrete ordinate method, which is the most popular one in literature.
(d) Monte Carlo methods

For more details we refer also to the lectures from A. Klar and M. Frank.

2.1.1 Fast Numerical Methods for Radiative Heat Transfer

The Rosseland approximation is widely used in industrial practice and describes the radiation as a correction of the heat conductivity. There exist different methods to derive the Rosseland approximation, for instance using asymptotic analysis. We refer also to the lecture of A. Klar or the PhD thesis [18]. Originally the method was derived in 1924 by S. Rosseland to investigate the radiation of stars. (see [9])

Here we will use the formal solution of the radiative transfer equation (5)

$$I(\mathbf{r},\Omega,\lambda) = I(\mathbf{r}_b,\Omega,\lambda)e^{-\kappa d(\mathbf{r},\Omega)} + \kappa \int\limits_0^{d(\mathbf{r},\Omega)} B(T(\mathbf{r}-s\Omega),\lambda)e^{-\kappa s}ds. \qquad (8)$$

$d(\mathbf{r},\Omega)$ is the distance from the point $\mathbf{r}$ to the boundary ∂D in direction $-\Omega$. Using a linear approximation of the Planck function in the point $\mathbf{r}$

$$B(T(\mathbf{r}-s\Omega),\lambda) \approx B(T(\mathbf{r}),\lambda) - s\frac{\partial B}{\partial T}(T(\mathbf{r}),\lambda)\Omega \cdot \nabla T(\mathbf{r}),$$

and assuming that the desired glass is optically thick

$$\kappa d(\mathbf{r},\Omega) \longrightarrow \infty, \qquad (9)$$

the integral on the right hand side of (8) can be calculated analytically

$$I(\mathbf{r},\Omega,\lambda) = B(T(\mathbf{r},t),\lambda) - \frac{1}{\kappa(\lambda)}\frac{\partial B}{\partial T}(T(\mathbf{r},t),\lambda)\Omega \cdot \nabla T(\mathbf{r},t). \qquad (10)$$

Using now the definition of the radiative flux vector (4) we obtain

$$\mathbf{q}_r = k_r \nabla T(\mathbf{r},t),$$

$$\text{with} \quad k_r = \frac{4\pi}{3} \int\limits_0^\infty \frac{1}{\kappa(\lambda)} \frac{\partial B}{\partial T}(T,\lambda) d\lambda. \tag{11}$$

Instead of the combination of heat transfer equation (1) and radiative transfer equation (5) one has to solve only the heat transfer equation with corrected heat conductivity

$$c_m \rho_m \frac{\partial T}{\partial t}(\mathbf{r},t) = \nabla \cdot ((k_h + k_r) \nabla T(\mathbf{r},t)) \quad (\mathbf{r},t) \in D \times (0,t^\star). \tag{12}$$

It is important to mention that the boundary condition (3) has to be changed too. Taking the energy balance over the whole wavelength region into account for the boundary condition yields (see [6])

$$\left(k_h + \frac{1}{2}k_r\right) \frac{\partial T}{\partial n}(\mathbf{r}_b,t) = \gamma[T_a - T(\mathbf{r}_b,t)] + \sigma\varepsilon[T_a^4 - T^4(\mathbf{r}_b,t)],$$

$$\mathbf{r}_b \in \partial D, \ 0 < t \leq t^\star. \tag{13}$$

The Rosseland approximation treats thermal radiation as a correction of heat conductivity. The method is very fast and easy to implement into commercial software packages. On the other side it is easy to realize from the derivation of the method that it is valid only for optically thick glasses. Especially near the boundary ∂D the Rosseland approximation will give wrong simulation results.

The reason for this behavior is given by the requirement (9). Due to that requirement the approximation for the spectral intensity (10) has lost almost all geometrical information included in the radiative transfer equation (5)–(6).

We want to present now two numerical methods developed at Fraunhofer ITWM and published first in [13] and [14].

These methods are similar to the Rosseland approximation but they will give much more accurate results because they conserve all geometrical information as we will see later on.

The starting point is once more the formal solution of the radiative transfer equation (8). As before we take the Taylor expansion to linearize the Planck function but we skip the requirement (9). We calculate the integral in (8) analytically

$$I(\mathbf{r},\Omega,\lambda) = I(\mathbf{r}_b,\Omega,\lambda)e^{-\kappa d(\mathbf{r},\Omega)} + B(T(\mathbf{r},t),\lambda)\left(1 - e^{-\kappa d(\mathbf{r},\Omega)}\right)$$

$$-\frac{1}{\kappa(\lambda)}\left[1 - (1 + \kappa d(\mathbf{r},\Omega))e^{-\kappa d(\mathbf{r},\Omega)}\right]\frac{\partial B}{\partial T}(T(\mathbf{r},t),\lambda)\Omega \cdot \nabla T(\mathbf{r},t). \tag{14}$$

In contrast to the classical Rosseland approximation the formal solution was taken without any assumption for the optical thickness. This means that all geometrical information of the spectral radiative intensity is preserved. Applying the definition of the radiative flux (4) to formula (14) for the divergence of the radiative flux vector yields

$$
\nabla \cdot \mathbf{q}_r(\mathbf{r}) = -\nabla \cdot \left[\left(\int_0^\infty \frac{1}{\kappa(\lambda)} \mathscr{A}(\mathbf{r}, \lambda) \frac{\partial B}{\partial T}(T, \lambda) d\lambda \right) \nabla T(\mathbf{r}, t) \right]
$$

$$
+ \int_0^\infty \kappa(\lambda) [B(T, \lambda) - B(T_a, \lambda)] \int_{S^2} (1 - \rho(\Omega)) e^{-\kappa d(\mathbf{r}, \Omega)} d\Omega d\lambda
$$

$$
- \int_0^\infty \frac{\partial B}{\partial T}(T, \lambda) \int_{S^2} (1 - \rho(\Omega)) \Omega \cdot \nabla T e^{-\kappa d(\mathbf{r}, \Omega)} d\Omega d\lambda, \tag{15}
$$

with a symmetric anisotropic diffusion tensor $\mathscr{A}(\mathbf{r}, \lambda)$. We approximate $\mathscr{A}(\mathbf{r}, \lambda)$ by an orthotropic diffusion tensor in form of a diagonal matrix

$$
\mathscr{A}(\mathbf{r}, \lambda) = \begin{pmatrix} a_1(\mathbf{r}, \lambda) & 0 & 0 \\ 0 & a_2(\mathbf{r}, \lambda) & 0 \\ 0 & 0 & a_3(\mathbf{r}, \lambda) \end{pmatrix}, \tag{16}
$$

$$
a_j(\mathbf{r}, \lambda) = \int_{S^2} \Omega_j^2 [1 - (1 + \kappa(\lambda) d(\mathbf{r}, \Omega)) e^{-\kappa(\lambda) d(\mathbf{r}, \Omega)}] d\Omega, \quad j = 1, 2, 3
$$

In contrast to the anisotropic diffusion tensor, such an orthotropic diffusion tensor can easily be used in commercial software packages.

The first term on the right hand side of (15) represents a correction to the heat conduction due to radiation. The second term contains the boundary condition and the third a thermal convection because it depends on the gradient of the temperature T.

According to the reference [13] we call the method (15)–(16) *Improved Diffusion Approximation*.

Another method similar to the Improved Diffusion Approximation was presented in [14] and used for the simulation of flat glass tempering. Starting point is as well as before the formal solution (8) of the radiative transfer equation, and as well as before we skip the requirement (9). Integrating the radiative transfer equation (5) over all directions the divergence of the radiative flux vector can be calculated as

$$
\nabla \cdot \mathbf{q}_r(\mathbf{r}) = \int_0^\infty \kappa(\lambda) \left(4\pi B(T, \lambda) - \int_{S^2} I(\mathbf{r}, \Omega, \lambda) d\Omega \right) d\lambda \tag{17}
$$

Together with the formal solution (8) we obtain the so-called *Formal Solution Approximation*

$$\nabla \cdot \mathbf{q}_r(\mathbf{r}) = \int_0^\infty \kappa(\lambda) \int_{S^2} [B(T,\lambda) - I(\mathbf{r},\Omega,\lambda)] e^{-\kappa d(\mathbf{r},\Omega)} d\Omega d\lambda +$$

$$+ \int_0^\infty \frac{\partial B}{\partial T}(T,\lambda) \int_{S^2} [1 - (1 + \kappa(\lambda) d(\mathbf{r},\Omega)) e^{-\kappa d(\mathbf{r},\Omega)}] \Omega \cdot \nabla T d\Omega d\lambda \quad (18)$$

We want to demonstrate the introduced two new methods for three examples typical for glass industry – the heating of a glass plate, the radiative heat transfer in a cavity with natural convection, and an application to flat glass tempering.

2.1.2   The Heating of a Glass Plate

A thin glass plate with thickness of 0.005 m and uniform initial temperature of $T_0(x) = 200°C$ is heated up from above and below by electrical heaters with $T_{above} = 800°C$ and $T_{below} = 600°C$. The glass is described by physical parameters listed in Table 1.

The considered glass is semitransparent in the region between 0.01 and 7.0 µm, what is realistic for optical glasses. The used absorption coefficients are collected in Table 2.

Figure 2 shows the steady state temperature profile along the thickness of the glass plate simulated using the discrete ordinate method (DOM), the improved diffusion approximation (ida) (15), and the formal solution method (fsa) (18). All

Table 1   Material parameters

density	$2,500 kg/m^3$
specific heat	$1,250 J/kg/K$
conductivity	$1 W/m/K$
refractive index	1.5

Table 2   Absorption coefficients

λ_{k-1} in μm	λ_k in μm	κ_k in $1/m$
0.01	0.2	0.40
0.2	3.0	0.50
3.0	3.5	7.70
3.5	4.0	15.45
4.0	4.5	27.98
4.5	5.5	267.98
5.5	6.0	567.32
6.0	7.0	7,136.06
7.0	∞	opaque

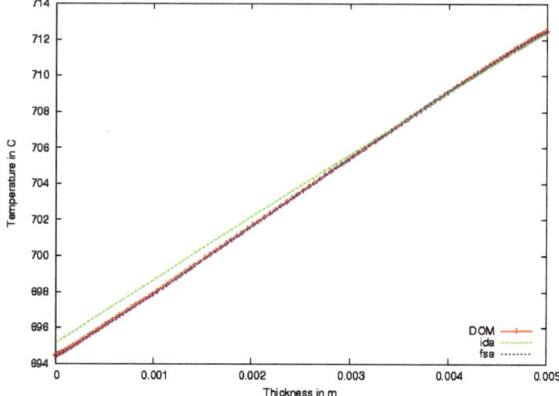

Fig. 2 Temperature distribution in a glass plate

three methods deliver comparable results. The maximum difference between the
discrete ordinate method and (15) is about 0.7°C, whereas the difference between
DOM and (18) is less than 0.03°C. From the computational point of view the two
"ITWM-approaches" are much faster than the discrete ordinate method.

2.1.3 Radiative Heat Transfer with Natural Convection

As a second example we simulate the radiative heat transfer with natural convection
in a two-dimensional cavity. The cavity has a hot right wall of 1,800°C, a colder
left wall with 1,300°C, and adiabatic walls at top and bottom. Gravity points
downwards. A buoyant flow develops because of thermally-induced density gradients. We compare our radiation model with DOM available in the commercial
software package FLUENT. We have implemented our improved diffusion method
into FLUENT. Therefore, the FLUENT radiation is turned off and the volume radiation is added by a user defined function developed at Fraunhofer ITWM into the
source term. The main parameters are given in Table 3.

The walls are taken as diffusely reflecting. The glassmelt is considered as a grey
material.

Figure 3 shows the temperature distribution inside the two-dimensional cavity
using the commercial software package FLUENT. The left picture was calculated
with the discrete ordinate method available in FLUENT and the right with the improved diffusion approximation developed by ITWM. The temperature distribution
is very similar but there are big differences in the computational time. FLUENT-
DOM needs more than 5,000 iterations to converge whereas the ITWM-UDF needs
only 86 iterations.

Table 3 Material parameters
of the cavity problem

density	$2,500\,kg/m^3$
specific heat	$1,200\,J/kg/K$
conductivity	$1\,W/m/K$
heat transfer coefficient	$0\,W/m^2/K$
absorption coefficient	$40\,1/m$
refractive index	1.5
viscosity	$30\,Pas$
thermal expansion coefficient	$7 \cdot 10^{-5}\,1/K$

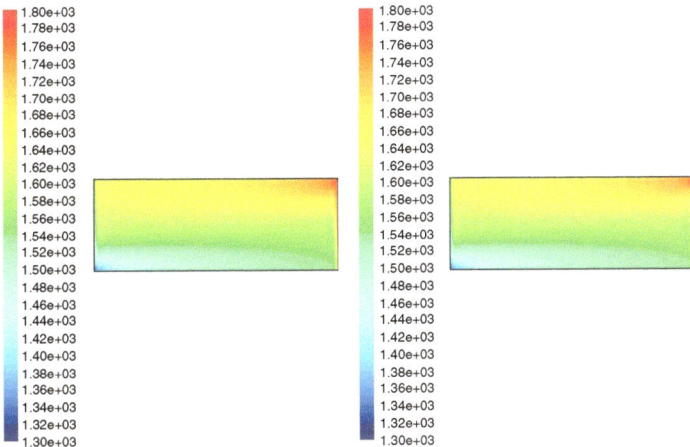

Fig. 3 Temperature profile in a two-dimensional cavity calculated with FLUENT. *Left*: using FLUENT-DOM. *Right*: using ITWM-Approaches

2.1.4 Application to Flat Glass Tempering

Thermal tempering of flat glass plates consists of two stages: First the glass is heated up at a temperature higher than the transition temperature (so-called "T_g"). In the second step it is cooled down rapidly by an air jet. This thermal treatment gives better mechanical and thermal strengthening to the glass by way of residual stresses generated along the thickness of the glass plate. The present example concerns the calculation of stresses inside a sodalime silicate flat glass of thickness 6 mm. For a detailed description especially for the used parameters we refer to [14].

The cooling of the glass melt depends on the temperature distribution in time and space. There exists a temperature range, where the glass changes from fluid to solid state. The essential property is the viscosity of the glass. The viscosity is high for low temperatures and the glass behaves like a linear-elastic material. High temperatures cause high viscosity of glass, which behaves in that case like an Newtonian flow. The viscosity changes the density depending on the temperature. A change in density influences the stress inside the glass. A numerical method for the calculation of transient and residual stresses in glass, including both structural relaxation and viscous stress relaxation, has been developed by Narayanaswamy. (See [14, 15])

In [14] we have investigated how the numerical calculation of the temperature influences the calculation of the thermal stress inside the flat glass plate. Calculating the formal solution of the one-dimensional radiative heat transfer with 200 different directions and 30 wavelength bands we have got a reference which was called to be exact. We have compared it with the Rosseland model, a model where we have skipped the radiation totally, and the formal solution approximation (18). The Figs. 4 and 5 show the evolution over time of the mid-plane and surface temperature

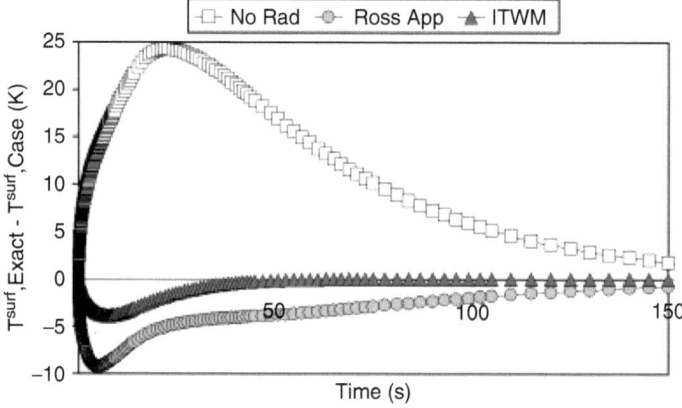

Fig. 4 Evolution over time of surface temperature difference for the Exact solution compared with Case (where Case is on Of No Rad, Ross App, or ITWM models)

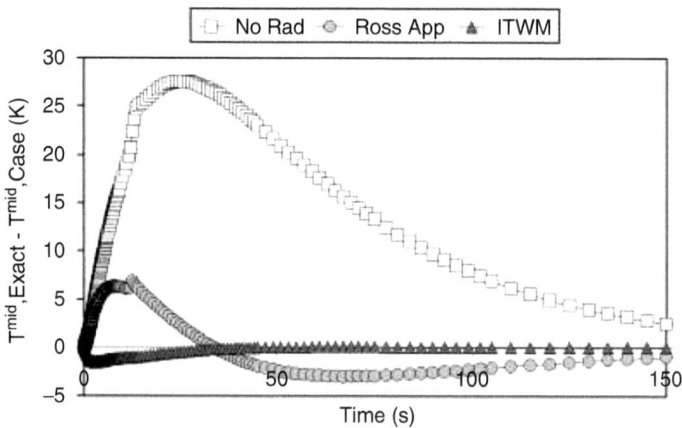

Fig. 5 Evolution over time of mid-plate temperature difference for the Exact solution compared with Case (where Case is on Of No Rad, Ross App, or ITWM models)

differences of the exact solution and the other mentioned methods. It turns out that
the ITWM approach gives the closest result to the exact temperature calculation.

Comparing the computational time the ITWM model is almost as fast as
the Rosseland approximation and about 100 times faster than the exact solution
procedure (Table 4).

Using Narayanaswamy's model [14] transient and residual stresses were computed along the glass thickness. The results are shown in Figs. 6 and 7. As for
the temperature itself the ITWM method gives the closest results to the exact

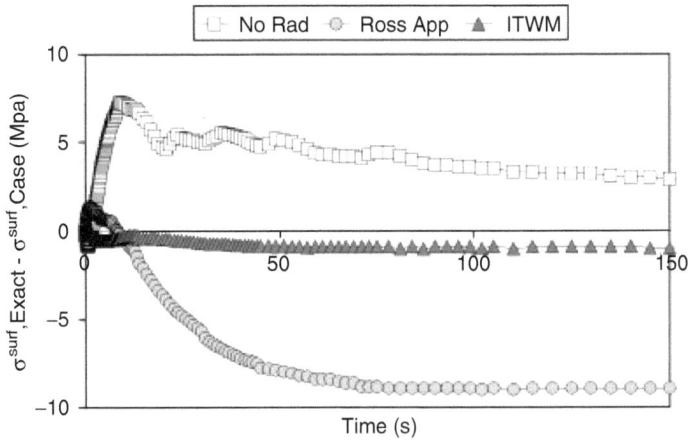

Fig. 6 Evolution over time of the stress differences on the glass surface between the Exact solution
and Case (where Case is on Of No Rad, Ross App, or ITWM models)

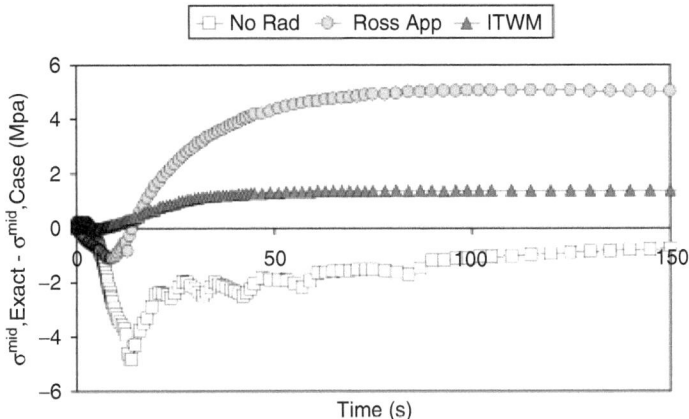

Fig. 7 Evolution over time of the stress differences in the mid-plane between the Exact solution
and Case (where Case is on Of No Rad, Ross App, or ITWM models)

Table 4 CPU time in s

Exact solution model	806.50 s
No radiation	0.65 s
Rosseland approximation	1.65 s
ITWM model	7.50 s

solution for transient and residual stress calculation. It is interesting to mention that for the residual stress it is better to exclude radiation than to use the Rosseland approximation.

2.1.5 Conclusions

Temperature is one of the most important parameters to make "good" glasses. To simulate the thermal behavior of glass radiation must be taken into account because of the fact that glass is a semitransparent material. From the mathematical point of view one has to solve a non-linear and high-dimensional system of partial differential equations, what can be done only numerically. Even the numerical solution of the radiative transfer equation is very complex. For a domain discretization with 20,000 space grid points, 60 different directions, and 10 wavelength bands one has to solve using the Discrete Ordinate Method, which is favored in literature, a linear system of equations with 12 million unknowns. That is the reason why one has to look for fast numerical methods. The above described Improved Diffusion Approximation and Formal Solution Approach are good alternatives, because they make a balanced compromise between exactness and quickness of the numerical solution. The methods are nearly as fast as the classical Rosseland approximation, which treats the thermal radiation as a correction of the thermal conductivity, and nearly as accurate as the Discrete Ordinate Method, which is asymptotically exact. Compared to the Rosseland approximation all geometrical information is conserved.

3 Indirect Temperature Measurement of Hot Glasses

3.1 Introduction

As discussed in the previous chapters temperature is the most important parameter in almost all stages of glass production and glass processing. To determine the temperature of hot glasses one can use numerical simulation or – what is the most conventional method – one measures the temperature. Measurement techniques are required that work in harsh environment of glass industry. The usage of thermocouples which continously are in touch with glass or penetrate it disturbs the manufacturing process. Thermal pyrometers are used to measure the surface temperature of hot glasses, their usage to measure the inside temperature leads to wrong measuremet results. On the other side glass is semitransparent, i.e. it radiates not

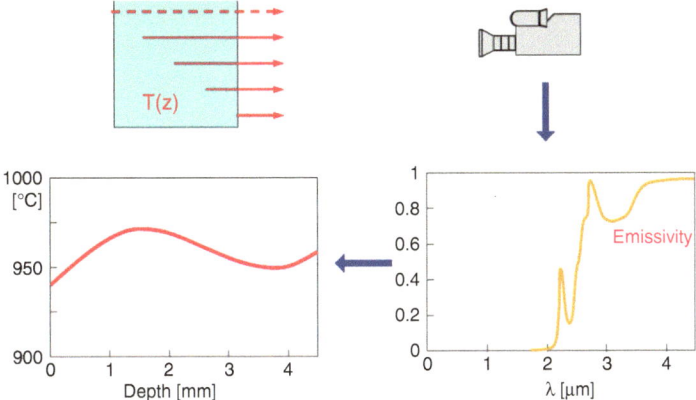

Fig. 8 The principle of remote spectral temperature sensing

only from the surface of a glass body but also from the inner parts. Hence one can use this glass property to measure the temperature in the following way: The optics of an instrument (spectrometer, spectral thermocamera) is directed towards a glass gob and records the wavelength dependent spectral intensity of the glass. From these signals the temperature distribution inside the glass gob has to be computed using a mathematical algorithm. Figure 8 illustrates the principle of the remote spectral temperature sensing. In the following we will develop and discuss a mathematical algorithm of the spectral remote sensing (see [5]).

3.2   The Basic Equation of Spectral Remote Temperature Sensing

The one-dimensional radiative transfer equation is the starting point for the indirect temperature determination

$$\mu \frac{\partial I}{\partial z}(z,\mu,\lambda) + \kappa(\lambda)I(z,\mu,\lambda) = \kappa(\lambda)B(T(z),\lambda), \tag{19}$$

where $I(z,\mu,\lambda)$ denotes the spectral radiative intensity depending on position z, on direction $\mu = \cos(\theta)$, and on wavelength λ. κ denotes the absorption coefficient, and $B(T,\lambda)$ Planck's function defined by (7).

At the boundaries of the one-dimensional domain we consider specular reflection

$$I(z_b,\mu,\lambda) = \rho(\mu)I(z_b,-\mu,\lambda) + (1-\rho(\mu))n_g^2 B(T_{ab},\lambda). \tag{20}$$

For $\mu > 0$ we have $z_b = 0$ and $T_{ab} = T_{a1}$ and for $\mu < 0$ $z_b = D$ and $T_{ab} = T_{a2}$. D is the length of the considered glass domain and n_g the refractive index of glass.

The formal solution $I(z, \mu, \lambda)$ of the equations (19), (20) can be calculated and is related to the measured spectral intensities $I_m(\lambda)$ by

$$I_m(\lambda) = (1 - \rho)I(0, -1, \lambda). \tag{21}$$

Therefore, for the measured intensities hold

$$I_m(\lambda) = \frac{1 - \rho}{1 - \rho^2 e^{-2\kappa D}} \left\{ (1 - \rho)n_g^2 e^{-\kappa D} [B(T_{a2}, \lambda) + \rho e^{-\kappa D} B(T_{a1}, \lambda)] \right. $$
$$\left. + \int_0^D \kappa B(T(s), \lambda)[e^{-\kappa s} + \rho e^{-2\kappa D + \kappa s}]ds \right\}. \tag{22}$$

For measured spectral intensities $I^m(\lambda)$ we have to determine the temperature profile $T(z)$ inside the glass body $0 \leq z \leq D$. Thus, (22) represents a non-linear Fredholm's integral equation of first kind. We have to solve an ill-posed inverse problem.

3.3 Some Basics of Inverse Problems

The field of inverse problems has become one of the most important and one of the fastest growing areas in applied mathematics during the last years driven by the needs of industry as well as sciences. There is a vast literature on inverse and ill-posed problems. We refer here to [3] and [4].

Inverse problems are concerned with finding causes for observed or desired effects. If one looks for the cause of an observed effect, we call it identification or reconstruction. If one looks for the cause of an desired effect, we call it control or design. At the beginning of this section we will show some examples of inverse problems arising in various fields of industrial application.

3.3.1   Example 1: Numerical Differentiation

Assume that an input signal I is transformed by a "black box" function "f" to an output signal g which can be measured. The transformation is described by the convolution

$$\int_0^x I(x - t)f(t)dt = g(x). \tag{23}$$

If we take $I \equiv 1$ we obtain Volterra's integral equation of first kind

$$\int_0^x f(t)dt = g(x),$$

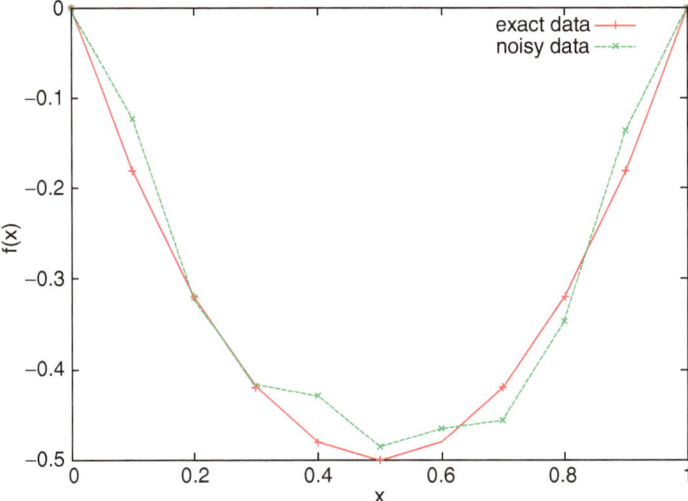

Fig. 9 Numerical differentiation with $h = 0.1$

and if, furthermore, we assume that $g(x)$ is continuously differentiable and $g(0) = 0$ we get the solution of the integral equation as the first derivative of $g(x)$:

$$f(x) = g'(x).$$

$g(x)$ represent data coming from measurement. In praxis the measured data are finite and noisy:

$$g_i^\delta = g(x_i) + \delta, \quad h = x_i - x_{i-1}, \quad i = 1, 2, ... n,$$

where we assume that the points x_i are equally distributed. The solution $f(x)$ is calculated for discrete points x_i by numerical differentiation using

$$f_i^\delta = D_h G_i^\delta = \frac{g_{i+1}^\delta - g_{i-1}^\delta}{2h}. \tag{24}$$

It is known (see [2]) that the discretization error of central differences is of second order. As a test example we take $g(x) = \frac{2}{3}x^3 - x^2 + 1$ and random noise of 1%. For a uniform step size $h = 0.1$ we obtain the solution shown in Fig. 9.

Normally a smaller step size $h = 0.01$ leads to a smaller discretization error, but for (24) that seams to be not true: Here a finer discretization leads to a bigger error for the reconstruction of $f(x)$ as can be seen from Fig. 10.

3.3.2 Example 2: Identification of Heat Conductivity

As a second example we consider the one-dimensional heat transfer equation

$$\frac{\partial}{\partial x}\left(a(x)\frac{\partial u}{\partial x}\right) = -f(x), \quad 0 < x < l, \quad a(0)\frac{\partial u}{\partial x}(0) = g_0, \quad u(l) = g_1. \tag{25}$$

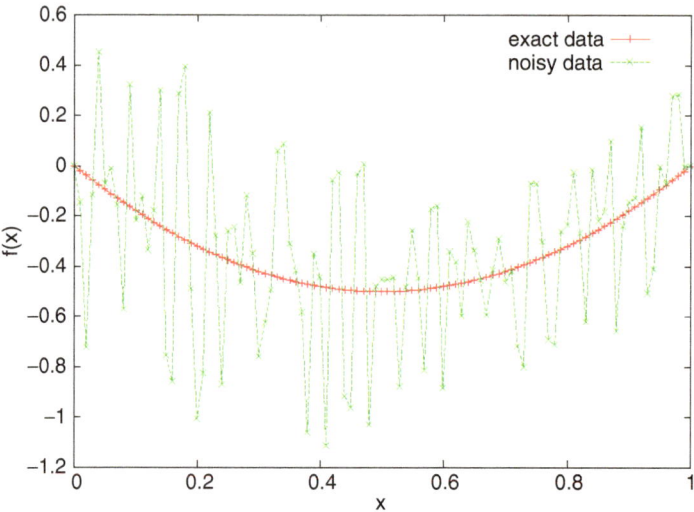

Fig. 10 Numerical differentiation with $h = 0.01$

Knowing the temperature $u(x)$ we want to determine the heat conductivity $a(x)$. If we assume that $\frac{\partial u}{\partial x} \geq u_0 > 0$ we can transform (25) into

$$a(x) = \frac{g_0 - \int_0^x f(y)dy}{\frac{\partial u}{\partial x}}. \tag{26}$$

For $f(x) = 4x - 2$, $g_0 = 4$, $g_1 = 4$, $l = 1$, and $u(x) = -x^2 + 4x + 1$ we obtain the exact value for $a(x)$

$$a(x) = x + 1.$$

Instead of exact measurement $u(x)$ we know $u^\delta(x)$ with 0.01% random noise. Figure 11 shows the exact and noisy values $u(x)$ and $u^\delta(x)$.

As can be seen from Fig. 12 in that case it is not possible to reconstruct the right value for the heat conductivity.

3.3.3 Example 3: An Other Heat Transfer Problem

As a third and last example we consider once more the heat transfer equation

$$\frac{\partial}{\partial x}\left(a(x)\frac{\partial u}{\partial x}\right) = -f(x), \ 0 < x < l, \quad a(0)\frac{\partial u}{\partial x}(0) = 0, \ -a(1)\frac{\partial u}{\partial x}(1) = u(1) \tag{27}$$

with Robbin-type boundary conditions and strongly varying heat conductivity $2 > a(x) \geq 10^{-8}$.

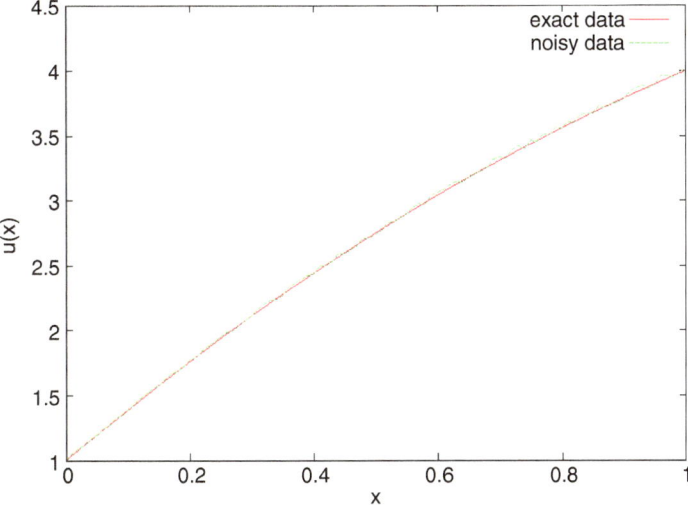

Fig. 11 Measurement with $\delta = 0.01\%$

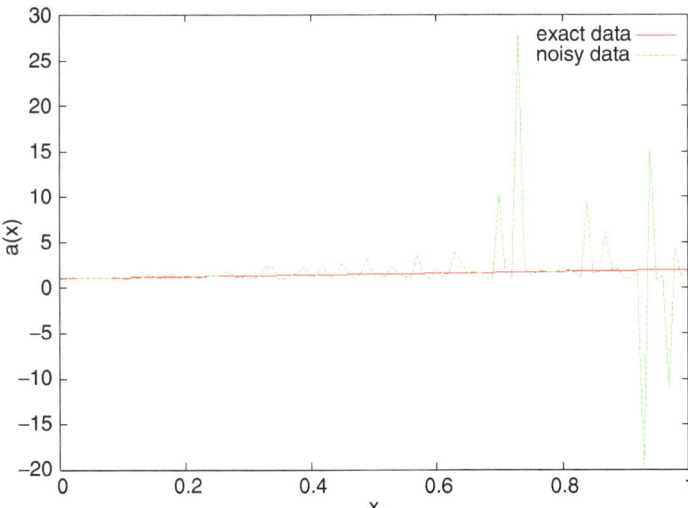

Fig. 12 Reconstructed heat conductivity with $\delta = 0.01\%$

Instead of (27) we consider the finite volume approximation for three grid points $x_0 = 0$, $x_1 = 0.5$, and $x_2 = 1$. The discretization of right hand side is given by $f = (8, -4, 0)^T$ and the conductivity should have the values $a = (1.99999801, 1.99 \cdot$

$10^{-6}, 1.0 \cdot 10^{-8})^T$. Then the discrete solution of (27) is given by the solution $u = (1, 0, 0)^T$ of the linear system

$$\begin{pmatrix} 1 & -1 & 0 \\ -1 & 1.000001 & -0.000001 \\ 0 & -0.000001 & 1.000001 \end{pmatrix} \begin{pmatrix} u_1 \\ u_2 \\ u_3 \end{pmatrix} = \begin{pmatrix} 1 \\ -1 \\ 0 \end{pmatrix}. \tag{28}$$

If we add a small disturbance $\delta = (0.01, 0.01, 0)^T$ to the right hand side vector f, so that $f^\delta = f + \delta$ we get as a new solution of the system (28) the vector $u^\delta = (20001.03, 20000.02, 0.000002)^T$. Once more we notice that, a small error in the measurement (right hand side) causes a big error for the reconstruction (solution vector).

What is the reason for that behavior?

A common property of all these three examples and of a vast majority of inverse problems is their ill-posedness. Hadamard (1865–1963) has given a definition of well-posedness.

A mathematical problem is well-posed, if

  (i) For all data, there exists a solution of the problem
 (ii) For all data, the solution is unique
(iii) The solution depends continuously on the data

A problem is ill-posed, if at least one of these three conditions is violated.

What is the reason for the ill-posedness of the considered examples?

3.3.4 Example 1: Numerical Differentiation

The function g was given with a small error

$$||g - g^\delta||_\infty \le \delta.$$

For the numerical differentiation we used the central difference quotient with step size h. If $g \in C^3[0, 1]$ we get from the Taylor expansion

$$D_h g = \frac{g(x_{i+1}) - g(x_{i-1})}{2h} = g'(x) + \frac{h^2}{6} g'''(\xi), \quad \xi \in [0, 1].$$

Thus the total error behaves like

$$||f - D_h g^\delta||_\infty \approx \frac{h^2}{6} ||g'''||_\infty + \frac{\delta}{h}. \tag{29}$$

For a fixed error level δ the first term on the right hand side is decreasing for decreasing h while the second term is increasing. If the step size h becomes too small

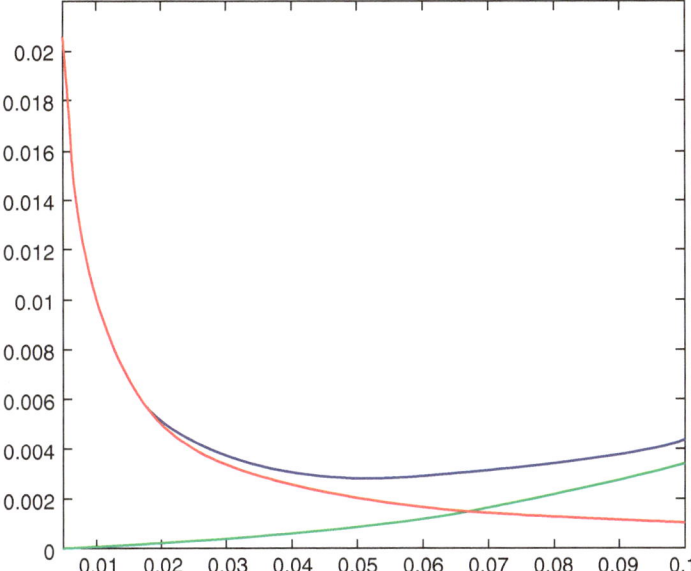

Fig. 13 Numerical error as function of step size. *Green*: Discretization error, *Red*: Measurement error, *Blue*: Resulting error

the total error is increasing. The behavior of the errors is plotted in Fig. 13. There exists an optimal step size h_{opt}, which can not be computed explicitly because it depends on the smoothness of the unknown exact data.

3.3.5 Example 2: Identification of Heat Conductivity

The numerical differentiation in (26) of noisy data $u(x_i)$ is ill-posed like shown for the first example. That is the reason why the second example is ill-posed too.

3.3.6 Example 3: An Other Heat Transfer Problem

Let us consider the third example, the linear system (28). It is easy to calculate (using for instance MATLAB) that the (numerical) eigenvalues of the system are

$$\lambda_1 = 0.5 \cdot 10^{-6}, \quad \lambda_2 = 1.000001, \quad \lambda_3 = 2.0000005. \tag{30}$$

Therefore, the condition number is

$$\kappa = \frac{\lambda_3}{\lambda_1} \approx 4 \cdot 10^6.$$

Let v_1, v_2, and v_3 be the eigenvectors with respect to the eigenvalues. The solution of (28) can be written as

$$u = \sum_{i=1}^{3} \lambda_i^{-1}(f, v_i) v_i, \tag{31}$$

where (f, v_i) denotes the scalar product of the vectors f and v_i. A small error in the right hand side $f^\delta = f + \delta$ leads to

$$u^\delta = \sum_{i=1}^{3} \lambda_i^{-1}(f, v_i) v_i + \sum_{i=1}^{3} \lambda_i^{-1}(\delta, v_i) v_i. \tag{32}$$

The interesting part is the second term in (32). The error is divided by the eigenvalues. For the smallest eigenvalue λ_1 we obtain

$$\frac{\delta}{\lambda_1} \approx \frac{10^{-2}}{0.5 \cdot 10^{-6}} = 20000. \tag{33}$$

A small error in the right hand side leads to large error in the solution.

What can be done to overcome the ill-posedness?

The answer to this question is regularization. In [3] is given a kind of definition:

Regularization is the approximation of an ill-posed problem by a family of neighboring well-posed problems.

There exist a lot of methods to construct such a family of neighboring well-posed problems. We want to describe here only some of the most important methods. For a more detailed overview we refer to [3] or [4].

3.3.7 Truncated Singular Value Decomposition

The simplest method to overcome the ill-posedness of a problem is to make a singular value decomposition and skip the smallest singular values, i.e. for a fixed (large enough) $\lambda^\star$ we skip all

$$\lambda_1 < \lambda_2 < ... \lambda_j < \lambda^\star.$$

Using that method in our third example we obtain for $\lambda^\star = 0.5$

$$u^\delta = \sum_{i=2}^{3} \lambda_i^{-1}(f, v_i) v_i = \begin{pmatrix} 0.5 \\ -0.5 \\ -0.5 \cdot 10^{-6} \end{pmatrix} \tag{34}$$

It can be shown that the truncated singular value decomposition is identical to the minimization problem

$$J(u) = ||Au - f||_{L_2} \rightarrow \min \tag{35}$$

and to take from all existing solutions of (35) that with minimal norm

$$||u||_{L_2} \to \min.$$

3.3.8 Tichonov (Lavrentiev) Regularization

We look for a problem which is near by the original one but well-posed. Thereto we increase the eigenvalues. Instead of (31) we obtain

$$u^\alpha = \sum_{i=1}^{3} \frac{1}{\lambda_i + \alpha}(f, v_i)v_i, \quad \alpha > 0 \tag{36}$$

For α we take a sequence $\alpha_n \to 0$. To decide which α_n gives the best results one can use different methods. Here we use the so-called L-curve method: We solve (36) for a sequence α_n and look at the solution norm $||u||_{L_2}$ versus residual norm $||Au - f||_{L_2}$ for different α. The result for the considered example is shown in Fig. 14. The graph

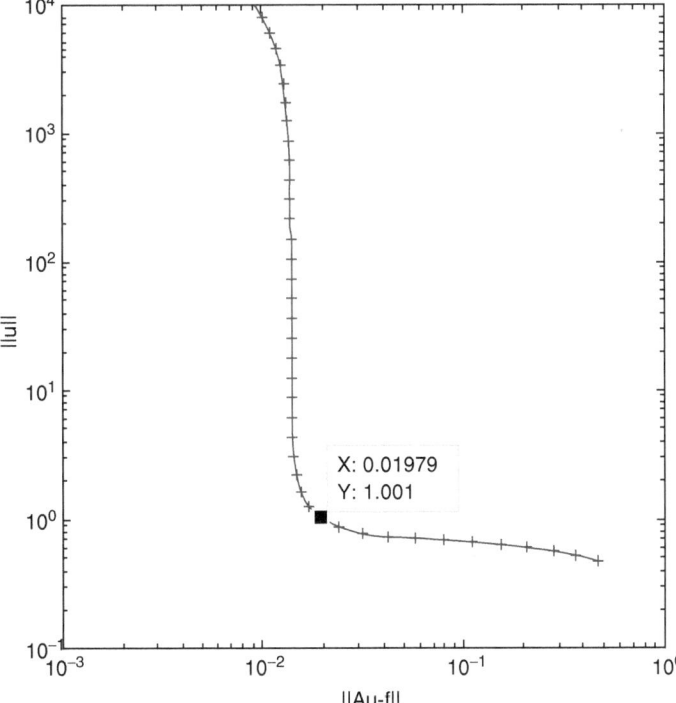

Fig. 14 The L-curve method

is "L-shaped". As the optimal regularization parameter $\alpha^\star$ we take the value in the corner. In the example $\alpha^\star = 0.01977$ for which we obtain

$$u^{\alpha^\star} = \begin{pmatrix} 1.0004 \\ 0.0120 \\ 1 \cdot 10^{-8} \end{pmatrix},$$

as the solution of (36), which is in good agreement to the exact solution of the problem.

3.3.9 Landweber Iteration

Instead of the system (28) we consider the normal equation

$$A^\star A u^\delta = A^\star f^\delta$$

which we solve by a fixed point iteration

$$u_{k+1}^\delta = u_k^\delta - \tau A^\star(A u_k^\delta - f^\delta), \quad u_0^\delta \text{ given.} \tag{37}$$

Here the iteration number k plays the role of the regularization parameter $\alpha = \frac{1}{k}$. As the stopping rule we use the so-called discrepancy principle:

$$k^\star(\delta, f^\delta) = \inf\left\{ k \in N : ||A u_k^\delta - f^\delta||_{L_2} < \eta\delta \right\}, \tag{38}$$

with some real value $\eta \geq \eta_0 > 1$.

For the third example we find after four iterations ($k^\star = 4$) the minimal norm solution $u_{k^\star}^\delta = (0.5 \ -0.5 \ -1 \cdot 10^{-7})^T$.

3.3.10 The classical Tichonov Regularization

The Russian mathematician A.N. Tichonov (1906–1993) proposed the regularization of the normalized system:

$$A^\star A u_\alpha^\delta + \alpha I u_\alpha^\delta = A^\star f^\delta. \tag{39}$$

It can be shown, that this procedure is equivalent to solve the minimization problem (see [3])

$$J(u_\alpha^\delta, \alpha) = ||A u_\alpha^\delta - f^\delta||_{L_2}^2 + \alpha ||u||_{L_2}^2 \rightarrow \min \tag{40}$$

Dealing with an ill-posed problem means to find the right balance between accuracy for which stands the first term on the right hand side and stability for which stands the second term.

Using the discrepancy or the L-curve principle one obtains for our example once more the minimal norm solution.

From all discussion above it follows that for a good regularization one has to include as much information as available. If we know, for instance, that the solution of (28) has the norm $||u||_{L_2}^2 = 1$ the functional (40) can be modified

$$J(u_\alpha^\delta, \alpha) = ||Au_\alpha^\delta - f^\delta||_{L_2}^2 + \alpha \left(||u||_{L_2}^2 - 1 \right) \to \min,$$

which gives approximately the exact solution

$$u_\alpha^\delta = \begin{pmatrix} 1.00104 \\ 0.00104 \\ -9 \cdot 10^{-9} \end{pmatrix}.$$

3.4 Spectral Remote Sensing

Now, we return to the integral equation (22). We write the non-linear equation as

$$G(T) = I_m \tag{41}$$

After linearization

$$(G'(T_k))^\star G'(T_k)(T_{k+1} - T_k) = (G'(T_k))^\star (I_m - G(T_k)) \tag{42}$$

we have to apply regularization. Assume that $G'(T_k) : \mathcal{X} \to \mathcal{Y}$, where $\mathcal{X}$ is the space of temperature functions and $\mathcal{Y}$ the image space. As we discussed earlier the classical Tichonov regularization can be written as a minimization problem:

$$\inf_{T_{k+1} \in \mathcal{U}} ||(G'(T_k))^\star G'(T_k)(T_{k+1} - T_k) - (G'(T_k))^\star (I_m - G(T_k))|| \tag{43}$$

with some constraint condition $\mathcal{U}$.

If we assume that the temperature in the glass results from combined heat transport by conduction and radiation, $T(z)$ fulfills the stationary equation

$$\frac{\partial}{\partial z} \left(k_h \frac{\partial T}{\partial z} - q_r(T(z)) \right) = 0. \tag{44}$$

$q_r(T(z))$ represents the temperature-dependent radiative flux vector. The influence of radiation can be calculated by the Rosseland approximation. Therefore, as a constraint we choose

$$\mathscr{U} = \left\{ T_{k+1} \in \mathscr{X} : \frac{\partial}{\partial z}\left((k_h + k_r)\frac{\partial T_{k+1}}{\partial z}\right) = 0, \right.$$

$$\left. k_r = \frac{4\pi}{3}\int\limits_{\lambda_0}^{\lambda_N}\frac{1}{\kappa(\lambda)}\frac{\partial B}{\partial T}(T_{k+1}(z),\lambda)d\lambda \right\}. \tag{45}$$

$[\lambda_0, \lambda_N]$ denotes the semitransparent wavelength region and κ the absorption coefficient depending on wavelength. The problem (43), (45) can be written as

$$||(G'(T_k))^\star G'(T_k)(T_{k+1} - T_k) - (G'(T_k))^\star (I_m - G(T_k))||^2$$

$$+\alpha||\frac{\partial}{\partial z}\left((k_h + k_r)\frac{\partial T_{k+1}}{\partial z}\right)||^2 \to \min. \tag{46}$$

We assume that the intensities are measured with some noise

$$||I_m - I_m^\delta|| = \left(\int\limits_{\lambda_0}^{\lambda_N}\left(I_m(\lambda) - I_m^\delta(\lambda)\right)^2 d\lambda\right)^{\frac{1}{2}} < \delta.$$

The regularization parameter α is chosen according to the discrepancy rule. We choose a sequence $\{\alpha_n\} \to 0$ and take this $\alpha^\star$ as the optimal one for which

$$||I_m^\delta - G(T_{k+1}^\alpha)|| < \eta\delta, \quad \eta > 1, \tag{47}$$

is satisfied.

We have tested the method for different applications from glass industry like the heating of a quartz glass rod in a electrically heated tube furnace or the measurement of the temperature distribution in a glass drop during hot forming.

Figure 15 shows the experimental set-up of a laboratory experiment (see [5]). A 280-mm-long quartz glass rod with diameter of 25 mm is placed in a tube furnace

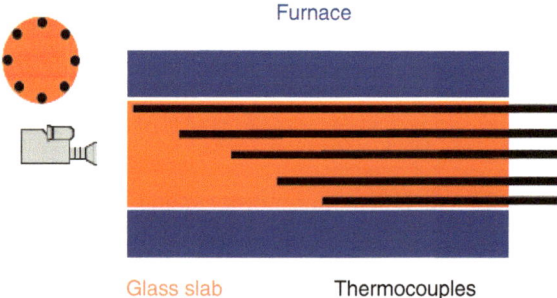

Fig. 15 Experimental set-up for the laboratory experiment

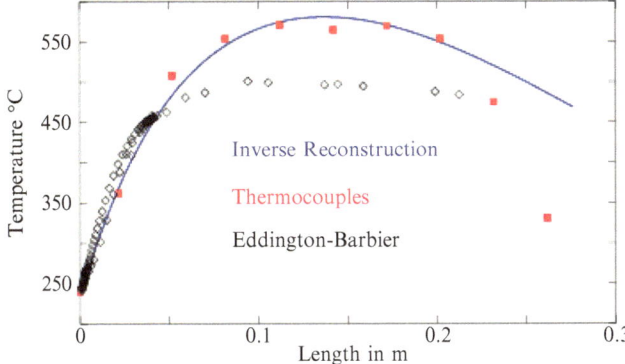

Fig. 16 Temperature inside the rod. Experimental measurements and reconstruction by spectral remote sensing

and heated up. The temperature profile of the furnace is transferred to the quartz rod during the heating process. The corresponding temperature profile $T(z)$ inside the glass was determined by means of spectral pyrometry. In six longitudinal grooves in the rod, eleven thermocouples were placed, whose sensor tips were equidistantly distributed along the furnace axis. Figure 16 represents the comparison between the measured temperature and the temperature calculated by spectral remote sensing. The thermocouple profile corresponds very well with the reconstructed profiles using the above described algorithm. Using spectral remote sensing the temperature inside the glass can be "measured" from the boundary to the hotest point inside the quartz rod.

For more details about the test examples we refer to [5].

3.5 Reconstruction of Initial Temperature

In the previous section we have discussed some examples of inverse problems. If initial temperature distribution, boundary conditions and all physical parameters are known, the forward problem can be solved by standard techniques. But in many industrial applications like glassmaking and glass processing some of the information is missing. In this section we will discuss the case where the initial temperature distribution T_0 is unknown. Instead of that one may be able to measure the temperature and the heat flux at the whole boundary or parts of it for some time $[t_1, t_2] \subseteq [0, t^\star]$, e.g. by infrared camera or pyrometer. The main question we want to investigate can be stated: How to reconstruct the initial temperature distribution from additional boundary measurements? Reconstructing the initial temperature from measured boundary data is another possibility of indirect temperature determination for the inner part of a hot glass body (Fig. 17).

Fig. 17 The boundary temperature of a body is measured over a certain time to "look into the body"

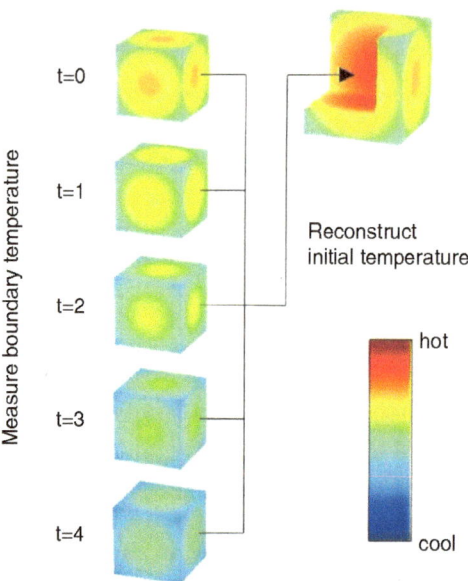

The material of this lecture was taken from [17] and [16]. Furthermor we refer to [11] and [12].

Like for the previous discussed problems we will show that the determination of the initial temperature distribution from boundary data is ill-posed, i.e. small measurement errors cause huge deviations in the result. Regularization techniques have to be used. We consider the three-dimensional (forward) heat transfer equation

$$\frac{\partial T}{\partial t} - \nabla \cdot (k \nabla T) = f, \quad \text{in } (0, t^\star) \times D,$$

$$T|_{t=0} = T_0, \quad \text{on } D, \tag{48}$$

$$k \frac{\partial T}{\partial n} + \gamma T = g, \quad \text{on } S_t := (0, t^\star) \times \partial D. \tag{49}$$

As said before, it may be possible to measure the boundary temperature on the whole boundary or parts of it. Let Γ be a subset of ∂D and $S_\Gamma := (0, t^\star) \times \Gamma$. We define the extended forward operator $\tilde{A}$ by

$$\tilde{A}(T_0, f, g) := T|_{S_\Gamma}.$$

The inverse problem states:

Given f, g and additionally (measured) boundary temperatures y on S_Γ. Find T_0 satisfying

$$\tilde{A}(T_0, f, g) = y. \tag{50}$$

The inverse problem (50) is non-linear, but due to the linearity of the forward problem (48) one can eliminate the inhomogeneities f and g and one obtains

$$\tilde{A}(T_0,f,g) = \tilde{A}(T_0,0,0) + \tilde{A}(0,f,g).$$

Evaluating $\tilde{A}(0,f,g)$ by solving a well-posed (forward) problem, instead of (50) we have to solve the following inverse problem

$$(AT_0) := \tilde{A}(T_0,0,0) = y - \tilde{A}(0,f,g) = y - v. \tag{51}$$

Without loss of generality we want to discuss a simplified one-dimensional heat transfer equation to analyze the (51) in more detail. Consider

$$\frac{\partial T}{\partial t}(x,t) - \frac{\partial^2 T}{\partial x^2}(x,t) = 0 \quad \text{in} \quad D_t = (0,t^\star] \times (0,\pi), \tag{52}$$

with boundary conditions

$$\frac{\partial T}{\partial x}(0,t) = 0, \quad \frac{\partial T}{\partial x}(\pi,t) = 0, \quad t \in (0,t^\star], \tag{53}$$

and additional measurements at the left boundary

$$T(0,t) = y(t), \quad t \in (0,t^\star]. \tag{54}$$

From these data we want to reconstruct the initial temperature distribution $T(x,0) = T_0(x)$, i.e. we want to solve

$$(AT_0)(t) = y(t). \tag{55}$$

Figure 18 illustrates the problem.

In literature one can find two heat conduction problems related to the here presented one, namely the backward heat conduction problem and the sideways heat equation.

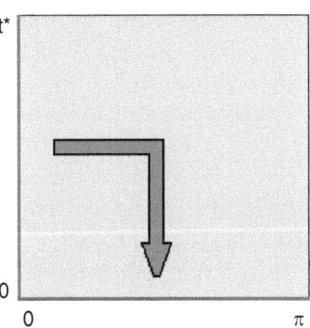

Fig. 18 Initial temperature reconstruction

Fig. 19 Backward heat
equation

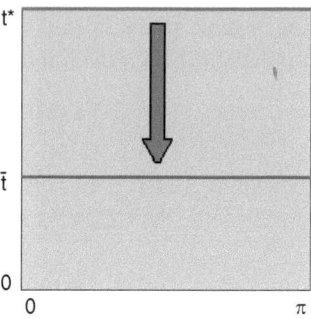

Fig. 20 Sideways heat
equation

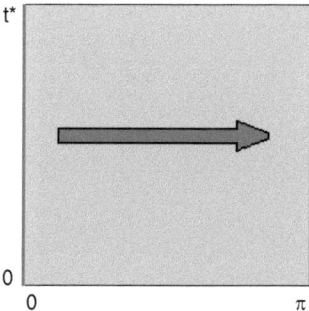

Consider the heat transfer equation (52) with boundary conditions (53). If it
is required to determine the temperature at some time $\bar{t} \in (0, t^*)$ from measurements at the end time t^*, i.e. $T(x, t^*)$, $x \in [0, \pi]$, then one has to solve the heat
equation backward in time – the so-called backward heat equation. This problem
is illustrated in Fig. 19. Figure 20 shows the second known inverse heat transfer
problem.

Consider the heat transfer equation (52) with initial condition $T(x, 0) = T_0(x)$
and boundary conditions at one side, for instance

$$T(0, t) = y(t), \quad \frac{\partial T}{\partial x}(0, t) = 0, \quad t \in (0, t^*].$$

We assume $t^* = \infty$. Now, it is required to find the temperature at $x = \pi$. This problem
is called sideways heat equation.

Both inverse problems – backward heat and sideways heat – are ill-posed. A
detailed discussion and several numerical schemes of solving these problems can be
found in [3] and [10].

Now, we return to the initial temperature reconstruction problem (55), which is
different to both – the backward and sideways heat equation. The solution of the

forward problem, i.e. with given initial condition $T(x,0) = T_0(x)$, can be calculated analytically by Fourier's series

$$T(x,t) = \sum_{n=0}^{\infty} e^{-n^2 t} (T_0(x), w_n(x))_{L_2(0,\pi)} w_n(x), \tag{56}$$

with

$$w_n(x) = \begin{cases} \sqrt{\frac{1}{\pi}}, & n = 0 \\ \sqrt{\frac{2}{\pi}} \cos(nx), & n > 0 \end{cases}$$

and

$$(u(x), v(x))_{L_2(0,\pi)} = \int_0^{\pi} u(x)v(x)dx.$$

Using (56) the temperature at the boundary $x = 0$ can be written as

$$T(0,t) = \frac{1}{\pi} \int_0^{\pi} T_0(x)dx + \frac{2}{\pi} \sum_{n=1}^{\infty} e^{-n^2 t} \int_0^{\pi} T_0(x) \cos(nx)dx$$

or in a more compact way as

$$T(0,t) = \int_0^{\pi} k(x,t) T_0(x)dx, \quad k(x,t) = \sum_{n=0}^{\infty} e^{-n^2 t} w_n(0) w_n(x), \tag{57}$$

with kernel $k(x,t)$ defined in (57). Therefore the inverse problem for calculating $T_0(x)$ using boundary measurements (54) states

$$(AT_0)(t) = y(t) = \int_0^{\pi} k(x,t) T_0(x)dx. \tag{58}$$

We have to solve an integral equation of the first kind.

It can be shown, that there exists a unique solution $T_0(x)$ for (58) (see [17]). On the contrary the continuous dependence on the right hand side is violated as can be seen from the following analytical consideration.

Let $T_0(x)$ be the unique solution of $(AT_0)(t) = y(t)$ and assume that instead of $y(t)$ we have measured

$$y^k(t) = y(t) + \sqrt{\frac{2k}{\pi}} e^{-k^2 t}, \quad k = 1, 2, ...$$

$y^k(t)$ deviates only slightly from $y(t)$

$$\|y^k(t) - y(t)\|_{L_2(0,t^*)}^2 = \frac{2k}{\pi} \int_0^{t^*} e^{-2k^2 t} dt = \frac{1}{\pi k} \left(1 - e^{-2k^2 t^*}\right) \xrightarrow{k \to \infty} 0.$$

The solution of $(AT_0^k)(t) = y^k(t)$ is given by

$$T_0^k(x) = T_0(x) + \sqrt{k} w_k(x), \tag{59}$$

what can be easily seen from

$$
\begin{aligned}
(AT_0^k)(t) &= (AT_0)(t) + \sqrt{k}(Aw_k)(t) \\
&= y(t) + \sqrt{k} \int_0^\pi k(x,t) w_k(x) dx \\
&= y(t) + \sqrt{\frac{2k}{\pi}} e^{-k^2 t} \\
&= y^k(t).
\end{aligned}
$$

From (59) one obtains

$$\|T_0^k(x) - T_0(x)\|_{L_2(0,\pi)}^2 = k \xrightarrow{k \to \infty} \infty$$

A small error in the measurements leads to a huge deviation in the reconstructed initial condition. That is the reason why the problem is ill-posed.

We now turn back to the more general three-dimensional problem (51). For the solution of ill-posed problems one can apply one of the regularization methods discussed before. For given noisy measurement data

$$\|y - y^\delta\|_{L_2(S_\Gamma)} < \delta$$

we use the Tichonov regularization to solve the discretized normalized system of equations

$$(A_h^\star A_h + \alpha I) T_{0,\alpha}^\delta = A_h^\star y^\delta,$$

where $T_{0,\alpha}^\delta$ is the regularized discrete solution. For the choose of α we apply the discrepancy rule.

The performance of the algorithm can be enhanced if an initial guess T_0^δ is known, which is already close to T_0. Let $y^\delta(0)$ be the measurement at $t = 0$ on Γ. As the initial guess for T_0^δ we use the solution of the well-posed forward problem

$$
\begin{aligned}
-\nabla \cdot (k \nabla T_0^\delta) &= f(0), &&\text{in } D, \\
T_0^\delta &= y^\delta(0), &&\text{on } \Gamma, \\
k \frac{\partial T_0^\delta}{\partial n} + \gamma T_0^\delta &= g(0), &&\text{on } \partial D \setminus \Gamma
\end{aligned}
\tag{60}
$$

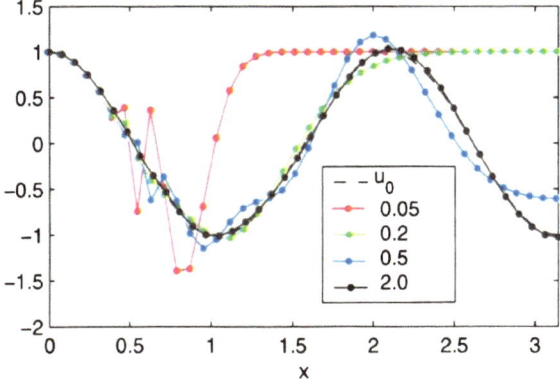

Fig. 21 Initial temperature reconstruction for different measured times

As a numerical example we show a figure from [17], where the inverse problem (55) for the simplified one-dimensional forward problem (52)–(54) was considered. The normalized equation $A^\star A T_0 = y$ was solved using a conjugate gradient algorithm for four different measurement times: $t^\star = 0.05$, $t^\star = 0.2$, $t^\star = 0.5$, and $t^\star = 2.0$. The number of iterations acts as regularization parameter. As a stopping rule the discrepancy criterion was used. The exact initial temperature was chosen as $T_0(x) = \cos(3x)$.

Near the boundary $x \in [0,0.4]$ all reconstructions with different $t^\star$ give satisfactory results. From Fig. 21 it can also be stated that the accuracy of the reconstruction depends on the number of measured data. Whereas for $t^\star = 0.05$ only the left part of the reconstructed curve fits to the exact solution, the reconstruction with $t^\star = 2.0$ fits for the whole interval $x \in [0,\pi]$.

In [16] and [11] the initial temperature reconstruction for a nonlinear heat equation including thermal radiation was investigated. As a test example the cooling of a hot glass plate was discussed:

$$c_m \rho_m \frac{\partial T}{\partial t}(z,t) = k_h \frac{\partial^2 T}{\partial z^2}(z,t) - \kappa \left(4\sigma T^4(z,t) - 2\pi \int_{-1}^{1} I(z,\mu)d\mu \right), \text{in}(0,t^\star] \times (0,D),$$

$$k_h \frac{\partial T}{\partial n}(z,t) = 0, \quad z \in 0,D, \; t \in (0,t^\star], \tag{61}$$

with measurements at the left boundary

$$T(0,t) = y(t), \quad t \in (0,t^\star]. \tag{62}$$

As before c_m denotes the specific heat, ρ_m the density of the glass, and k_h the glass conductivity. $\sigma = 5.67051 \cdot 10^{-8} \frac{W}{m^2 K^4}$ is the Stefan–Boltzmann constant. The radiative intensity $I(z,\mu)$ depending on position z and direction $\mu \in [-1,1]$ is defined as the solution of the radiative transfer equation

$$\mu \frac{\partial I}{\partial z}(z,\mu) = \kappa \left(\frac{\sigma}{\pi} T^4(z,t) - I(z,\mu) \right), \quad z \in (0,D),$$

$$I(0,\mu) = \frac{\sigma}{\pi} T_a^4, \; \mu > 0, \quad I(1,\mu) = \frac{\sigma}{\pi} T_a^4, \; \mu < 0. \tag{63}$$

From (61) to (63) we have to identify the initial temperature distribution $T(z,0) = T_0(z)$. The inverse problem is denoted as

$$\tilde{A}(T_0, f, g) = y, \tag{64}$$

where in this case $g = 0$ and $f = \kappa \left(4\sigma T^4(z,t) - 2\pi \int\limits_{-1}^{1} I(z,\mu) d\mu \right)$ is a non-linearity depending via radiative intensity I on temperature T and hence on the initial temperature distribution T_0.

Similar to (51) we make a decomposition of the non-linear equation

$$\tilde{A}(T_0, f, 0) = \tilde{A}(T_0, 0, 0) + \tilde{A}(0, f(T_0), 0) = y$$

and use now a fixed-point iteration to determine the initial condition

$$(AT_{0,k+1}) := \tilde{A}(T_{0,k+1}, 0, 0) = y - \tilde{A}(0, f(T_{0,k}), 0) := y - v(T_{0,k}). \tag{65}$$

Instead of exact measurements y we consider $y^\delta : ||y - y^\delta||_{L_2(S_\Gamma)} < \delta$ with some noise level δ.

In [11, 16] the problem was solved using the Tichonov regularization

$$(A^\star A + \alpha_i I) T_{0,k+1}^{\delta,\alpha_i} = A^\star (y^\delta - v(T_{0,k}^{\delta,\alpha_i})) \tag{66}$$

taking a series of decreasing regularization parameters $\alpha_i = \alpha_0 q^i$, $q < 1$, $i = 1, ..., m$. For each α_i we solve (66) and obtain $T_{0,k+1}^{\delta,\alpha_i}$. Among all $\left\{ T_{0,k+1}^{\delta,\alpha_i} \right\}_{i=0}^{m}$ we choose $T_{0,k+1}^{\delta,\alpha_j}$ such that

$$||T_{0,k+1}^{\delta,\alpha_j} - T_{0,k+1}^{\delta,\alpha_{j-1}}|| = \min \left\{ ||T_{0,k+1}^{\delta,\alpha_i} - T_{0,k+1}^{\delta,\alpha_{i-1}}||_{L_2(0,D)}, i = 1, 2, ..., m \right\}. \tag{67}$$

(67) is called quasi-optimality criterion, which does not depend on the noise level δ.

Figures 22 and 23 show the reconstruction of the initial temperature profile, where we have taken a typical situation of uniform cooling from both boundaries (see [11]) with random noise of 0.1% and 1.0%.

Both figures show satisfactory results of solving the inverse problem.

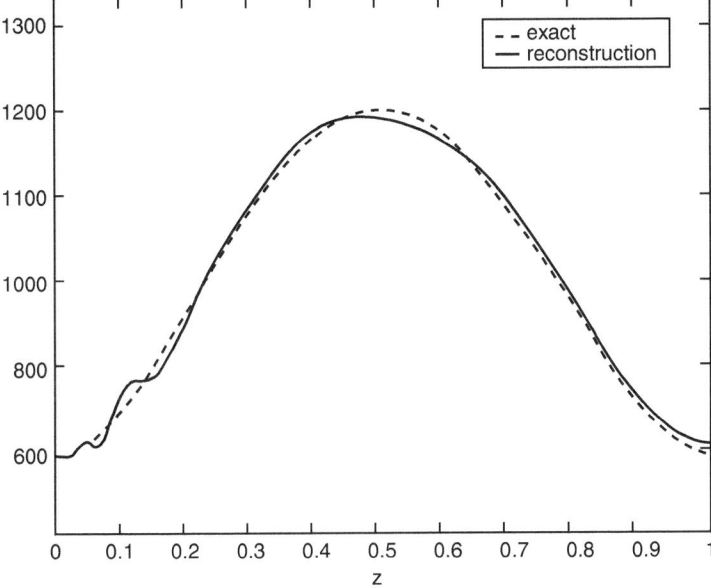

Fig. 22 Initial temperature reconstruction with 0.1% noisy data

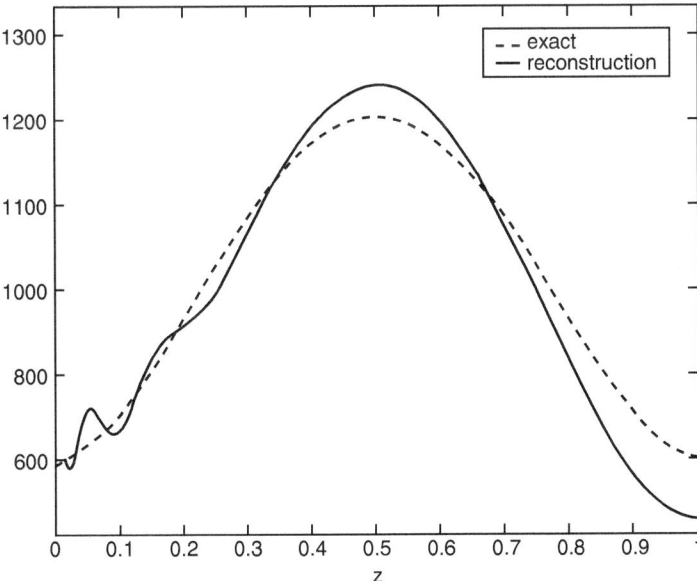

Fig. 23 Initial temperature reconstruction with 1.0% noisy data

3.6 Conclusions

Inverse problems are concerned with finding causes for an observed or desired effect. A common property of a vast majority of inverse problems is their ill-posedness. Very often the solution does not depend continuously on the data. To solve an ill-posed problem one has to use regularization techniques, that is, replace the ill-posed problem by a family of neighboring well-posed problems. The regularization has to be taken in accordance with the problem one wants to solve. The spectral remote sensing leads to an ill-posed integral equation. We use the method to reconstruct the inner temperature of a hot glass body. That is why the usage of the heat transfer operator with Rosseland approximation is an excellent way to regularize the problem. The reconstruction of the initial temperature distribution for a glass cooling process offers another possibility to reconstruct the temperature profile inside the hot glass. This ill-posed heat transfer problem is different to those from literature – backward head and sideways heat equation. We have discussed a solution procedure even in that case when heat radiation has to be taken into account. We solved the ill-posed problems with Tichonov regularization.

References

1. Modest, M.F.: Radiative Heat Transfer. Academic, San Diego (2003)
2. LeVeque, R.J.: Finite difference methods for ordinary and partial differential equations: Steady-state and time-dependent problems. SIAM (2007)
3. Engl, H.W., Hanke, M., Neubauer, A.: Regularization of Inverse Problems. Kluwer Academic, Dordrecht (2000)
4. Louis, A.K.: Inverse und schlecht gestellte Probleme, Teubner, Stuttgart (1989)
5. Brinkmann, M., Siedow, N., Korb, T: Remote Spectral Temperature Profile Sensing. In: Loch, H., Krause, D. (ed): Mathematical Simulation in Glass Technology. Springer, Berlin (2002)
6. Neunzert, H., Siedow, N., Zingsheim, F.: Simulation of the Temperature Behaviour of Hot Glass during Cooling. In: Cumberbatch, E., Fitt, A. (ed): Mathematical Modelling. Case Studies from Industry, pp. 181–198. Cambridge Univerity Press, Cambridge (2001)
7. Viskanta, R., Anderson, E.E.: Heat transfer in semitransparent solids. In: Irvine, T.F. Jr., Harnett, J.P. (eds.) Advances in Heat Transfer, vol. 11, pp. 317–441. Academic, New York (1975)
8. Choudhary, M.K., Huff, N.T.: Mathematical modeling in glass industry: An overview of status and needs. Glastech. Ber. Glass Sci. Technol **70**, 363–370 (1997)
9. Rosseland, S.: Note on the absorption of radiation within a star. M.N.R.A.S. **84**, 525 (1924)
10. Berntsson, F.: Numerical Methods for Solving a Non-Characteristic Cauchy Problem for a Parabolic Equation. Technical report LiTH-MAT-R-2001-17, Department of Mathematics, Linköping University (2001)
11. Pereverzyev, S.S. Jr., Pinnau, R., Siedow, N.: Regularized fixed-point iterations fotr nonlinear inverse problems. Inverse Probl. **22**,1–22 (2006)
12. Pereverzyev, S.S. Jr., Pinnau, R., Siedow, N.: Initial temperature reconstruction for a nonlinear heat equation: application to radiative and convective heat transfer. Inverse Problems in Science and Engineering. **16**, 55–67 (2008)
13. Lentes, F.T., Siedow, N.: Three-dimensional radiative heat transfer in glass cooling processes. Glastech. Ber. Glass Sci. Technol. **72**(6), 188–196 (1999)

14. Siedow, N., Grosan, T., Lochegnies, D., Romero, E.: Application of a New Method for Radiative Heat Transfer to Flat Glass Tempering. J. Am. Ceram. Soc. **88**(8), 2181–2187 (2005)
15. Narayanaswamy, O. S.: A model of structural relaxation in glass. J. Am. Ceram. Soc. **54**(10), 491–498 (1971)
16. Pereverzyev, S.S. Jr.: Method of Regularized Fixed-Point and its Application. PhD Thesis, Technical University Kaiserslautern (2006)
17. Justen, L.: An Inverse Heat Conduction Problem with Unknown Initial Condition. Diploma Thesis, Technical University Kaiserslautern (2002)
18. Zingsheim, F.: Numerical Methods for Radiative Transfer in Semitransparent Media. PhD Thesis, Technical University Kaiserslautern (1999)

Non-Isothermal Flow of Molten Glass: Mathematical Challenges and Industrial Questions

Angiolo Farina, Antonio Fasano, and Andro Mikelić

Abstract With specific reference to the process of glass fibers drawing we review the models proposed to describe the various stages of the flow of molten glass from the furnace to the winding spool: the slow flow in the die, the jet formation under rapid cooling, the terminal fiber profile. In the course of our exposition we will present a general model for non-isothermal flows of mechanically incompressible but thermally expansible fluids (the basic model here assumed for glass), and the Oberbeck–Boussinesq limit is discussed. Both the modelling and the mathematical aspects will be illustrated in detail. An appendix is devoted to the question of stability analysis.

Keywords Glass fiber drawing · Isochoric viscous fluids · Heat conducting · Navier-Stokes equations · Justification of the Boussinesq approximation · Singular perturbation · Non-isothermal elongational free boundary flow

1 Introduction

In this chapter we present a mathematical theory for molten glass flow, with specific reference to the industrial process of glass fibers drawing, roughly depicted in Fig. 1.

Molten glass is kept at a high temperature (1,200–1,400°C) and goes by gravity through an array of hundreds of dies. From each drop a filament is pulled down which become thinner and thinner during its motion, reaching a final diameter which

A. Farina and A. Fasano (✉)
Università degli Studi di Firenze, Dipartimento di Matematica "Ulisse Dini", Viale Morgagni 67/A, I-50134 Firenze, Italy
e-mail: farina@math.unifi.it; fasano@math.unifi.it

A. Mikelić
Université de Lyon, Lyon, F-69003, FRANCE; Université Lyon 1, Institut Camille Jordan, UMR 5208 CNRS, Bât. Braconnier, 43, Bd du onze novembre 1918 69622 Villeurbanne Cedex, FRANCE
e-mail: Andro.Mikelic@univ-lyon1.fr

A. Fasano (ed.), *Mathematical Models in the Manufacturing of Glass*,
Lecture Notes in Mathematics 2010, DOI 10.1007/978-3-642-15967-1_4,
© Springer-Verlag Berlin Heidelberg 2011

Fig. 1 A single glass fiber drawing

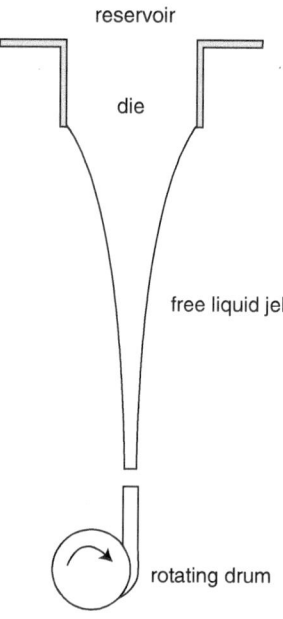

can be three orders of magnitude less than the diameter of the die (various shapes have been adopted for the die, but here we refer to a simple cylindrical geometry). As an obvious consequence of mass conservation the terminal speed of the glass can be up to six orders of magnitude the one at which glass enters the die (typical values can be: a fraction of 1 mm/s for the entrance velocity, and 30 m/s for the terminal phase). The final product is as flexible as a textile fiber and is wound around a rotating drum. The latter device provides the traction driving the entire process. From the die to the spool the flow conditions change in a dramatic way. While only moderate temperature changes take place within the die, at the exit of the die heat is removed at a high rate, both by radiation, and by an intense transversal air flow. Consequently, density will increase and, above all, viscosity will go very rapidly to quite large values (a phenomenon which makes the fiber traction possible). The shape of the boundary will be determined by surface tension (also sensitive to temperature changes) and the flow conditions (strongly influenced by the rapidly varying viscosity). Roughly speaking, we can distinguish four stages (see Fig. 1):

(a) The flow of molten glass at high temperature in the reservoir, feeding the fiber production system.
(b) The non-isothermal flow through the die, with rigid lateral boundaries.
(c) The viscous jet flow with rapidly changing physical parameters, owing to the fast cooling, up to the formation of a "fiber" (high viscosity, small variation of the axial velocity and very small radial velocity).
(d) The motion of the glass fiber, drawn down by a device called spinner (or spool).

In our analysis we will not deal with stage (a), which is treated elsewhere in this book. Stage (b) terminates at the (unknown) point at which the jet is formed. To be more precise, depending on the glass temperature and on the pulling speed (ultimately determining the discharge), the glass may form the jet while it is still inside the die, or it may fill completely the die region and even flow over its external rim (an undesired effect) before forming the jet. In order to describe the flow in the die we have adopted the ideal scheme of a mechanically incompressible, but thermally expansible fluid, which is illustrated in detail in Sect. 2. Such an approach has been heavily criticized in the past (see [2, 3]) on the basis of a linear stability analysis, showing that under such conditions the rest state is linearly unstable. Putting aside the philosophical question of how significant linear instability can be, we will just emphasize that in the context of the approximation $\mathbf{Ec} \approx 0$ ($\mathbf{Ec}$ stands for Eckert number, which basically compares a typical kinetic energy of the unit mass with its typical heat content) such unpleasant property disappears (see Appendix 1). In our case it will be clear that $\mathbf{Ec}$ is really negligible. Existence and uniqueness of a solution to the stationary fluid dynamical problem formulated in Sect. 2 is proved in Sect. 3, supposing that inflow and outflow temperature and velocity are known. In this same section we discuss the question of the Oberbeck–Boussinesq limit, a recurrent subject in non-isothermal flows that has been treated in a number of papers in different contexts on the basis of the choice of a small parameter. In our case (flow in the die, where temperature excursion is small) the most sensible choice is a parameter combining the thermal expansion coefficient and the maximum temperature variation. We will be able to obtain not only the corresponding O–B limit, but also to estimate the higher order correction.

The third stage, i.e. stage (c), is the one of the jet flow in rapidly changing conditions. It starts at the point where the jet is formed (which is unknown) and its endpoint is usually established according to some thumb rule. Of course one of the main difficulty of the problem as a whole is to make the solutions at the three stages agree with each other. This requires a sequence of iterations, but the complete free boundary problem is still open. The shape of the jet results from the action of surface tension (via the surface curvature), creating the confining force to be balanced by the stress tensor. The temperature dependence of both quantities is clearly a source of great difficulty.

The fourth stage is the terminal phase of the fiber motion, characterized by the fact that viscosity has already reached a very large value, the fiber is already considerably thin and further thinning is slow. The particular geometry allows to look for the expansion of all relevant quantities in power series of the ratio between two typical lengths in the transverse and in the longitudinal directions. We will derive the so called Matovich–Pearson approximation on a rigorous basis and prove an existence and uniqueness theorem for the limit problem.

One of the most delicate aspects of the whole theory is the selection of the confine between stage (c) and stage (d). Here we present an approach based on the Matovich–Pearson model (Sect. 4.1 and Appendix 2).

2 Mathematical Modelling

2.1 *Definitions and Basic Equations*

We start by recalling the mass balance, the momentum and energy equations, as
well as the Clausius–Duhem inequality in the *Eulerian formalism*. Following an approach similar to the one presented in [25], pages 51–85, we denote by $\{\rho, \mathbf{v}, e, T, s\}$
the *density*[1], *velocity, specific internal energy, absolute temperature and specific entropy*, satisfying the following system of equations

$$\frac{D\rho}{Dt} = -\rho \operatorname{div} \mathbf{v},\tag{1}$$

$$\rho \frac{D\mathbf{v}}{Dt} = -\rho g \mathbf{e}_3 + \operatorname{div} \mathbf{T},\tag{2}$$

$$\rho \frac{De}{Dt} = -\operatorname{div} \mathbf{q} + \mathbf{D}(\mathbf{v}) : \mathbf{T},\tag{3}$$

$$\rho \frac{Ds}{Dt} + \operatorname{div}\left(\frac{\mathbf{q}}{T}\right) \geq 0,\tag{4}$$

where:

- $\dfrac{D}{Dt} = \dfrac{\partial}{\partial t} + \mathbf{v} \cdot \nabla$ denotes the *material derivative*.
- $\mathbf{D}(\mathbf{v}) = \dfrac{1}{2}\left(\nabla \mathbf{v} + (\nabla \mathbf{v})^T\right)$ is the *rate of strain tensor*.
- $\mathbf{T}$ is the *Cauchy stress tensor*. Further, $\mathbf{D}(\mathbf{v}) : \mathbf{T} = \operatorname{tr}\left(\mathbf{D}(\mathbf{v})\mathbf{T}^T\right) = \sum_{i,j} \mathbf{D}_{ij}\mathbf{T}_{ij}$.
- $\mathbf{q}$ is the *heat flux vector*.
- $-g\mathbf{e}_3$ is the *gravity acceleration*. Indeed $\mathbf{e}_3$ is the unit vector relative to the x_3
  axis directed upward.
- In (3) the internal heat sources are disregarded.

Introducing now the specific *Helmholtz free energy*

$$\psi = e - Ts,\tag{5}$$

inequality (4) becomes

$$\rho\left(\frac{D\psi}{Dt} + s\frac{DT}{Dt}\right) - \mathbf{T} : \mathbf{D}(\mathbf{v}) + \mathbf{q} \cdot \frac{\nabla T}{T} \leq 0.\tag{6}$$

---

[1] Density, in the Eulerian formalism, means mass per unit volume of the current configuration.

2.2 Fluids Physical Properties and Constitutive Equations

Key point of the model is to select suitable constitutive equations for

$$e, \mathbf{T}, \mathbf{q}, \text{ and } s,$$

in term of the primary dependent variables $\mathbf{v}$, T, and ρ. Of course the selection of the constitutive equations must be consistent with the physical properties of the fluids we are considering. For our purposes molten glass can be reasonably described as follows:

1. The fluid is *mechanically incompressible* but *thermally dilatable* (i.e. the fluid can sustain only isochoric[2] motion in isothermal conditions).
2. The fluid behaves as a *linear viscous fluid (Newtonian fluid)*, i.e. the shear stress is proportional to the shear rate.

Focusing on the first aspect (i.e. thermally dilatable fluid), we can say that thermal dilation is described by the *thermal expansion coefficient*, defined as follows

$$\beta = -\frac{1}{\rho}\frac{d\rho}{dT}, \quad [\beta] = {}^{\circ}K^{-1}. \tag{7}$$

The coefficient β may depend on temperature, i.e. $\beta = \beta(T)$, but, since the material is mechanically incompressible, β does not depend on pressure.

Remark 2.1. Considering a material particle denoted by the Lagrangian coordinate $\mathbf{X}$ and $\beta(T)$ given, we have

$$\rho(\mathbf{X},t) = \rho_o(\mathbf{X})\exp\left\{-\int_{T_o(\mathbf{X})}^{T(\mathbf{X},t)}\beta(s)\,ds\right\}, \text{ i.e. } \rho = \rho(T),$$

where $\rho_o(\mathbf{X})$ and $T_o(\mathbf{X})$ are density and temperature in the reference configuration, respectively.

We now introduce the Lagrangian density, ρ_L, defined as mass per reference configuration unit volume. We have

$$\rho_L = \rho J,$$

with J determinant of the deformation gradient, $J = \det\mathbf{F}$. Now, since ρ_L does not vary, we have

$$\frac{d\rho_L}{dT} = 0, \quad \Rightarrow \quad \frac{1}{J}\frac{dJ}{dT} + \underbrace{\frac{1}{\rho}\frac{d\rho}{dT}}_{-\beta} = 0,$$

---

[2] Volume preserving.

namely

$$\beta(T) = \frac{1}{J}\frac{dJ}{dT}, \tag{8}$$

or, $J = \exp\{\int_{T_o}^{T} \beta(s)\,ds\}$, i.e. $J = J(T)$.

Equation (1), because of (7), reads as a constraint linking temperature variations with the divergence of the velocity field

$$-\beta\frac{DT}{Dt} + \mathbf{D}(\mathbf{v}) : \mathbf{I} = 0, \quad \Leftrightarrow \quad -\beta\frac{DT}{Dt} + \operatorname{div}\mathbf{v} = 0. \tag{9}$$

Once again, if T is uniform and constant, $\operatorname{div}\mathbf{v} = 0$, the flow is isochoric.

Thus, the constraint $\rho = \rho(T)$ models the fluid as being incompressible under isothermal conditions, but with density changing in response to changes in temperature. In other words, given a temperature T the fluid is capable to exert any force for reaching the corresponding density. This fact is, of course, not physical and as we shall see in Appendix 1 will introduce a structural instability in the model. Nevertheless this has no serious consequences in our case, as we shall see later.

Let us now turn to a discussion of the constraint. We proceed applying classical procedure, see [12], which tours out to be feasible in our situation. We modify $\mathbf{T}$, $\mathbf{q}$, ψ and s by adding a term (the so called *constraint response*) due to the constraint itself. We consider

$$\mathbf{T} = \mathbf{T}_c + \mathbf{T}_r, \quad \mathbf{q} = \mathbf{q}_c + \mathbf{q}_r, \quad \psi = \psi_c + \psi_r, \quad s = s_c + s_r, \tag{10}$$

where the suffix "$_c$" denotes the *constraint response* while the suffix "$_r$" denotes the *constitutive part*, which is function of the primary state variables.

Concerning the constraint responses, following standard practice, we require that:

1. $\mathbf{T}_c$, $\mathbf{q}_c$, ψ_c, and s_c have no dependence on the state variables.
2. The constraint responses do not dissipate energy, namely recalling (6),

$$\rho(T)\left(\frac{D\psi_c}{Dt} + s_c\frac{DT}{Dt}\right) - \mathbf{T}_c : \mathbf{D}(\mathbf{v}) + \mathbf{q}_c \cdot \frac{\nabla T}{T} = 0, \tag{11}$$

in any thermo–mechanical process that fulfills (9).

Now, considering various special subsets of the set of all allowable thermo–mechanical processes, the above equation leads to $\mathbf{q}_c = 0$ and $\frac{D\psi_c}{Dt} = 0$, implying $psi_c \equiv 0$ if it vanishes in the reference configuration. Hence, (11) reduces to

$$\rho\,s_c\frac{DT}{Dt} - \mathbf{T}_c : \mathbf{D}(\mathbf{v}) = 0, \quad \text{when} \quad -\beta\frac{DT}{Dt} + \mathbf{I} : \mathbf{D}(\mathbf{v}) = 0.$$

The latter is therefore verified if $\rho\, s_c$ and $\mathbf{T}_c$ are proportional to $-\beta$ and to $\mathbf{I}$, respectively. We thus set

$$
\begin{cases}
\rho\left(T\right) s_c = -p\beta, \quad \Rightarrow \quad s_c = -p\dfrac{\beta}{\rho\left(T\right)}, \\[2mm]
\mathbf{T}_c = -p\mathbf{I},
\end{cases}
$$

where p is a function of position and time, referred to as *mechanical pressure*. Recall that p is not the thermodynamic pressure, defined through an equation of state. In this framework there is *no constitutive equation* for p.

Remark 2.2. We may reach the same conclusions in a more standard way considering the theory of a compressible (i.e. unconstrained) fluid, introducing the Helmholtz free energy $\psi = \psi\left(\rho, T\right)$, and we develop the standard theory eventually considering the limit $\rho = \rho\left(T\right)$, which gives $\psi = \psi\left(T\right) = \psi\left(\rho\left(T\right), T\right)$. We deduce that:

- There is no equation of state defining the pressure p.
- $s = s(T) - p\dfrac{\beta}{\rho}$.

Now, going back to (10), we have

$$
\mathbf{T} = \mathbf{T}_r - p\mathbf{I}, \quad \mathbf{q} = \mathbf{q}_r, \quad \psi = \psi_r, \quad s = s_r - p\frac{\beta}{\rho\left(T\right)}. \tag{12}
$$

So, as usual, we have now to select appropriate constitutive equations for $\mathbf{T}_r$, $\mathbf{q}_r$, ψ_r, and s_r requiring that inequality (6) is fulfilled, namely

$$
\rho\left(\frac{D\psi_r}{Dt} + s_r\frac{DT}{Dt}\right) - \mathbf{T}_r : \mathbf{D}(\mathbf{v}) + \mathbf{q}_r \cdot \frac{\nabla T}{T} \leq 0. \tag{13}
$$

The *first constitutive assumption* is the following:

A1. $\psi = \psi(T)$.

Inserting the latter into (13), yields

$$
\rho\left(\frac{d\psi}{dT} + s_r\right)\frac{DT}{Dt} - \mathbf{T}_r : \mathbf{D}(\mathbf{v}) + \mathbf{q}_r \cdot \frac{\nabla T}{T} \leq 0, \tag{14}
$$

which must hold true for all thermo-mechanical processes. Thus, we deduce

$$
\begin{cases}
s_r = -\dfrac{d\psi}{dT}, \\[3mm]
\mathbf{T}_r : \mathbf{D}(\mathbf{v}) \geq 0, \\[3mm]
\mathbf{q}_r \cdot \nabla T \leq 0.
\end{cases}
\tag{15}
$$

In particular, $(15)_2$ and $(15)_3$ play the role of restrictions on the constitutive assumptions. Recalling that we are considering a Newtonian fluid, we *assume*:

A2. The constitutive part of the Cauchy stress tensor is

$$\mathbf{T}_r = 2\mu \mathbf{D}(\mathbf{v}) + \eta \, (\mathbf{D}(\mathbf{v}) : \mathbf{I}) \, \mathbf{I}, \tag{16}$$

where $\mu = \mu(T) > 0$ is the so–called *dynamic viscosity* and $\eta = \eta(T)$ the so–called *bulk viscosity*.

Now setting

$$\widehat{\mathbf{D}}(\mathbf{v}) = \mathbf{D}(\mathbf{v}) - \frac{1}{3} (\mathbf{D}(\mathbf{v}) : \mathbf{I}) \, \mathbf{I}, \quad \text{so that } \mathbf{I} : \widehat{\mathbf{D}}(\mathbf{v}) = 0, \tag{17}$$

we may write

$$\mathbf{T}_r = 2\mu \widehat{\mathbf{D}}(\mathbf{v}) + \left(\frac{2}{3}\mu + \eta \right) (\mathbf{D}(\mathbf{v}) : \mathbf{I}) \, \mathbf{I}. \tag{18}$$

Hence, inserting $(15)_1$ and (18) into (14), we are left with

$$-2\mu \widehat{\mathbf{D}}(\mathbf{v}) : \widehat{\mathbf{D}}(\mathbf{v}) - 3 \left(\frac{2}{3}\mu + \eta \right) (\mathbf{D}(\mathbf{v}) : \mathbf{I})^2 + \mathbf{q}_r \cdot \frac{\nabla T}{T} \leq 0. \tag{19}$$

We thus *stipulate*:

A3. $\dfrac{2}{3}\mu + \eta = 0, \Rightarrow \eta = -\dfrac{2}{3}\mu$, *Stokes assumption.*

A4. The heat flux vector is

$$\mathbf{q}_r = -\lambda \nabla T, \tag{20}$$

where λ is the *thermal conductivity*, $\lambda = \lambda(T) > 0$.

Then inequality (19) reduces to

$$-2\mu \widehat{\mathbf{D}}(\mathbf{v}) : \widehat{\mathbf{D}}(\mathbf{v}) - \frac{|\nabla T|^2}{T} \leq 0,$$

which is always fulfilled.

Before proceeding further, let us recall the results obtained so far. Starting form (12) and applying the classical continuum mechanics procedure as well as assumptions **A1.–A4.** we have obtained

$$\begin{cases} \mathbf{T} = -p\mathbf{I} + 2\mu \mathbf{D}(\mathbf{v}) - \dfrac{2}{3}\mu \, (\mathbf{D}(\mathbf{v}) : \mathbf{I}) \, \mathbf{I}, \\[2mm] \mathbf{q} = -\lambda \nabla T, \\[2mm] \psi = \psi(T), \\[2mm] s = -\left(\dfrac{d\psi}{dT} + p \dfrac{\beta}{\rho(T)} \right). \end{cases} \tag{21}$$

2.3 The General Model

Let us start with the internal energy. From (5), considering (21)$_3$ and (21)$_4$, we have

$$e = e(T,p) = \psi(T) - T\left(\frac{d\psi(T)}{dT} + p\frac{\beta}{\rho}\right),$$

and

$$\frac{De}{Dt} = \left(-T\frac{d^2\psi}{dT^2} + Tp\underbrace{\left(\frac{\beta}{\rho^2}\frac{d\rho}{dT}\right)}_{-\frac{\beta^2}{\rho}} - \frac{pT}{\rho}\frac{d\beta}{dT}\right)\frac{DT}{Dt}$$

$$-\underbrace{\left(\frac{\beta p}{\rho}\frac{DT}{Dt}\right)}_{\frac{p}{\rho}\,\mathrm{div}\,\mathbf{v}} - \frac{\beta T}{\rho}\frac{Dp}{Dt}.$$

Concerning the term $\mathbf{D}(\mathbf{v}) : \mathbf{T}$, recalling also (17), we obtain

$$\mathbf{T} : \mathbf{D}(\mathbf{v}) = \left(-p\mathbf{I} + 2\mu\underbrace{\left[\mathbf{D}(\mathbf{v}) - \frac{1}{3}(\mathbf{D}(\mathbf{v}) : \mathbf{I})\,\mathbf{I}\right]}_{\widehat{\mathbf{D}}(\mathbf{v})}\right) : \mathbf{D}(\mathbf{v})$$

$$= -p\,\mathrm{div}\,\mathbf{v} + 2\mu\,\widehat{\mathbf{D}}(\mathbf{v}) : \widehat{\mathbf{D}}(\mathbf{v}),$$

where the term $-p\,\mathrm{div}\,\mathbf{v}$ represents the mechanical work per unit time operated by the system during expansion or compression. Of course such a term is necessarily compensated by internal energy gain (or loss) associated with dilation (or compression). Indeed we have developed the theory assuming that the constraint response does not dissipate energy.

Thus, taking (21)$_2$ into account, the energy equation (3) can be rewritten as

$$\rho\left(-T\frac{d^2\psi}{dT^2} - \frac{\beta^2}{\rho}Tp - \frac{pT}{\rho}\frac{d\beta}{dT}\right)\frac{DT}{Dt} = \mathrm{div}\left(\lambda\nabla T\right) + \beta T\frac{Dp}{Dt} + 2\mu|\widehat{D}(\mathbf{v})|^2, \quad (22)$$

where the term $2\mu|\widehat{D}(\mathbf{v})|^2$ represents mechanical energy converted into heat by the internal friction.

We remark that the coefficient in front of $\dfrac{DT}{Dt}$ is, from the physical point of view, the isobaric specific heat c_p. Actually, (22) is exactly the corresponding general

energy equation in the specific enthalpy formulation from [25], pages 51–85, but
with the particular choice

$$c_p(T,p) = c_{p1}(T) - p\frac{T}{\rho}\left(\beta^2 + \frac{d\beta}{dT}\right), \quad \text{with} \quad c_{p1}(T) = -T\frac{d^2\psi(T)}{dT^2}.$$

Indeed, the fluid we are modelling admits only the isobaric specific heat, since,
because of (8), any change of body's temperature implies a change in volume. Hence
it is *not possible* to work with the isochoric specific heat c_v. Consequently the form
of the energy equation here adopted is necessarily different from the theory developed in [29] and [30].

Next, experiments show that the variations of c_p with respect to pressure are
generally very small. We impose that c_p is constant with respect to the pressure p,
requiring

$$\frac{\partial c_p}{\partial p} = 0, \Rightarrow \beta^2 + \frac{d\beta}{dT} = 0,$$

that implies

$$\beta(T) = \frac{\beta_R}{1 + \beta_R(T - T_R)},$$

with T_R reference temperature and $\beta_R = \beta(T_R)$.

As a consequence, from (7), we have the following law for the density

$$\rho = \frac{\rho_R}{1 + \beta_R(T - T_R)}, \tag{23}$$

with $\rho_R = \rho(T_R)$ reference density. In particular, we will consider the linearized
version of (23), namely

$$\rho(T) = A_\rho - B_\rho T.$$

with

$$\beta_R = \frac{B_\rho}{\rho_R} \quad \text{and} \quad \rho_R = A_\rho - B_\rho T_R. \tag{24}$$

We remark that, from the mathematical point of view such a simplification is not
crucial and it is consistent with the data reported in the experimental literature.

Therefore, once assumptions **A1, A2, A3** and **A4** are stated, the mathematical
model (1)–(3) rewrites as

$$\beta\frac{DT}{Dt} = \text{div } \mathbf{v}, \tag{25}$$

$$\rho\frac{D\mathbf{v}}{Dt} = -\rho g\mathbf{e}_3 - \nabla p + \text{div}\left\{2\mu D(\mathbf{v}) - \frac{2\mu}{3}\text{div } \mathbf{v}I\right\}, \tag{26}$$

$$\rho c_{p1}(T)\frac{DT}{Dt} = \text{div}(\lambda\nabla T) + \beta T\frac{Dp}{Dt} + 2\mu|\widehat{D}(\mathbf{v})|^2, \tag{27}$$

$$\rho(T) = A_\rho - B_\rho T. \tag{28}$$

We then introduce the *rescaled dimensionless temperature*, that is

$$\vartheta = \frac{T - T_R}{T_w - T_R}, \quad \Leftrightarrow \quad T = T_R\left[1 + \vartheta\left(\widetilde{T}_w - 1\right)\right], \quad \text{with } \widetilde{T}_w = \frac{T_w}{T_R} > 1, \tag{29}$$

where T_w, $T_w > T_R$, is another characteristic temperature, as we shall see in next section. Note that (28) reads now

$$\rho(\vartheta) = \rho_R - \beta_R \rho_R T_R \left(\widetilde{T}_w - 1\right)\vartheta. \tag{30}$$

Next, we introduce the *hydraulic head*

$$P = p + \rho_R g x_3. \tag{31}$$

Hence $-\rho g \mathbf{e}_3 - \nabla p$ in (26) becomes

$$-\rho g \mathbf{e}_3 - \nabla p = (\rho_R - \rho)\,g\mathbf{e}_3 - \nabla P = \left[\rho_R \beta_R T_R\left(\widetilde{T}_w - 1\right)\vartheta\right]g\mathbf{e}_3 - \nabla P.$$

We note that, after "eliminating" the hydrostatic pressure, the effect of the temperature field on the flow is more easy to observe.

Concerning the shear viscosity μ we assume the well known Vogel–Fulcher–Tamman's (VFT) formula

$$\log \mu(\vartheta) = -C_\mu + \frac{A_\mu}{T_R - B_\mu + T_R\left(\widetilde{T}_w - 1\right)\vartheta}, \quad A_\mu, B_\mu, C_\mu > 0. \tag{32}$$

In particular, μ is monotonically decreasing with ϑ. For more details we refer e.g. to [26], Chap. 6. We note that the temperature in the problems we are considering is such that the denominator in (32) is always positive. Consequently, μ is a given strictly positive and C^∞ function of the temperature. In the polymer physics, formula (32) is known as Williams–Landau–Ferry (WLF) relation.

Finally, we suppose c_{p1} and λ to be smooth functions of ϑ, bounded from above and from below by positive constants.

2.4 Scaling and Dimensionless Formulation

The scaling of model (25)–(27) will be operated paying particular attention to the specific problem we are interested in.

We recall that we are considering a gravity driven flow of molten glass through a nozzle in the early stage of a fiber manufacturing process. At this stage we consider the inlet and outlet temperatures of the fluid as prescribed quantities. In particular, referring to Fig. 2, the fluid temperature on Γ_{in} is higher than the one on Γ_{out},

$$T\left(\Gamma_{in}\right) = T_w, \quad T\left(\Gamma_{out}\right) = T_R < T_w.$$

We introduce the following dimensionless quantities

$$\begin{cases} \tilde{\mathbf{x}} = \dfrac{\mathbf{x}}{H}, \ \tilde{\mathbf{v}} = \dfrac{\mathbf{v}}{V_R}, \ \tilde{t} = \dfrac{t}{t_R} \ \text{with} \ t_R = \dfrac{H}{V_R}, \\[2mm] \tilde{\mu} = \dfrac{\mu}{\mu_R}, \ \tilde{\lambda} = \dfrac{\lambda}{\lambda_R}, \ \tilde{c}_{p1} = \dfrac{c_{p1}}{c_{pR}}, \\[2mm] \tilde{\rho} = \dfrac{\rho}{\rho_R}, \ \tilde{P} = \dfrac{P}{P_R}, \ \tilde{\psi} = \dfrac{\psi}{T_R c_{pR}}, \end{cases}$$

H playing the role of a length scale (see again Fig. 2). The choice of the reference velocity V_R has to be made according to the particular flow conditions we deal with[3]. We will return to this point later on. As the reference pressure P_R we take the point of view that flows of glass or polymer melts are essentially dominated by viscous effects. Accordingly we set[4]

$$P_R = \frac{\mu_R}{H} V_R. \tag{33}$$

Concerning the reference quantities μ_R, λ_R and c_{pR} we identify them with the values taken by the respective quantities μ, λ and c_p for $P = P_R$ and $T = T_R$.

Suppressing tildes to keep notation simple, model (25)–(27) rewrites[5]

$$\operatorname{div} \mathbf{v} = \frac{|K_\rho|}{\rho(\vartheta)}(T_w - 1)\frac{D\vartheta}{Dt}, \tag{34}$$

$$\rho(\vartheta)\frac{D\mathbf{v}}{Dt} = \left(\frac{|K_\rho|(T_w - 1)}{\mathbf{Fr}^2}\right)\vartheta \mathbf{e}_3 - \left(\frac{P_R}{\rho_R V_R^2}\right)\nabla P$$

$$+ \frac{1}{\mathbf{Re}} \operatorname{Div}\left\{2\mu(\vartheta)D(\mathbf{v}) - \frac{2\mu(\vartheta)}{3}\operatorname{div}\mathbf{v}I\right\}, \tag{35}$$

$$\rho(\vartheta)c_{p_1}(\vartheta)\frac{D\vartheta}{Dt} = \left(\frac{|K_\rho|P_R}{\rho_R c_{pR}T_R(T_w - 1)}\right)\frac{1 + (T_w - 1)\vartheta}{\rho(\vartheta)}\left[\frac{DP}{Dt} - \frac{\rho_R g H}{P_R}v_3\right]$$

$$+ \frac{1}{\mathbf{Pe}}\operatorname{div}(\lambda\nabla\vartheta) + 2\frac{\mathbf{Ec}}{\mathbf{Re}(T_w - 1)}\mu(\vartheta)\left(|D(\mathbf{v})|_2^2 - \frac{1}{3}(\operatorname{div}\mathbf{v})^2\right), \tag{36}$$

---

[3] This makes our approach different from the ones presented in [22] and in [11] where there is no velocity scale defined by exterior conditions.

[4] Notice that $P_R \to 0$ as V_R tends to 0 and, as a consequence p tends to the hydrostatic pressure. This is consistent with the fact that P represents the deviation of the pressure from the hydrostatic–one due to the fluid motion.

[5] In the phenomena we are considering $T_w - 1$ is small but not negligible. Typically $T_w - 1$ is of order 10^{-1}.

with

$$c_{p1}(\vartheta) = -\frac{1 + (T_w - 1)\vartheta}{(T_w - 1)^2}\frac{d^2\psi}{d\vartheta^2}. \tag{37}$$

We list the non-dimensional characteristic numbers appearing in (34)–(36):

$K_\rho = -\beta_R T_R$, *thermal expansivity number* (negative, according to the usual notation) and we may write (30) as

$$\rho(\vartheta) = 1 - |K_\rho|(T_w - 1)\vartheta. \tag{38}$$

Fr $= \dfrac{V_R}{\sqrt{gH}}$, *Froude's number.*

$v_R = \dfrac{\mu_R}{\rho_R}$, *kinematic viscosity.*

Re $= \dfrac{V_R H}{v_R}$, *Reynolds' number*, which, because of (33), can be also written as

$$\mathbf{Re} = \frac{\rho_R V_R^2}{P_R}.$$

Pe $= \mathbf{Re \cdot Pr} = \dfrac{V_R H \rho_R c_{pR}}{\lambda_R}$, *Peclet's number*, with $c_{pR} = \psi_R / T_R$.

Pr $= \dfrac{\mu_R c_{pR}}{\lambda_R}$, *Prandtl's number.*

Ec $= \dfrac{V_R^2}{c_{pR} T_R}$, *Eckert's number.*

As mentioned, we are interested in studying vertical slow flows of very viscous heated fluids (molten glass, polymers, etc.) which are thermally dilatable. So, introducing the so–called *expansivity coefficient* (or *thermal expansion coefficient*)

$$\alpha = |K_\rho|(T_w - 1). \tag{39}$$

we will consider the system (34)–(36) in the realistic situation in which the parameter α is small. Typically (e.g. for molten glass) $10^{-3} \lesssim \alpha \lesssim 10^{-2}$. Equation (38) can be rewritten as

$$\rho(\vartheta) = 1 - \alpha\vartheta.$$

Next, we define the *Archimedes' number*

$$\mathbf{Ar} = \frac{\alpha}{\mathbf{Fr}^2} = \frac{|K_\rho|(T_w - 1)}{V_R^2} gH, \Rightarrow \alpha = \mathbf{Ar}\,\mathbf{Fr}^2.$$

So, system (34)–(36) rewrites as[6]

---

[6] We remark that

$$\text{div } \mathbf{v} = \frac{\alpha}{\rho(\vartheta)} \frac{D\vartheta}{Dt}, \tag{40}$$

$$\rho(\vartheta)\frac{D\mathbf{v}}{Dt} = \mathbf{Ar}\,\vartheta\,\mathbf{e}_3 + \frac{1}{\mathbf{Re}}\left[-\nabla P + \text{Div}\left(2\mu(\vartheta)D(\mathbf{v}) - \frac{2\mu(\vartheta)}{3}\,\text{div } \mathbf{v}I\right)\right], \tag{41}$$

$$\rho(\vartheta)c_{p_1}(\vartheta)\frac{D\vartheta}{Dt} = \frac{1+(T_w-1)\vartheta}{\rho(\vartheta)}\frac{\mathbf{Ec}\,\mathbf{Ar}}{(T_w-1)^2}\left[\frac{\mathbf{Fr}^2}{\mathbf{Re}}\frac{DP}{Dt} - v_3\right]$$
$$+\frac{1}{\mathbf{Pe}}\,\text{div}\,(\lambda\nabla\vartheta) + 2\frac{\mathbf{Ec}}{\mathbf{Re}(T_w-1)}\mu(\vartheta)\left(|D(\mathbf{v})|^2 - \frac{1}{3}(\text{div } \mathbf{v})^2\right). \tag{42}$$

Notice that the ratios $\dfrac{\mathbf{Ec}}{\mathbf{Re}}$ and $\dfrac{\mathbf{Fr}^2}{\mathbf{Re}}$ which appear naturally in the equation above can be interpreted as ratios of characteristic times

$$\frac{\mathbf{Ec}}{\mathbf{Re}} = \left(\frac{v_R}{c_{pR}T_R}\right)\frac{1}{t_R}, \quad \frac{\mathbf{Fr}^2}{\mathbf{Re}} = \left(\frac{v_R}{gH}\right)\frac{1}{t_R}.$$

Note also that

$$\mathbf{Ec}\,\mathbf{Ar} = \alpha\frac{gH}{c_{pR}\,T_R} = \alpha\frac{\mathbf{Ec}}{\mathbf{Fr}^2}.$$

We consider a flow regime such that $\mathbf{Ar} = \mathcal{O}(1)$, $\mathbf{Re} = \mathcal{O}(1)$, $\mathbf{Pe} = \mathcal{O}(1)$ and $\mathbf{Ec} \ll 1$ (e.g. $\mathbf{Ec} \lesssim 10^{-9}$). The terms in energy equation (42) containing the Eckert are dropped. So (40) and (41) remain unchanged while (42) reads as follow

$$\rho(\vartheta)c_{p1}(\vartheta)\frac{D\vartheta}{Dt} = \frac{1}{\mathbf{Pe}}\,\text{div}\,(\lambda(\vartheta)\nabla\vartheta). \tag{43}$$

We remark that in the limit α small, while $\mathbf{Ar}$ and $\mathbf{Re}$ order 1, the buoyancy and the viscous forces cannot be negligible. We will show that in such a case the system (40), (41) and (43) can be approximated by a system similar to the Oberbeck–Boussinesq system. Indeed our choice of the parameters takes us close to the conditions of the formal derivation of the Oberbeck–Boussinesq system for bounded domain.

Remark 2.3. Let us note that the model (40)–(42) is criticized in [2] and [3], on the basis of the fact that giving the density as a function of the temperature contradicts the Gibbs convexity inequalities thus generating a rest state which is not stable. In order to overcome such a difficulty, they propose to replace the constraint $\rho = \rho(\vartheta)$

$$\frac{\alpha}{\rho(\vartheta)} = \frac{\alpha}{1-\alpha\vartheta}.$$

Hence, if order α^2 terms are neglected, we may approximate $\frac{\alpha}{\rho(\vartheta)}$ by α.

with the constraint $\rho = \rho(s)$. However, we emphasize that the system (40), (41), (43) (i.e. with $\mathbf{Ec} = 0$) has a rest state (in absence of any body force) linearly stable. We have also to remark that, when Eckert's number is very small, the rest state instability of the system (40)–(42) would show up after such a long time to be of no interest in technical problems. Clearly, for some other problems (e.g. atmospheric flows) the presence of instabilities, pointed out in [2] and [3], is a real difficulty and the meaning of the stationary problems is not clear. Furthermore, we neglected some other terms from the full compressible Navier–Stokes system, whose influence, in the authors' knowledge, on linear stability of the rest state has not been investigated.

In Appendix 1 we show a direct analysis of the linear stability of the rest state according to model (25), (26), (27), (28) and model (25), (26), (43), (28).

3 Study of the Stationary Non-Isothermal Molten Glass Flow in a Die

In Sect. 2 a mathematical model which describes the motion of a *mechanically incompressible, but thermally expansible viscous fluid* was derived.

Such model can be widely used in industrial simulations of flows of hot melted glass, polymers etc.

The model for a mechanically incompressible, but thermally expansible viscous fluid could be thought as a particular case of the compressible heat-conducting Navier–Stokes system.

Nevertheless, in the mechanically incompressible but thermally expansible case the pressure is not linked any more to the density and to the temperature, very much the same as in the incompressible case. Consequently, we had to be careful with the thermodynamical modelling.

An important quantity that we have introduced is the *thermal expansion coefficient* $\beta = -\frac{d}{dT}\log\rho$, where ρ is the molten glass density and T is the absolute temperature, and its typical value β_R.

In this section we study the stationary flow within the die, i.e. the stationary flow in stage (b).

The flow domain (see Fig. 2) is the set $\Omega = \{\{r < R(x_3, \phi)\} \times [0, 2\pi] \times (0, H)\}$, where $R : [0, H] \times [0, 2\pi] \to [R_{min}, R_{max}]$ is a C^∞-map. For the reference problem we have in mind R and H are lengths of comparable size.

The boundary of Ω contains 3 distinct parts (see again Fig. 2):

- Lateral boundary $\Gamma_{lat} = \{r = R(x_3, \phi), x_3 \in (0, H), \phi \in [0, 2\pi]\}$.
- Inlet boundary or upper boundary $\Gamma_{in} = \{x_3 = H \text{ and } r \leq R(H, \phi), \phi \in [0, 2\pi]\}$.
- Outlet boundary $\Gamma_{out} = \{x_3 = 0 \text{ and } r \leq R(0, \phi), \phi \in [0, 2\pi]\}$.

The stationary problem in non–dimensional form, for the velocity $\mathbf{v}$, the dimensionless hydraulic head $P = (p + \rho_R g H x_3)/P_R$ and the rescaled temperature ϑ reads as follows:

Problem A. Find $(\mathbf{v}, P, \vartheta)$ such that

Fig. 2 The domain Ω

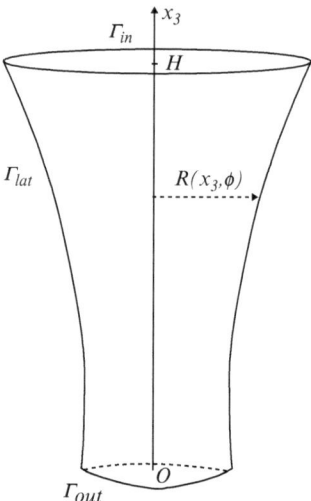

$$\text{div } (\rho(\vartheta)\mathbf{v}) = 0, \qquad \rho(\vartheta) = 1 - \alpha\vartheta, \tag{44}$$

$$\rho(\vartheta)(\mathbf{v}\nabla)\mathbf{v} = \mathbf{Ar}\ \vartheta\mathbf{e}_3 - \frac{1}{\mathbf{Re}}\left\{\nabla\left(P + \frac{2\mu(\vartheta)}{3}\text{div }\mathbf{v}\right) - \text{div } (2\mu(\vartheta)D(\mathbf{v}))\right\}, \tag{45}$$

$$c_{p1}(\vartheta)\rho(\vartheta)\mathbf{v}\cdot\nabla\vartheta = \frac{1}{\mathbf{Pe}}\text{div } (\lambda(\vartheta)\nabla\vartheta), \tag{46}$$

$$\mathbf{v} = v_1\mathbf{e}_3, \quad \vartheta = 1, \quad \text{on } \Gamma_{in}, \qquad \mathbf{v} = v_2\mathbf{e}_g, \quad \vartheta = 0, \quad \text{on } \Gamma_{out}, \tag{47}$$

$$\mathbf{v} = 0, \quad -\frac{1}{\mathbf{Pe}}\lambda(\vartheta)\nabla\vartheta\cdot\mathbf{n} = q_0\vartheta + \mathscr{S}, \quad \text{on } \Gamma_{lat}, \tag{48}$$

where the fluid's *thermal expansivity* $\alpha \geq 0$ is defined by formula (39) and c_{p1} is a function of temperature which can be expressed by means of Helmholtz free energy, as shown in (37). We refer the reader to Sect. 2 for the notations which, on the other hand, are quite standard. In (48) $\mathscr{S}$ is a non–negative prescribed function.

The plan of this section is the following:

- In Sect. 3.1 we will explain the basic steps of an existence proof for the boundary value problem (44)–(48), which does not require small data. Furthermore, we will give a hint on uniqueness for small data and prove that solution is infinitely differentiable, if the data are so.
- In Sect. 3.2 we exploit the uniqueness and regularity results concerning system (44)–(48) to obtain a *rigorous mathematical justification* of the Oberbeck–Boussinesq approximation (in the limit of small expansivity parameter), a long debated question, and also to compute the next order approximation (the correction). We note that this is another singular limit, not linked at all to the zero Mach number limit (see [19] or [28]).

3.1 Existence and Uniqueness Result for the Stationary Problem

In order to study the existence of a solution to the problem (44)–(48), we list the assumptions on the data:
We suppose that

$$\begin{cases} v_1 \in C_0^1(\Gamma_{in}) \cap C^\infty(\bar{\Gamma}_{in}), \\[2mm] v_2 \in C_0^1(\Gamma_{out}) \cap C^\infty(\bar{\Gamma}_{out}), \\[2mm] \mathcal{S} \in C^\infty(\bar{\Gamma}_{lat}), \quad \mathcal{S} \geq 0, \end{cases} \tag{49}$$

and that

$$\int_{\Gamma_{in}} \rho(1)v_1 \, rdrd\phi = \int_{\Gamma_{in}} (1-\alpha)v_1 \, rdrd\phi = \int_{\Gamma_{out}} v_2 \, rdrd\phi, \tag{50}$$

since $\rho(0) = 1$.

Concerning the coefficients, it is natural that μ corresponds to Vogel–Fulcher–Tamman's VFT formula (32), that we rewrite

$$\log \mu(\vartheta) = -C_\mu + \frac{A_\mu}{B_\mu - \vartheta}, \quad C_\mu > 0. \tag{51}$$

Next, as mentioned we suppose that

$$c_{p1}, \mu, \rho, \lambda \in C^\infty(\mathbb{R}), \tag{52}$$

and take values between two positive constants. We note that the coefficients are defined only on an interval and we may need to extend them on $\mathbb{R}$ in some obvious way, not affecting the final estimates.

3.1.1 Existence Result

The existence is proved by constructing an iterative procedure, where we first calculate the temperature for given velocity; then for given temperature we calculate the velocity and the pressure. Procedure is repeated until getting a fixed point. Our idea is to prove that such iterations give a uniformly bounded sequence in appropriate Sobolev spaces. Then elementary Sobolev compactness leads to the result. We explain here the main steps of the proof. For details we refer to the article [9].

Step 1. We start by studying *the energy equation* for a given $\mathbf{w} = \rho \mathbf{v} \in \mathcal{H}(\Omega)$, where $\mathcal{H}(\Omega) = \{\mathbf{z} \in L^3(\Omega)^3 \mid \operatorname{div} \mathbf{z} = 0 \text{ in } \Omega, \ \mathbf{z} \cdot \mathbf{n} = 0 \text{ on } \Gamma_{lat}, \ \mathbf{z} \cdot \mathbf{n}|_{\Gamma_{in}} = \rho(1)v_1, \mathbf{z} \cdot \mathbf{n}|_{\Gamma_{out}} = \rho(0)v_2\}$.

Our nonlinearities c_{p1} and λ are defined only for ϑ such that the density is not negative. We extend it on $\mathbb{R}$ by setting

$$c_{p1}(\vartheta) = \begin{cases} c_{p1}(1)/\vartheta^2 & \text{for } \vartheta > 1 \\ (\dfrac{\|\mathscr{S}\|_{L^\infty(\Gamma_{lat})}}{q_0\vartheta})^2 c_{p1}(-\dfrac{\|\mathscr{S}\|_{L^\infty(\Gamma_{lat})}}{q_0}) & \text{for } \vartheta < -\dfrac{\|\mathscr{S}\|_{L^\infty(\Gamma_{lat})}}{q_0}, \end{cases} \tag{53}$$

$$\lambda(\vartheta) = \begin{cases} \lambda(1) & \text{for } \vartheta > 1 \\ \lambda(-\dfrac{\|\mathscr{S}\|_{L^\infty(\Gamma_{lat})}}{q_0}) & \text{for } \vartheta < -\dfrac{\|\mathscr{S}\|_{L^\infty(\Gamma_{lat})}}{q_0}. \end{cases} \tag{54}$$

We will prove that $\vartheta \in [-\dfrac{\|\mathscr{S}\|_{L^\infty(\Gamma_{lat})}}{q_0}, 1]$ and the result is independent of the extension. Now for a given

$$\begin{cases} \mathbf{w} \in \mathscr{H}(\Omega), \ \mathscr{S} \in L^\infty(\Gamma_{lat}), \ \mathscr{S} \geq 0 \\ \lambda, c_{p1} \in W^{1,\infty}(\mathbb{R}), \ \lambda \geq \lambda_0 \text{ and constants } q_0, \ \text{Pe} > 0, \end{cases} \tag{55}$$

we consider the problem

Problem Θ: Find $\vartheta \in H^1(\Omega) \cap L^\infty(\Omega)$ such that

$$c_{p1}(\vartheta)\mathbf{w}\nabla\vartheta = \frac{1}{\text{Pe}} \text{div}\left(\lambda(\vartheta)\nabla\vartheta\right), \tag{56}$$

$$\vartheta = 1 \text{ on } \Gamma_{in} \quad \text{and} \quad \vartheta = 0 \text{ on } \Gamma_{out} \tag{57}$$

$$-\frac{1}{\text{Pe}}\lambda(\vartheta)\nabla\vartheta \cdot \mathbf{n} = q_0\vartheta + \mathscr{S} \quad \text{on } \Gamma_{lat} \tag{58}$$

We list some results for **Problem Θ**, whose prove is given in [9].

Proposition 3.1. *Under the stated assumptions,* **Problem Θ** *has at least one variational solution in $H^1(\Omega)$, satisfying the estimate*

$$\frac{\lambda_0}{2\text{Pe}}\|\nabla\vartheta\|^2_{L^2(\Omega)^3} + \frac{q_0}{4}\|\vartheta\|^2_{L^2(\Gamma_{lat})} \leq A_0 + B_0\|\mathbf{w}\|_{L^1(\Omega)^3}, \tag{59}$$

where B_0 and A_0 are positive constants.

Corollary 3.2. *Under the assumptions of Proposition 3.1, we have the following estimate*

$$\|\vartheta\|^2_{L^2(\Omega)} \leq 8\max\left\{\frac{\text{Pe}}{\lambda_o}, \frac{1}{q_0}\|\partial_{x_3}R\|_\infty\right\}\left(A_0 + B_0\|\mathbf{w}\|_{L^1(\Omega)^3}\right). \tag{60}$$

Lemma 3.3. *Any variational solution $\vartheta \in H^1(\Omega)$ to Problem Θ, satisfying the a priori estimate (59), satisfies also*

$$-\frac{\|\mathscr{S}\|_{L^\infty(\Gamma_{lat})}}{q_0} \leq \vartheta \leq 1 \quad \text{a.e. on } \Omega. \tag{61}$$

Remark 3.4. Let us now recall a very general result for a class of elliptic problems, including Θ (see [1]). Under a suitable condition on the interior angle between the lateral boundary $r = R(x_3, \phi)$ and the upper and lower surfaces, $\ell_3(\Omega)$ is the elliptic regularity constant, the solution u for the mixed problem

$$\text{div} \, (\nabla u - \mathbf{f}) = 0 \quad \text{in} \ \Omega \tag{62}$$

$$\nabla u \cdot \mathbf{n} = g \quad \text{on} \ \Gamma_{lat} \quad \text{and} \quad u = 0 \ \text{on} \ \Gamma_{out} \cup \Gamma_{in} \tag{63}$$

is estimated as

$$\|\nabla u\|_{L^3(\Omega)^3} \leq \ell_3(\Omega) \left\{ \|g\|_{L^3(\Gamma_{lat})} + \|\mathbf{f}\|_{L^3(\Omega)^3} \right\}. \tag{64}$$

Hence, applying such a result, we have

$$\|\nabla \vartheta\|_{L^3(\Omega)^3} \leq E_{00} + E_{01} \|\mathbf{w}\|_{L^3(\Omega)}, \tag{65}$$

where

$$E_{00} = \frac{|\Gamma_{lat}|^{1/3} \, \ell_3(\Omega) \, \mathbf{Pe}}{\lambda_0} \left(\|\mathscr{S}\|_{L^\infty(\Gamma_{lat})} + q_o \max\left\{ 1, \frac{\|\mathscr{S}\|_{L^\infty(\Gamma_{lat})}}{q_0} \right\} \right)$$
$$+ \frac{\|\lambda\|_\infty}{\lambda_0} \left(|\Omega|^{1/3} + \ell_3(\Omega)|\Gamma_{lat}|^{1/3} \right), \tag{66}$$

and

$$E_{01} = \frac{\ell_3(\Omega)}{\lambda_0} \mathbf{Pe} \, \|c_{p1}\|_\infty \max\left\{ 1, \frac{\|\mathscr{S}\|_{L^\infty(\Gamma_{lat})}}{q_0} \right\}. \tag{67}$$

Step 2. Now we turn to the *continuity and momentum equations*, determining the impulse $\mathbf{w} = \rho \mathbf{v}$ and the pressure p.

Our system reads as follows:

For given ϑ determine $\{\mathbf{w}, p\}$ satisfying

$$\text{div} \, \mathbf{w} = 0 \quad \text{in} \ \ \Omega \tag{68}$$

$$(\mathbf{w}\nabla)\frac{\mathbf{w}}{\rho} = -\mathbf{Ar}\vartheta \mathbf{e}_3 - \nabla(p + \frac{2\mu}{3\mathbf{Re}} \, \text{div} \, \frac{\mathbf{w}}{\rho})$$
$$+ \frac{1}{\mathbf{Re}} \, \text{Div} \, (2\mu(\vartheta)D(\frac{\mathbf{w}}{\rho})) \quad \text{in} \ \ \Omega \tag{69}$$

$$\mathbf{w} = 0 \ \text{on} \ \Gamma_{lat}, \ \mathbf{w} = (1 - \alpha)v_1 \mathbf{e}_3 \ \text{on} \ \Gamma_{in} \ \text{and} \ \mathbf{w} = v_2 \mathbf{e}_3 \ \text{on} \ \Gamma_{out}, \tag{70}$$

Next, there exists

$$\zeta \in H^2(\Omega)^3, \nabla\zeta \in L^3(\Omega)^9, \quad \text{with curl } \zeta \text{ satisfying (70)} \tag{71}$$

Now we adapt the well-known Hopf construction to our non-standard nonlinearities. we have

Proposition 3.5. *(see [9]) Let us suppose that $1/\rho \in L^\infty(\Omega)$, $1/\rho \leq 1/\rho_{min}$. Then, for every $\gamma > 0$ there is a ξ, depending on γ and ρ_{min}, such that*

$$\xi \in H^1(\Omega)^3, \ div \, \xi = 0 \ in \ \Omega, \ \xi = \zeta \ on \ \partial\Omega \ and \tag{72}$$

$$\left| \int_\Omega \frac{1}{\rho} (\phi\nabla)\phi \cdot \xi \, dx \right| \leq \gamma \|\phi\|_{H^1(\Omega)^3}^2, \quad \forall \phi \in H_0^1(\Omega)^3 \tag{73}$$

Let us introduce now some useful constants:

$$\begin{cases} \gamma = \dfrac{2\mu_{min}}{\rho_{max}\mathbf{Re}}, \quad A_{00} = 4\dfrac{\sqrt{6}}{\rho_{min}^2}, \quad B_{00} = \dfrac{\mu_{min}}{\rho_{max}}, \quad C_{01} = 2\cdot 6^{1/6}|\Omega|^{5/6} \\[2mm] B_{01} = 2\cdot 6^{1/3}\dfrac{1}{\rho_{min}^2}\|\xi\|_{L^6(\Omega)^3}, \quad B_{02} = \dfrac{4\mu_{max}6^{1/6}}{\rho_{min}^3} \\[2mm] C_{00} = \dfrac{1}{\rho_{min}}\left(\|\xi\|_{L^6(\Omega)^3}^2 + \dfrac{2\mu_{max}}{\mathbf{Re}}\|D(\xi)\|_{L^2(\Omega)^9}\right). \end{cases} \tag{74}$$

Our basic result is the following:

Theorem 3.6. *(see [9]) Let us suppose that*

$$\Delta_B = \frac{1}{\mathbf{Re}}\left(B_{00} - \alpha B_{02}\|\nabla\vartheta\|_{L^3(\Omega)^3}\right) - \alpha B_{01}\|\nabla\vartheta\|_{L^2(\Omega)^3} > 0 \tag{75}$$

$$and \ \Delta_{det} = \Delta_B^2 - 4A_{00}\alpha\|\nabla\vartheta\|_{L^2(\Omega)^3}(C_{00} + \mathbf{Ar}C_{01}\|\nabla\vartheta\|_{L^2(\Omega)^3}) > 0, \tag{76}$$

Then there is a solution $\mathbf{w} \in H^1(\Omega)^3$ for the problem (68)–(70), satisfying the estimate

$$\|D(\mathbf{w})\|_{L^2(\Omega)^9} \leq \|D(\xi)\|_{L^2(\Omega)^9} + \frac{C_{00} + \mathbf{Ar}C_{01}\|\nabla\vartheta\|_{L^2(\Omega)^3}}{\sqrt{\Delta_{det}}} \tag{77}$$

Step 3. Next we define our *iterative procedure*:

Let $\gamma = \dfrac{\mu_{min}}{\rho_{max}\mathbf{Re}}$ and ξ be the corresponding vector valued function from Hopf's construction.

For a given $\mathbf{w}^m = \mathbf{W}^m + \xi$, such that $\mathbf{W}^m \in B_R = \{\mathbf{z} \in H_0^1(\Omega)^3: \ div \, \mathbf{z} = 0 \ in \ \Omega$ and $\|D(\mathbf{W}^m)\|_{L^2(\Omega)^9} \leq R\}$, we calculate ϑ^m, a solution to (56)–(58).

Next, with this ϑ^m, we determine a solution $\mathbf{w}^{m+1} = \mathbf{W}^{m+1} + \xi$ for the problem (68)–(70), satisfying the estimate (77).

The natural question arising in the iterative process is *does $\mathbf{W}^{m+1}$ remains in B_R* ? Of course R has to be selected in a suitable way.

We have the following result

Proposition 3.7. *(see [9]) Let the constants A_0 and B_0 be given as in Proposition 3.1 and let E_{00} and E_{01} be given by (66)–(67). Let the constants $B_{00}, B_{01}, B_{02}, A_{00}, C_{00}$ and C_{01} be given by formula (74). Let ξ be generalized Hopf's lift, given by Proposition 3.5 and corresponding to γ. Let R be given by*

$$R = \sqrt{2}\mathbf{Re}\frac{\rho_{max}}{\mu_{min}}\left(2\mathbf{Ar}|\Omega|^{5/6}6^{1/6}\max\left\{1, \frac{\|\mathscr{S}\|_{L^\infty(\Gamma_{lat})}}{q_0}\right\}\right.$$
$$\left. +\frac{1}{\rho_{min}}\left(\|\xi\|^2_{L^6(\Omega)^3} + \frac{2\mu_{max}}{\mathbf{Re}}\|D(\xi)\|_{L^2(\Omega)^9}\right)\right). \qquad (78)$$

Then for all $\alpha > 0$ such that

$$\Delta_1 = \frac{1}{\mathbf{Re}}\left(B_{00} - \alpha B_{02}\left(E_{00} + E_{01}(\|\xi\|_{L^3(\Omega)^3} + 48^{1/12}|\Omega|^{1/6}R)\right)\right)$$
$$-\alpha B_{01}\sqrt{\frac{\mathbf{Pe}}{\lambda_0}}\left(\sqrt{2A_0} + \sqrt{2B_0}(\|\xi\|^{1/2}_{L^1(\Omega)^3} + |\Omega|^{1/4}\sqrt{H}R^{1/2})\right) > 0 \quad \text{and} \quad (79)$$

$$\Delta_2 = \Delta_1^2 - 4A_{00}\alpha\sqrt{\frac{2\mathbf{Pe}}{\lambda_0}}\left[\sqrt{A_0} + \sqrt{B_0}(\|\xi\|^{1/2}_{L^1(\Omega)^3} + |\Omega|^{1/4}\sqrt{R})\right]\cdot$$
$$\left[C_{00} + \mathbf{Ar}C_{01}\max\{1, \frac{\|\mathscr{S}\|_{L^\infty(\Gamma_{lat})}}{q_0}\}\right] > \frac{B_{00}^2}{2\mathbf{Re}^2}, \qquad (80)$$

$\mathbf{W}^m \in B_R$ *implies* $\mathbf{W}^{m+1} \in B_R$.

After this invariance result, a simple compactness argument gives the existence of at least one solution.

The result could be summarize in the following theorem:

Theorem 3.8. *(see [9]) There is a weak solution $\{\vartheta, \mathbf{v}\} \in W^{1,3}(\Omega) \times H^1(\Omega)^3$ for* **Problem A**, *such that*

$$\begin{cases}\|\nabla\vartheta\|_{L^2(\Omega)^3} \leq \sqrt{\frac{2\mathbf{Pe}}{\lambda_0}}\left(\sqrt{A_0 + B_0\|\xi\|_{L^1(\Omega)}} + \sqrt{B_0|\Omega|^{1/2}R}\right), \\ -\frac{\|\mathscr{S}\|_{L^\infty(\Gamma_{lat})}}{q_0} \leq \vartheta \leq 1, \quad \text{and} \quad \|D(\mathbf{v} - \xi)\|_{L^2(\Omega)^9} \leq R,\end{cases} \qquad (81)$$

where ξ is given by Proposition 3.5 with $\gamma = 2\mu_{min}/(\rho_{max}\mathbf{Re})$.

3.1.2 Regularity and Uniqueness

Regularity of solutions and the uniqueness are also studied in [9]. Getting uniqueness is quite technical. After proving the regularity, one obtains that for small data there is a unique weak solution $\{\mathbf{v}, \vartheta\} \in H^1(\Omega)^3 \times (W^{1,3}(\Omega) \cap C(\bar{\Omega}))$ for the problem (44)–(48), satisfying the bounds (81). Detailed calculations are in [9].

Let us say a bit more concerning the regularity of solutions:

Lemma 3.9. *Let c_{p0} and $\lambda \in C^\infty(\mathbb{R})$. Furthermore let $\Gamma_{lat} \in C^\infty$ and $\mathcal{S} \in C^\infty(\bar{\Gamma}_{lat})$. Then $\vartheta \in W^{2,6}(\Omega) \subseteq C^{1,1/2}(\bar{\Omega})$.*

Lemma 3.10. *Let $\vartheta \in W^{2,6}(\Omega)$, let $\Gamma_{lat} \in C^\infty$ and let $v_j \in C^\infty$, $j = 1, 2$ satisfy (49). Then $\{\mathbf{w}, p\} \in W^{2,q}(\Omega) \times W^{1,q}(\Omega)$, $\forall q < +\infty$.*

Theorem 3.11. *Let the assumptions on the data from two previous Lemmata hold true. Then every weak solution $\{\mathbf{v}, p, \vartheta\} \in H^1(\Omega)^3 \times L^2(\Omega) \times (H^1(\Omega) \cap L^\infty(\Omega))$ for the **Problem A** is an element of $W^{2,q}(\Omega)^3 \times W^{1,q}(\Omega) \times W^{2,q}(\Omega)$, $\forall q < \infty$. Furthermore $\{\mathbf{v}, p, \vartheta\} \in C^\infty(\Omega)^5$.*

3.2 Oberbeck–Boussinesq Model

Now we are in position to pass to the limit when the expansivity parameter α tends to zero.

First we remark that the a priori estimates from the previous section are *independent* of α, $|\alpha| \le \alpha_0$, where α_0 is the maximal positive α satisfying (79)–(80). Consequently, by a simple weak compactness argument, we have

Theorem 3.12. *(see [9]). Let $\{\vartheta(\alpha), \mathbf{v}(\alpha)\}$, $\alpha \in (0, \alpha_0)$, be a sequence of weak solutions to Problem **A**, satisfying the bounds (81). Then there exists $\{\vartheta^{OB}, \mathbf{v}^{OB}\} \in W^{1,3}(\Omega) \times H^1(\Omega)^3$ and a subsequence $\{\vartheta(\alpha_k), \mathbf{v}(\alpha_k)\}$ such that*

$$\begin{cases} \vartheta(\alpha_k) \to \vartheta^{OB}, \text{ uniformly on } \bar{\Omega} \\ \vartheta(\alpha_k) \rightharpoonup \vartheta^{OB}, \text{ weakly in } W^{1,3}(\Omega) \\ \mathbf{v}(\alpha) \rightharpoonup \mathbf{v}^{OB}, \text{ weakly in } H^1(\Omega)^3. \end{cases}$$

Furthermore, $\{\vartheta^{OB}, \mathbf{v}^{OB}\}$ is a weak solution for the equations

$$div \, \mathbf{v}^{OB} = 0, \tag{82}$$

$$(\mathbf{v}^{OB}\nabla)\mathbf{v}^{OB} = \mathbf{Ar}\vartheta^{OB}\mathbf{e}_3 - \nabla P^{OB} + \frac{1}{\mathbf{Re}} \, div \, \{2\mu(\vartheta^{OB})D(\mathbf{v}^{OB})\}, \tag{83}$$

$$c_{p1}(\vartheta^{OB})\mathbf{v}^{OB}\nabla\vartheta^{OB} = \frac{1}{\mathbf{Pe}} \, div \, (\lambda(\vartheta^{OB})\nabla\vartheta^{OB}), \tag{84}$$

satisfying the boundary conditions (47)–(50) and the bounds (81).

For the mathematical theory of the system (82)–(84), subject to various boundary conditions, but with constant viscosity, we refer to [27], pages 129–137. The existence and uniqueness for the complete non–stationary problem (82)–(84) is in the article [8].

3.2.1   Discussion of the Formal Limit

For the glass molt in the die typical values of the thermal expansivity number K_ρ are of order $0.5 \cdot 10^{-2}$, typical temperature oscillation is 10^{-1}. The buoyancy forces (mixed forced and natural convection) are proportional to $\dfrac{1}{\mathbf{Fr}^2} = gH/V_R^2$ and their typical order of magnitude is 10^4. Therefore it is important to consider a flow regime such that $\mathbf{Ar} = \mathscr{O}(1)$, $\mathbf{Re} = \mathscr{O}(1)$ and $\mathbf{Pe} = \mathscr{O}(1)$. So we have $\alpha = K_\rho(1 - T_w) \to 0$ with the Archimedes number remaining constant. We have $\rho(\vartheta) \to 1$, as $\alpha \to 0$ and the continuity equation (44) becomes the incompressibility condition (82). In (45), the density $\rho(\vartheta)$ becomes 1 and we get the (83). Note that extracting the static pressure and working with the hydraulic head was crucial for correct passing to the limit. Also, in the energy equation (46) the density $\rho(\vartheta)$ becomes equal to 1 and we get the (84). This limit was justified rigorously in Theorem 3.12.

Therefore, in practical situations of industrial importance the system (44), (45) and (46) can be approximated by a system similar to the Oberbeck–Boussinesq system. Indeed our choice of the parameters takes us close to the conditions of the formal derivation of the Oberbeck–Boussinesq system for bounded domain. Such conditions (see [25], pages 86–91) are

$$V_R \approx 0, \quad \mathbf{Ma} \approx 0, \quad \mathbf{Re} = \mathscr{O}(1) \quad \text{and} \quad \mathbf{Ar} = \mathscr{O}(1), \tag{85}$$

with $\mathbf{Ma}$ being *Mach's number*, which in our situation is of order 10^{-6}. Under conditions (85), the non–isothermal compressible Navier–Stokes system can be approximated (formally) by the Oberbeck–Boussinesq system (82)–(84).

We note that the same reasoning applies to the situations in which other boundary conditions are given, just modifying the definition of ϑ and of Archimedes' number $\mathbf{Ar}$. For details we refer again to [25], page 89.

In the case of natural convection flows, differently from gravity driven flows, there is no characteristic velocity and one takes $V_R = \mu_R/(H\rho_R)$. In this case the relevant characteristic number is the *Grashof number*, $\mathbf{Gr} = \mathbf{Ar}\,\mathbf{Re}^2$. Instead of imposing $\mathbf{Ar} = \mathscr{O}(1)$, one requires $\mathbf{Gr} = \mathscr{O}(1)$ and the condition $V_R \approx 0$ is replaced by $\mu_R/\rho_R \approx 0$.

Another approximation for the natural convection in the case of a layer of fluid of thickness H, with the top and bottom surfaces held at constant temperature, is derived in [22]. The authors take the Chandrasekhar's velocity $V_R = \sqrt{gL\beta(T_w - T_R)}$ (linked to buoyancy) as characteristic. Then they introduced the small parameter $\varepsilon = \left(\dfrac{V_R^3 \rho_R}{g\mu_R}\right)^{1/3}$, obtaining formally the Oberbeck–Boussinesq equations at order $\mathscr{O}(\varepsilon^3)$. This approach was developed further in [16] in order to get a modified Oberbeck–Boussinesq system, with important viscous heating.

A similar derivation of the Oberbeck–Boussinesq approximation for the natural convection in the case of a layer of fluid, can be found in [11], pages 43–53. There, for the sake of definiteness, a perfect gas is considered but the scalings and the orders of magnitude involved are similar to [22].

Next we mention a number of papers by R. Kh. Zeytounian on the thermocapillary problems with or without a free boundary. They are reviewed in [29] and [30]. He considered the natural convection in an infinite horizontal layer of viscous, thermally conducting and weakly expansible fluid, heated from below. The small parameter is α and for

$$\alpha \approx 0, \quad \mathbf{Fr} \approx 0, \quad \mathbf{Ma} \approx 0, \quad \text{and} \quad \mathbf{Gr} = \frac{\alpha \mathbf{Re}^2}{\mathbf{Fr}^2} = \mathscr{O}(1), \qquad (86)$$

he got the Oberbeck–Boussinesq approximation. In his derivation, the viscous dissipation term is absent if

$$1 \text{ mm} \approx \left(\frac{\mu_R^2}{g \rho_R^2} \right)^{1/3} << H \approx C_0 / (g(T_w - T_R)). \qquad (87)$$

The formal analysis of Zeytounian carries over to the free boundary cases.

Rigorous justification of the limit when the Mach number tends to zero was the subject of intense research in recent years. We mention in this direction papers [4, 6, 15, 19] and references therein. In the recent paper [10] the Oberbeck–Boussinesq system is obtained in the low Mach number limit.

Therefore we established a new singular limit for the isochoric Navier–Stokes-Fourier system. Nevertheless, the expansivity parameter α is small but not zero (mostly of order $0.5 \cdot 10^{-3}$. It makes sense to take the approximation with the next order correction. In the subsection which follows we undertake the rigorous derivation of the next order correction.

3.2.2 Correction to the Oberbeck–Boussinesq Model

In this section we reconsider the limit when the expansivity parameter α tends to zero. Having justified the Oberbeck–Boussinesq system as the limit equations when the expansivity parameter α tends to zero, the next question is: *What is the accuracy of the approximation ?*

The answer relies on the uniqueness and regularity results from the previous sections.

As usual for the expansions, we need the equations for the derivatives with respect to α. For simplicity we suppose that

$$v_2 \text{ is independent of } \alpha \text{ and } v_1 = V_1/(1 - \alpha) \text{ with } V_1 \text{ independent of } \alpha. \qquad (88)$$

Then we have

Proposition 3.13. *(see [9]). Let us suppose that data fulfill smallness conditions, from [9], ensuring existence, uniqueness and regularity of a solution lying inside* *the ball defined by the bounds (81). Furthermore let the solution $\{\mathbf{v}, p, \vartheta\}$ satisfy the inequalities*

$$\mathscr{N} = \frac{\lambda_0}{\mathbf{Pe}} - \frac{C_6(\Omega)}{2}(1 + \frac{H}{\sqrt{2}}) \left(\|c_{p1}\|_{L^\infty(\mathbb{R})} \|\mathbf{v}\|_{L^3(\Omega)^3} \rho_{max} \right.$$
$$\left. + \frac{\|\lambda'\|_{L^\infty(\mathbb{R})}}{\mathbf{Pe}} \|\nabla \vartheta\|_{L^3(\Omega)^3} \right) > 0 \qquad (89)$$

$$\frac{2}{\mathbf{Re}} \frac{\mu_{min}}{\rho_{max}} > 4(\frac{3}{2})^{1/4} \sqrt{H} \|D(\mathbf{v})\|_{L^2(\Omega)^9} + H\mathscr{L}_\Theta \left\{ \frac{2\alpha}{\mathbf{Re}} \frac{\mu_{max}}{\rho_{min}} \|\mathbf{v}\|_{L^\infty(\Omega)^3} \right.$$
$$+ \mathbf{Ar} \frac{H^2}{\sqrt{2}} + \frac{2}{\mathbf{Re}} \|\mu'\|_{L^\infty(\mathbb{R})} \|D(\mathbf{v})\|_{L^3(\Omega)^9} C_6(\Omega) + \left. \alpha H \|\mathbf{v}\|_{L^\infty(\Omega)^3}^2 \right\}, \qquad (90)$$

where

$$\mathscr{L}_\Theta = \frac{H\|c_{p1}\|_{L^\infty(\mathbb{R})} \|\vartheta\|_{L^\infty(\Omega)}}{\sqrt{2}\mathscr{N}} \qquad (91)$$

Then derivatives of the solution, with respect to α, exist at all orders as continuous functions of α.

With this result, we are ready to state the error estimate for Boussinesq's limit.

First, we write the 1st order correction, i.e. the system defining the first derivatives $\{\mathbf{w}^0, \tilde{\pi}^0, \theta^0\} = \frac{d}{d\alpha}\{\mathbf{v}, \tilde{p}, \vartheta\}|_{\alpha=0}$:

$$\operatorname{div}\left\{\mathbf{w}^0 - \vartheta^{OB}\mathbf{v}^{OB}\right\} = 0 \quad \text{in} \quad \Omega \qquad (92)$$

$$-\vartheta^{OB}(\mathbf{v}^{OB}\nabla)\mathbf{v}^{OB} + \left\{(\mathbf{w}^0\nabla)\mathbf{v}^{OB} + (\mathbf{v}^{OB}\nabla)\mathbf{w}^0\right\} = -\mathbf{Ar}\,\theta^0\mathbf{e}_3 - \nabla\tilde{\pi}^0$$
$$+ \frac{2}{\mathbf{Re}}\operatorname{Div}\left\{\mu(\vartheta^{OB})D(\mathbf{w}^0) + \mu'(\vartheta^{OB})\theta^0 D(\mathbf{v}^{OB})\right\} \text{ in } \Omega \qquad (93)$$

$$\operatorname{div}\left\{ -\frac{\lambda(\vartheta^{OB})}{\mathbf{Pe}}\nabla\theta^0 + (\mathbf{v}^{OB}c_p(\vartheta^{OB}) - \frac{\lambda'(\vartheta^{OB})}{\mathbf{Pe}}\nabla\vartheta^{OB})\theta^0 \right.$$
$$\left. -\vartheta^{OB}\mathbf{v}^{OB}C_p(\vartheta^{OB}) + \mathbf{w}^0 C_p(\vartheta^{OB}) \right\} = 0 \quad \text{in} \quad \Omega \qquad (94)$$

$$\theta^0 = 0, \mathbf{w}^0 = 0 \text{ on } \Gamma_{out}; \quad \theta^0 = 0, \mathbf{w}^0 = V_2\mathbf{e}_3 \text{ on } \Gamma_{in} \qquad (95)$$

$$\mathbf{w}^0 = 0 \text{ and } -\frac{1}{\mathbf{Pe}}(\lambda(\vartheta^{OB})\nabla\theta^0 + \lambda'(\vartheta^{OB})\theta^0\nabla\vartheta^{OB}) \cdot \mathbf{n} = q_0\theta^0 \text{ on } \Gamma_{lat}, \quad (96)$$

Under the conditions of the preceding Proposition, with $\alpha = 0$, the system (92)–(96) has a unique smooth solution. Hence we have established rigorously the $\mathscr{O}(\alpha^2)$ approximation for **Problem A**. Clearly, one could continue to any order.

The result is given by the following theorem, which is a straightforward corollary of Proposition 3.13.

Theorem 3.14. *(see [9]). Let us suppose the assumptions of Proposition 3.13. Then we have*

$$\|\mathbf{v} - \mathbf{v}^{OB} - \alpha \mathbf{w}^0\|_{W^{1,\infty}(\Omega)^3} + \|\vartheta - \vartheta^{OB} - \alpha\theta^0\|_{W^{1,\infty}(\Omega)} \le C\alpha^2 \qquad (97)$$

$$\inf_{\mathscr{C} \in \mathbb{R}} \|p - p^{OB} - \alpha p^0 + \mathscr{C}\|_{L^\infty(\Omega)^3} \le C\alpha^2 \qquad (98)$$

where $p^0 = \tilde{\pi}^0 - 2\mu(\vartheta^{OB})\, div\, \mathbf{w}^0/(3\,\mathbf{Re})$.

4 Modelling the Viscous Jet at the Exit of the Die

In this section we discuss briefly the stage (c) (recall Sect. 1), that is the free boundary flow in the air. Our goal is to derive a mathematical model, making some
simplifying assumptions, in order to discuss the influence of some physical parameters involved. In this stage one meets several difficulties. The first is that we deal
with the non-isothermal flow of a thermally expansible fluid. Next, when the fluid
leaves the die it experiences strong cooling. It is observed that, in a suitable range
of temperature and of pulling velocity, defining the so called "cold breakdown", the
jet detaches from the solid wall inside the die. Of course, phenomena of wetting of
the outer surface of the die are also observed if temperature is large enough, making
the molten glass less viscous, and/or if the pulling velocity is sufficiently small.

Our approach could describe the flow in stage (c) for a general geometry of the
die. Nevertheless, since the main difficulties are not really linked with the form of
the die and in order to simplify our exposition, we consider cylindrical geometry
with azimuthal symmetry. In particular, x will denote the axial coordinate, r the
radial–one and the symmetry axis coincides with the x axis.

The dependence of the fluid viscosity μ, on the temperature T plays a major role
in the process. Therefore, in order to simplify modelling at this stage we neglect
variations of the fluid density ρ and of the surface tension σ. Hence, within the
framework of incompressibility, the fluid constitutive model becomes

$$\mathbf{T} = -p\mathbf{I} + 2\mu\mathbf{D},$$

and the continuity equation, in the stationary case, writes

$$\frac{\partial(v_1)}{\partial x} + \frac{1}{r}\frac{\partial(rv_2)}{\partial r} = 0, \qquad (99)$$

where v_1 is the fluid axial velocity and v_2 the radial–one, that is $\mathbf{v} = v_1(x,r,t)\,\mathbf{e}_x + v_2(x,r,t)\,\mathbf{e}_r$.

We consider a die which is a cylinder of radius R and high L_1. After entering the
die the molten glass wets only a part of the die walls and then produces a viscous jet.
We confine our analysis to the case in which the jet starts at some point $x_T \in (0,L_1)$
and then we have a *free boundary* $\mathscr{S}$, $r = h(x,t)$, for $x_T < x < L$, separating the fluid

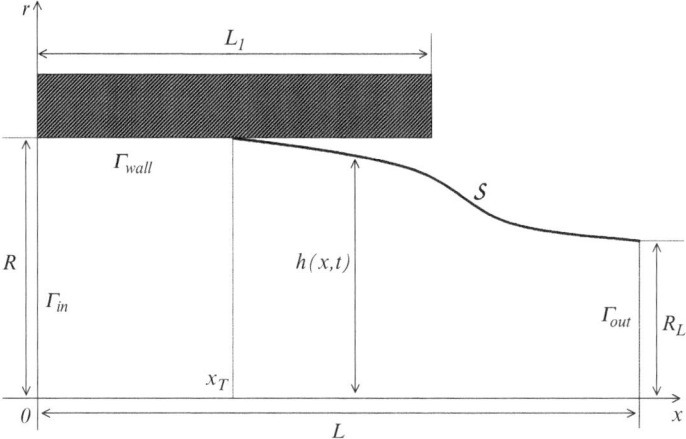

Fig. 3 A schematic of stage (c). The x-axis coincides with the cylinder symmetry line. $\mathbf{e}_x$, $\mathbf{e}_r$ denote the unit vectors, $\mathbf{x} = x\mathbf{e}_x + r\mathbf{e}_r$. The velocity field has the form $\mathbf{v}(\mathbf{x},t) = v_1(\mathbf{x},t)\mathbf{e}_x + v_2(\mathbf{x},t)\mathbf{e}_r$. The gravity acceleration is parallel to the x axis, $\mathbf{g} = g\mathbf{e}_x$

from the surrounding air (see Fig. 3). The fiber is cooled down by a stream of air, whose interaction with the fibers is also important, but we will not deal with this aspect here.

Referring to Fig. 3, the flow domain is the set

$$\Omega = \Omega^{die} \cup \Omega^{jet}, \text{ with}$$

$$\Omega^{die} = \{0 < x < x_T, \, 0 < r < R\},$$

$$\Omega^{jet} = \{x_T < x < L, \, 0 < r < h(x,t)\},$$

where, as mentioned, $r = h(x,t)$ is the unknown boundary of the jet. The boundary of Ω contains four distinct parts

- Inlet boundary $\Gamma_{in} = \{x = 0, \, 0 \le r \le R\}$.
- Wall boundary $\Gamma_{wall} = \{0 < x \le x_T, \, r = R\}$.
- Free boundary $\mathscr{S} = \{r = h(x,t), \, x \in (x_T, L)\}$.
- Outlet boundary $\Gamma_{out} = \{x = L, \, 0 \le r \le R_L < R\}$, where $R_L = h(L)$, fiber radius at $x = L$.

We observe that the whole process is governed by the following *easily accessible quantities* (besides the temperature of the cooling stream of air):

- Spool radius, R_{sp}.
- Spool angular velocity, ω.
- Power dissipated by the spinning device, W. Indeed, for estimating W it is enough to measure the intensity of the electric current I entering the electric engine and

the electric potential difference ΔV at the poles of the engine, so that $W = I\Delta V - W_o$, where W_o is the power consumption of the engine.
- Discharge, Q_{out}, i.e. rate of increase of the volume of collected fiber.

From these directly measured quantities, one can deduce:

- Spinning speed, $V_{sp} = \omega R_{sp}$, i.e. fibers velocity entering the spinner.
- Fiber radius, $R_f = (Q_{out}/\pi V_{sp})^{1/2}$.
- Torque applied to the spinner, $M = W/\omega$.
- Traction force applied to the fiber, $F_{sp} = M/R_{sp} = W/V_{sp}$.
- Mean stress applied to the fiber surface by the spinning device, $\Phi_{sp} = F_{sp}/\pi R_f^2 = W/Q_{out}$.

We denote by V_L the *mean fluid velocity on* Γ_{out}

$$V_L = \frac{2}{R_L^2} \int_0^{R_L} \mathbf{v}(L,r) \cdot \mathbf{e}_x r dr. \tag{100}$$

Now, since the discharge Q_{out} is prescribed, we have

$$\pi V_L R_L^2 = Q_{out}, \quad \Rightarrow \quad R_L = \sqrt{\frac{Q_{out}}{\pi V_L}}. \tag{101}$$

which links together V_L and R_L.

From the analysis of the simple Matovich–Pearson model for stage (d), that will be performed in Appendix 2, it will be clear that, once L has been fixed, quantities such V_L (the mean fluid velocity on Γ_{out}) and the mean traction stress on Γ_{out}

$$\Phi_L = \frac{2}{R_L^2} \int_0^{R_L} \mathbf{T}(L,r) \mathbf{e}_x \cdot \mathbf{e}_x \, r dr = \frac{2}{R_L^2} \int_0^{R_L} (-p(L,r) + 2\mu(L)\mathbf{D}\mathbf{e}_x \cdot \mathbf{e}_x) \, r dr, \tag{102}$$

as well as the fiber radius R_L can be explicitly related with the terminal quantities V_{sp}, R_{sp} and Q_{out}.

We note that the outlet boundary $x = L$ is *an artificial mathematical boundary*, because the jet continues beyond that point. In this section, we confine our attention to the portion of jet in the interval $x_T < x < L$, defining L as a location at which the fiber is cooled down to a point such that the fluid viscosity is large enough to prevent any further important fiber radius variations. On the other hand, the temperature should still be larger than the glass transition temperature. The value of L is an unknown of the problem and in fact after the point $x = L$ one is likely to use the averaged equations presented in the next section. Actually, giving a definition of L is a delicate point of the model and depends on how μ varies with the temperature.

Although there are authors [13] who, on the basis of empirical observations, suggest to take $L = 12R$, here we proceed in a rather different way, analyzing carefully the jet profile at the end of stage (c). Such an issue will be discussed in Sect. 4.1.

Concerning the boundary conditions to be imposed on the jet boundary $\mathscr{S}$ (see Fig. 3), we first assume a purely kinematic condition: the fluid velocity is tangent to the free surface, namely

$$\mathbf{v}\cdot\mathbf{n} = \mathbf{w}_{\mathscr{S}}\cdot\mathbf{n} = 0. \tag{103}$$

where $\mathbf{w}_{\mathscr{S}}$ is the free boundary velocity. Next, we impose a dynamic condition: equilibrium of the forces acting on $\mathscr{S}$, that is

$$\begin{cases} -p + 2\mu\,(\mathbf{D}\mathbf{n})\cdot\mathbf{n} = -\sigma\kappa, \\ 2\mu\,(\mathbf{D}\mathbf{n})\cdot\tau = (\tau\cdot\nabla)\sigma, \end{cases} \tag{104}$$

where:

- $2\kappa = \dfrac{1}{h\left(1+(h')^2\right)^{1/2}} - \dfrac{h''}{\left(1+(h')^2\right)^{3/2}}.$
- The air pressure, p_∞ (assumed to be constant), has been rescaled to 0. In other words, in place of p we consider $p - p_\infty$.

Concerning the boundary conditions on Γ_{in} and Γ_{wall}, we have:

- On Γ_{in}, we prescribe pressure, temperature and no radial velocity

$$\begin{cases} p(0,r) = P_{in}, & 0 \le r \le R, \\ v_2(0,r) = 0, & 0 \le r \le R. \end{cases}$$

- On Γ_{wall}, we assume no–slip $\mathbf{v}(x,R) = 0$, $0 < x < x_T$. Of course, some slipping occurs near to x_T, but it is generally admitted that the slipping length ℓ_s is extremely small.

The abscissa of the triple point, x_T, is not a priori given and has to be determined as a part of problem (as well as R_L). In most of the references, one imposes the contact angle (e.g. equal to 0) and uses this additional condition for having a totally determined problem.

The boundary conditions that have to be specified on Γ_{out} required coupling with the stage (d). This will be the subject of next section.

In this section we will suppose that the temperature is a given function of x and r. Actually, the main difficulties will remain, but our calculations will be less lengthy.

4.1 Definition of L and Jet's Profile at the End of Stage (c)

In this section, still considering *stationary conditions*, we will focus on these issues:

1. To define local boundary conditions on Γ_{out} to be used when studying stage (c).
2. To formulate a rational procedure allowing to determine L, replacing the heuristic rule $L = 12R$.

Table 1 Typical order of magnitude of the
main physical quantities at $x = L$. Notice that
the traction force F_{sp} corresponds to a weight
of $3\,gr$!

V_L	R_L	Φ_L	μ	σ
$10\,m/s$	$10^{-5}\,m$	$10^8\,Pa$	$10^7\,Pa\,s$	$10^{-1}\,N/m$

We point out that, for our purposes, L can be chosen in some range of values. Therefore we look for some criterion which has a physical motivation, but is based just on
the analysis of magnitude orders. So for the purposes we have in mind, i.e. locating
L within a reasonable approximation, we may confound V_L and Φ_L with their terminal values V_{sp} and Φ_{sp}, neglecting the small variations that will occur from $x = L$ to
the collecting spool. Therefore, in place of expressions (188), we set

$$V_L = V_{sp} \quad \text{and} \quad \Phi_L = \Phi_{sp}. \tag{105}$$

A full justification of this approach will be provided in Appendix 2. Indeed we are
mainly interested to the orders of magnitude, listed in Table 1, below:

Remark 4.1. After many studies (see e.g. [23] and references therein) and experimental observations (see e.g. [13] or [21]) we know that even for $x > L$ fiber could
exhibit a complicated behavior, oscillations and instabilities are possible. Here we
suppose a stabilized behavior when $x > L$ and the asymptotic analysis, which follows in this subsection applies to such situation. In general, one should couple
together the stages *(c)* and *(d)*, through an iterative procedure. Here we do not enter
this issue.

For the determination of a physically acceptable value of L we proceed considering
a characterization of L in terms of the dimensionless quantity h'. Hence we define
the endpoint of Ω, i.e. of stage (c), in the following way:

Definition 4.2. Given a sufficiently small number ς, typically $\varsigma = \mathcal{O}\left(10^{-6}\right)$, L is
chosen so that

$$\left|h'(L)\right| = \varsigma. \tag{106}$$

We will also make use of the following relationships, that will be checked
a–posteriori

$$\left|h''(L)\right| \lesssim \frac{\varsigma^2}{R_L} \quad \Leftrightarrow \quad \left|h''(L)\right|R_L \lesssim \varsigma^2. \tag{107}$$

Our strategy is the following:

(i) To define local boundary conditions such that (100) and (102) are fulfilled
    within a tolerance of order ς.
(ii) To determine L according to Definition 4.2.

Next, we define $p_L = p(L, R_L)$. Of course, p_L is unknown at this stage. In order to
determine p_L and L let us write explicitly (103) and (104) on $x = L$, $r = R_L$

$$\begin{cases} v_1\left(L,R_L\right)h'\left(L\right) - v_2\left(L,R_L\right) = 0, \\[2ex] p_L + \dfrac{2\mu\left(L\right)}{1+\left(h'\right)^2}\left[h'\left(\dfrac{\partial v_1}{\partial r} + \dfrac{\partial v_2}{\partial x}\right) - \dfrac{\partial v_2}{\partial r} - \dfrac{\partial v_1}{\partial x}\left(h'\right)^2\right] = \dfrac{\sigma}{2R_L\left(1+\left(h'\right)^2\right)^{1/2}} \\[2ex] \hspace{6cm} - \dfrac{\sigma h''}{2\left(1+\left(h'\right)^2\right)^{3/2}}, \\[2ex] \left(\dfrac{\partial v_1}{\partial r} + \dfrac{\partial v_2}{\partial x}\right)\left(1-\left(h'\right)^2\right) - 2\left(\dfrac{\partial v_1}{\partial x} - \dfrac{\partial v_2}{\partial r}\right)h' = \dfrac{\left(\tau\cdot\nabla\right)\sigma}{\mu\left(L\right)}\left(1+\left(h'\right)^2\right). \end{cases}$$

$$(108)$$

Focusing on equations $(108)_2$ and $(108)_3$ we neglect $\mathscr{O}\left(\varsigma^2\right)$ terms (recall Definition 4.2) and[7] $\left(\tau\cdot\nabla\right)\sigma/\mu$. Hence, $(108)_2$ and $(108)_3$ can be rewritten as

$$\begin{cases} p_L + 2\mu\left(L\right)\left[h'\left(\dfrac{\partial v_1}{\partial r} + \dfrac{\partial v_2}{\partial x}\right) - \dfrac{\partial v_2}{\partial r}\right] = \dfrac{\sigma}{2R_L}, \\[2ex] \left(\dfrac{\partial v_1}{\partial r} + \dfrac{\partial v_2}{\partial x}\right) - 2\left(\dfrac{\partial v_1}{\partial x} - \dfrac{\partial v_2}{\partial r}\right)h' = 0. \end{cases}$$

$$(109)$$

Next, (109) can be further simplified. Indeed, neglecting once more $\mathscr{O}\left(\varsigma^2\right)$ terms, we have

$$p_L - 2\mu\left(L\right)\frac{\partial v_2\left(L,R_L\right)}{\partial r} = \frac{\sigma}{2R_L}. \tag{110}$$

We now assume:

A.1 The horizontal fluid velocity on $x = L$, $r = R$ is V_L, namely

$$v_1\left(L,R_L\right) = V_L, \tag{111}$$

Hence, going back to $(108)_1$, we obtain

$$v_2\left(L,R_L\right) = -\left|h'\left(L\right)\right|V_L. \tag{112}$$

We also recall that by symmetry $v_2\left(L,0\right) = 0$. Thus, following the results of Sect. 5, we also assume:

A.2 The profile of the radial velocity on Γ_{out} is

$$v_2\left(L,r\right) = -\frac{\left|h'\left(L\right)\right|V_L}{R_L}r = -\frac{\varsigma V_L}{R_L}r, \tag{113}$$

---

[7] If σ is just a function of temperature, $\nabla\sigma = \dfrac{d\sigma}{dT}\nabla T$ and $x = L$ is located in a region where the temperature variations are very small, i.e. $\sigma'\left(T\left(L\right)\right) \ll 1$.

yielding

$$\frac{\partial v_2}{\partial r} = -\frac{|h'(L)|V_L}{R_L}.$$

Hence, from (110)

$$p_L = -2\mu(L)\frac{|h'(L)|V_L}{R_L} + \frac{\sigma}{2R_L}, \tag{114}$$

or

$$p_L = -2\mu(L)\frac{|h'(L)|V_L}{R_L} = -2\mu(L)\frac{\varsigma V_L}{R_L}, \tag{115}$$

since the second term on the r.h.s. of (114) is negligible with respect to the first–one. Indeed $\frac{\sigma}{2R_L}\left/\frac{\mu(L)\varsigma V_L}{R_L}\right. \approx 10^{-3}$, according to Table 1.

Of course, we still have to determine L. At this point, consistently with the approximations already introduced, we take

$$p(L,r) = p_L, \quad 0 \le r \le R_L.$$

Then we exploit (102) and (99), obtaining

$$\Phi_L = -p_L + \frac{4\mu(L)}{R_L^2}\int_0^{R_L} \underbrace{\frac{\partial v_1}{\partial x}}_{-\frac{1}{r}\frac{\partial(rv_2)}{\partial r}} r\,dr = -p_L - \frac{4\mu(L)}{R_L}v_2(L,R_L),$$

which, recalling (113) and (115), gives[8]

$$\Phi_L = 6\mu(L)\frac{\varsigma V_L}{R_L}, \quad \text{or} \quad \Phi_L = 6\mu(L)\frac{\varsigma V_L}{\pi R_L^3}Q_{out}. \tag{116}$$

This formula allows us to identify L according to the profile of viscosity, which in turns is known (with a reasonable approximation) in terms of the profile of temperature. Thus L is defined via

$$\mu(L) = \frac{\Phi_L R_L}{6\varsigma V_L}. \tag{117}$$

According to Table 1, the r.h.s. of (117) is $\sim 10^7$ Pa s, consistently with the expected value of $\mu(L)$. Hence we conclude that L is the distance at which the glass has cooled down to the temperature corresponding to $\mu \approx 10^7$ Pa s.

We can proceed further obtaining the information on p_L and the expected profile of the fiber in the vicinity of $x = L$. Concerning p_L, from (116) and (115) we have

$$p_L = -\frac{1}{3}\Phi_L,$$

---

[8] It is easy to verify that $\varsigma \sim 10^{-6}$ is compatible with the order of magnitude of the quantities entering in formula (116).

so that the viscous stress on Γ_{out} is

$$\frac{4\mu(L)}{R_L^2}\int_0^{R_L}\frac{\partial v_1}{\partial x}r\,dr = \frac{2}{3}\Phi_L,$$

Next, we write also $v_2(L,r)$ in terms of Φ_L using (113) and (116), namely

$$v_2(L,r) = -\frac{\Phi_L}{6\mu(L)}r, \tag{118}$$

Summarizing, the boundary conditions on Γ_{out} read as follows

$$\begin{cases} p = -\dfrac{1}{3}\Phi_L, \\ v_2(r) = -\dfrac{\Phi_L}{6\mu(L)}r, \end{cases} \quad \text{on} \quad \Gamma_{out}, \tag{119}$$

with Φ_L expressed in terms of Φ_{sp}.

Going back to (113), we just suppose that it is valid for x from a neighborhood of $x = L$:

$$v_2(x,r) \approx -\frac{|h'(x)|}{h(x)}V_L r, \quad 0 \le r \le h(x), \quad \frac{|L-x|}{L} \ll 1. \tag{120}$$

Consistently with this assumption we have that, in the vicinity of $x = L$, $v_2(x,r)$ can be also expressed exploiting (118), namely

$$v_2(x,r) \approx -\frac{\Phi_L}{6\mu(x)}r, \quad 0 \le r \le h(x), \quad \frac{|L-x|}{L} \ll 1. \tag{121}$$

In this way, comparing (120) and (121), the position of the free boundary $r = h(x)$ is given by

$$\frac{\Phi_L}{6\mu(x)} = -\frac{h'(x)}{h(x)}V_L,$$

so that the following Cauchy problem is obtained

$$\begin{cases} h'(x) = -\dfrac{1}{6\mu(x)}\dfrac{\Phi_L}{V_L}h(x), \quad \dfrac{|L-x|}{L} \ll 1, \\ \\ h(L) = R_L, \end{cases} \tag{122}$$

which provides a sufficiently accurate information on the local profile of the fiber.

Problem (122) can be integrated, getting the following profile

$$h(x) = R_L \exp\left\{\frac{\Phi_L}{6V_L}\int_x^L \frac{1}{\mu(s)}ds\right\}, \tag{123}$$

referred to as *"transition profile"*, in the sequel denoted by $h_{TP}(x)$.

We can now check that h_{TP} fulfills condition (107), i.e. if our result is consistent with assumption (107). Evaluating h'' we have

$$\frac{h''}{h} = \left(\frac{\Phi_L}{6\mu V_L}\right)^2 \left(1 + \frac{6V_L}{\Phi_L}\mu'\right).$$

Since, close to $x = L$ we have $\dfrac{6V_L}{\Phi_L}|\mu'| \ll 1$, that yields

$$\frac{h''}{h} = \left(\frac{\Phi_L}{6\mu V_L}\right)^2 = \left(\frac{h'}{h}\right)^2, \quad \Rightarrow \quad h''h = (h')^2 = \mathcal{O}(\varsigma^2). \tag{124}$$

Hypothesis (107) is thus fulfilled.

As final check of our procedure we show that, considering the fiber profile (123), the assumption (111), the approximation (121) for v_2 and boundary conditions (119), we can evaluate $v_1(L,r)$, $0 \le r \le R_L$, so that the estimated volumetric discharge differs from the actual one, i.e. Q_{out}, within a tolerance $\mathcal{O}(\varsigma^2)$. We thus use the above mentioned approximations in

$$\begin{cases} v_1(x, h(x))\, h'(x) - v_2(x, h(x)) = 0, \\[2mm] \left(\dfrac{\partial v_1}{\partial r} + \dfrac{\partial v_2}{\partial x}\right) - 2\left(\dfrac{\partial v_1}{\partial x} - \dfrac{\partial v_2}{\partial r}\right) h' = \dfrac{(\tau \cdot \nabla)\sigma}{\mu(x)}, \end{cases} \tag{125}$$

and look for an estimate of $\partial v_1/\partial r$, since we approximate $v_1(L,r)$ in the following way

$$v_1(L,r) = V_L - \left.\frac{\partial v_1}{\partial r}\right|_{x=L, r=R_L}(R_L - r) + \mathcal{O}((R_L - r)^2).$$

Evaluating $(125)_2$ at $x = L$, $r = R_L$, we have

$$\left[\frac{\partial v_1}{\partial r} + \frac{\partial v_2}{\partial x}\right]_{x=L, r=R_L} = -\frac{\Phi_L}{\mu(L)}|h'(L)| = -6\frac{h'(L)^2 V_L}{R_L}, \tag{126}$$

where we have used (99) and (118) for expressing $\partial v_1/\partial x$, namely

$$\frac{\partial v_1}{\partial x} = -\frac{\partial v_2}{\partial r} - \frac{v_2}{r} = \frac{1}{3}\frac{\Phi_L}{\mu(L)} = -2\frac{|h'| V_L}{R_L},$$

and where $(\tau \cdot \nabla)\sigma/\mu$ has been neglected.

Differentiating $(125)_1$ w.r.t. x we get

$$\frac{\partial v_1}{\partial x}h' + \frac{\partial v_1}{\partial r}h'^2 + v_1 h'' - \frac{\partial v_2}{\partial x} - \frac{\partial v_2}{\partial r}h' = 0.$$

We now use again (99) and then exploit (118) and (111), obtaining

$$\left[\frac{\partial v_1}{\partial r}h'(L)^2 - \frac{\partial v_2}{\partial x}\right]_{x=L,r=R_L} = 3\frac{h'(L)^2 V_L}{R_L} - V_L h''(L).$$

Next, using (124) for evaluating $h''(L)$, i.e. $h''(L) = h'(L)^2/R_L$ and (126) for $\partial v_2/\partial x$, we have[9]

$$\left(1+h'(L)^2\right)\frac{\partial v_1}{\partial r}\bigg|_{x=L,r=R_L} = -4\frac{h'(L)^2 V_L}{R_L}.$$

We are now in position to estimate $\Delta v_1 = \max|v_1(r,L) - V_L| \approx \left|\frac{\partial v_1}{\partial r}(L,R_L)\right| R_L$, namely

$$\Delta v_1 = \mathcal{O}\left(\varsigma^2 V_L\right),$$

thus estimating the "relative error" on the outlet discharge

$$\frac{\Delta Q_{out}}{Q_{out}} = \frac{1}{\pi R_L^2 V_L}\left|\pi R_L^2 V_L - 2\pi\int_0^{R_L} v_1(L,r)\,r\,dr\right| = \mathcal{O}\left(\varsigma^2\right),$$

which is thus shown to respect the desired tolerance.

5 Terminal Phase of the Fiber Drawing

It is considered that the stage (d) starts when the fiber is sufficiently "cooled down."

This means that the viscosity is sufficiently large (larger than 10^5 Pa s) and the fiber radius is already rather small (smaller than hundred micrometers). Contrary to the stage (c), where one should treat the 3D Navier–Stokes system, with the free boundary, coupled with the nonlinear conduction of heat, here we have a long (several meters) and tiny filament (radii of hundred micrometers) of molten glass. In the stage (c) one needs an industrial code to solve the corresponding partial differential equations. Here it is possible to employ the asymptotic analysis and come out with a simplified effective model (Fig. 4).

Fundamental equations, describing the stage (c), are the temperature dependent incompressible Navier–Stokes equations with free boundary coupled with the energy equations. Their derivation from the first principles is in the article [9], where also the Oberbeck–Boussinesq approximation is justified rigorously, by passing to the singular limit when expansivity parameter goes to zero. For more details see Sects. 2 and 3.2.

---

[9] Notice that $\partial v_1/\partial r$ is negative, as one would have expected.

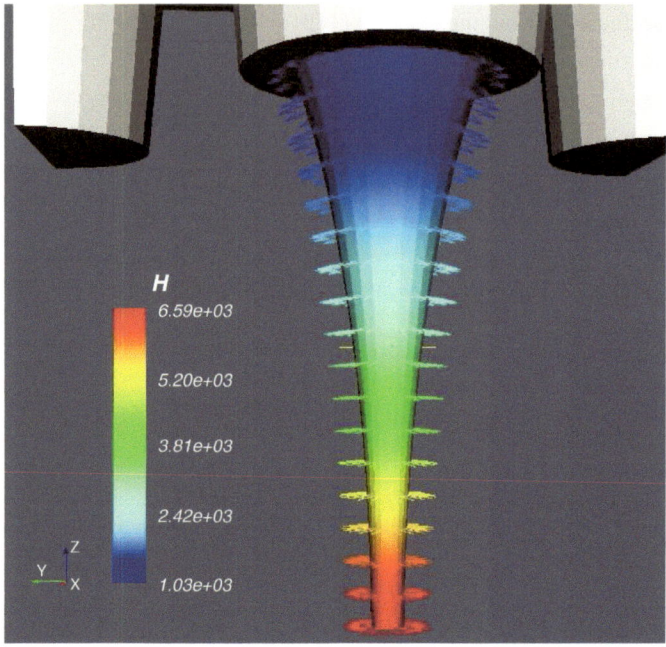

Fig. 4 A simulated glass fiber being drawn down

In the engineering literature on the fiber drawing, this system is frequently approximated by a quasi 1D approximation for viscous flows, in which the radius of the free boundary $r = R(z, \alpha, t)$, axial speed w and the temperature ϑ are independent of the radial variable and depend only on the axial coordinate z and of the time t. This approximation, summarized in the so–called the "equations of Matovich–Pearson" was introduced in the papers by Kase and Matsuo [17, 18] and Matovich and Pearson [20]. The model was obtained heuristically and reads

$$\frac{\partial \mathbf{A}}{\partial t} + \frac{\partial (w\mathbf{A})}{\partial z} = 0; \qquad \frac{\partial}{\partial z}\left(3\mu(T)\mathbf{A}\frac{\partial w}{\partial z}\right) + \frac{\partial(\sigma(T)\sqrt{\mathbf{A}})}{\partial z} = 0, \qquad (127)$$

where $\mathbf{A} = \mathbf{A}(z, t)$ is the area of fibre section, $w = w(z, t)$ is effective axial velocity, 3μ is Trouton's viscosity and σ is the surface tension. μ and σ depend on the temperature, and it is necessary to add the equation for the temperature $T = T(z, t)$.

The equations of Matovich and Pearson were obtained under the assumptions (H), listed below:

(a) The viscous forces dominate inertia
(b) Effects of the surface tension are in balance with the normal stress at the free boundary
(c) The heat conduction is small compared with the heat convection in the fiber
(d) All the phenomena are axially symmetrical and the fiber is nearly straight

5.1 Derivation of the Model of Matovich–Pearson for the Thermal Case

The derivation presented in [13] and in [14] surprisingly neglects the fact that the viscosity changes over several magnitude orders performing asymptotic expansions in ε as it was of order one, ε being the ratio of the characteristic thickness R_E in the radial direction with the characteristic axial length of fibre L. The correct formal derivation is in [5] and it confirms the model announced in [14]. We will present here the ideas and results. Since we deal with an axially symmetric and long fiber, we slightly change the notation and denote with subscript z the axial component and with the subscript r the radial one. Instead of x we use z to denote the axial variable.

Our unknowns are the following:

- Velocity $\mathbf{v} = v_z \mathbf{e}_z + v_r \mathbf{e}_r$
- Hydrodynamic pressure p
- Temperature T
- Fiber radius (being the distance from the symmetry axis) $R = R(t,z)$

Moreover the following physical quantities are given function of T:

- Specific heat $c_p = c_p(T)$
- Density $\rho = \rho(T)$
- Surface tension $\sigma = \sigma(T)$
- Viscosity $\mu = \mu(T)$
- Thermal conductivity $\lambda = \lambda(T)$

We will suppose that the fibre is long and thin, so that the principal flow is directed along the axis z and the velocity is basically one dimensional. In particular, the property of "small thickness" leads to the "lubrication approximation" of Reynolds. The strategy of the "lubrication approximation" is to expand the velocity field with respect to ε. The zero order terms should be sufficient to describe the motion.

Effective equations are then derived starting from the compatibility conditions for the solvability of the problems. This idea is traditionally applied to flows through thin domains. Treating the flows with a free boundary is much more complicated.

Initially, the radius r describes the position of the free boundary $r = R(x,t)$. The smallness of the expansion parameter $\varepsilon = \max_{x,t} R(x,t) / \{$ fiber length $\}$ will depend on the solution itself.

In the second place, it is not obvious which forces in the equations must be taken into account (for example torsion can be important or negligible).

In order to correctly model the heat exchange at the free boundary, we adopt for the heat transfer coefficient the empirical formula of Kase–Matsuo

$$h = \frac{\lambda_\infty}{R(z,t)} C \left(\frac{2\rho_\infty v_z(z,t) R(z,t)}{\mu_\infty} \right)^m, \qquad (128)$$

where the subscript ∞ denotes the parameters of the surrounding air and $C > 0$ and $m, 0 < m < 0.4$, are constants determined from the experimental data.

Concerning characteristic length, in [13] the authors take $L = \dfrac{\rho_E R_E c_{pE} v_E}{2h_E}$ $(1 - T_g/T_E)$, where T_g is the glass transition temperature and T_E is temperature at the orifice. Other possibility is to take the distance between the spooler and extrusion die. This will be our choice.

We will restrict our considerations to the stationary case, even if the generalization to the non-stationary case is straightforward. Our first difficulty is that viscosity changes over several orders of magnitude. This motivates us to write the Oberbeck–Boussinesq equations in $\Omega = \{z \in (0, L); \quad 0 \leq r < R(z)\}$, in the following form:

$$\nabla \left(\frac{p}{\mu(T)} + \frac{2}{3} \operatorname{div} \mathbf{v} \right) - \nabla^2 \mathbf{v} + \left(\left(\frac{p}{\mu(T)} + \frac{2}{3} \operatorname{div} \mathbf{v} \right) I - 2D(\mathbf{v}) \right)$$

$$\times \nabla \log \mu(T) - \nabla \operatorname{div} \mathbf{v} = \frac{\rho(T)}{\mu(T)} g \mathbf{e}_z - \frac{\rho(T)}{\mu(T)} \left\{ (\mathbf{v}\nabla)\mathbf{v} + \frac{\partial \mathbf{v}}{\partial t} \right\} \quad \text{in} \quad \Omega \ (129)$$

$$\frac{\partial \rho(T)}{\partial t} + \operatorname{div} (\rho(T)\mathbf{v}) = 0 \quad \text{in} \quad \Omega \tag{130}$$

$$\rho(T)c_p(T) \left(\frac{\partial T}{\partial t} + \mathbf{v}\nabla T \right) = \operatorname{div} (\lambda(T)\nabla T) \quad \text{in} \quad \Omega. \tag{131}$$

The stress tensor is now

$$\Sigma = -\mu(T) \left(\left(\frac{p}{\mu(T)} + \frac{2}{3} \operatorname{div} \mathbf{v} \right) I - 2D(\mathbf{v}) \right). \tag{132}$$

The natural small parameter is $\varepsilon = \dfrac{R_E}{L}$. We introduce the following dimensionless quantities:

$$r = R_E \tilde{r}, t = \frac{L}{v_E}\tilde{t}, \, v_z = v_E \tilde{v}_z, \, v_r = \varepsilon v_E \tilde{v}_r, \, \frac{T - T_\infty}{\Delta T} = \tilde{T},$$

$$\mu = \mu_E \tilde{\mu}, \sigma = \sigma_E \tilde{\sigma}, h = h_E \tilde{h}, \, \frac{p}{\mu(T)} + \frac{2}{3} \operatorname{div} \mathbf{v} = \frac{v_E \tilde{p}}{L}, \, \lambda = \lambda_E \tilde{\lambda}.$$

Equations in non-dimensional form will contain the following dimensionless numbers:

$\mathbf{Re} = \dfrac{\rho_E v_E L}{\mu_E}$ is Reynolds number; $\quad \mathbf{Ca} = \dfrac{\mu_E v_E}{\sigma_E}$ is capillary number;

$\mathbf{Pe} = \dfrac{c_{pE} \rho v_E L}{\lambda_E}$ is Peclet's number; $\quad \mathbf{Bi} = \dfrac{h_E R_E}{\lambda_E}$ is Biot's number;

$\mathbf{Fr} = \dfrac{v_E}{\sqrt{gL}}$ is Froude's number; $\quad \mathbf{Bo} = \dfrac{\mathbf{Re}}{\mathbf{Fr}^2}$ is Bond's number

and draw ratio is $E = \dfrac{V_f}{v_E}$, where V_f is the outlet fiber velocity. $\alpha = \log E$ is usually known as Hencky strain.

In the text, which follows, we will omit the wiggles. The axially symmetric generalized Oberbeck–Boussinesq system takes the following form:

$$\frac{\partial v_r}{\partial r} + \frac{v_r}{r} + \frac{\partial v_z}{\partial z} = -v_z \partial_z \log \rho(T) - v_r \partial_r \log \rho(T) - \partial_t \log \rho(T), \quad \text{in } \Omega; \quad (133)$$

$$\varepsilon^2 \mathbf{Re} \frac{\rho(T)}{\mu(T)} \left(\partial_t v_r + v_r \frac{\partial v_r}{\partial r} + v_z \frac{\partial v_r}{\partial z} \right) = -\frac{\partial p}{\partial r} + \left(\frac{\partial^2 v_r}{\partial r^2} + \frac{1}{r} \frac{\partial v_r}{\partial r} \right)$$

$$- \frac{v_r}{r^2} + \varepsilon^2 \frac{\partial^2 v_r}{\partial z^2} \right) - \frac{\partial \log \mu}{\partial r} \left(p - 2 \frac{\partial v_r}{\partial r} \right) + \frac{\partial \log \mu}{\partial z} \left(\varepsilon^2 \frac{\partial v_r}{\partial z} + \frac{\partial v_z}{\partial r} \right)$$

$$- \frac{\partial}{\partial r} (v_z \partial_z \log \rho(T) + v_r \partial_r \log \rho(T) + \partial_t \log \rho(T)), \quad \text{in } \Omega; \quad (134)$$

$$\varepsilon^2 \mathbf{Re} \frac{\rho(T)}{\mu(T)} \left(\partial_t v_z + v_r \frac{\partial v_z}{\partial r} + v_z \frac{\partial v_z}{\partial z} \right) = -\varepsilon^2 \frac{\partial p}{\partial z} + \left(\frac{\partial^2 v_z}{\partial r^2} + \frac{1}{r} \frac{\partial v_z}{\partial r} + \varepsilon^2 \frac{\partial^2 v_z}{\partial z^2} \right)$$

$$- \varepsilon^2 \frac{\partial \log \mu}{\partial z} \left(p - 2 \frac{\partial v_z}{\partial z} \right) + \frac{\partial \log \mu}{\partial r} \left(\varepsilon^2 \frac{\partial v_r}{\partial z} + \frac{\partial v_z}{\partial r} \right) + \varepsilon^2 \frac{\mathbf{Re}}{\mathbf{Fr}^2} \frac{\rho(T)}{\mu(T)}$$

$$- \varepsilon^2 \frac{\partial}{\partial z} (v_z \partial_z \log \rho(T) + v_r \partial_r \log \rho(T) \partial_t \log \rho(T)), \quad \text{in } \Omega; \quad (135)$$

$$\varepsilon^2 \mathbf{Pe} \rho(T) c_p(T) \left(\partial_t T + v_r \frac{\partial T}{\partial r} + v_z \frac{\partial T}{\partial z} \right) = \frac{1}{r} \frac{\partial}{\partial r} \left(r \lambda(T) \frac{\partial T}{\partial r} \right)$$

$$+ \varepsilon^2 \frac{\partial}{\partial z} \left(\lambda(T) \frac{\partial T}{\partial z} \right), \quad \text{in } \Omega. \quad (136)$$

Next we have

$$\frac{\partial R(z,t)}{\partial t} + v_z \frac{\partial R(z)}{\partial z} = v_r \quad \text{on } r = R(z) \quad \text{(the kinematic condition)} \quad (137)$$

and the dynamic conditions at the free boundary read:

$$\mu(T) \mathbf{Ca} \left(2\varepsilon^2 \frac{\partial R(z)}{\partial z} \left(\frac{\partial v_r}{\partial r} - \frac{\partial v_z}{\partial z} \right) + \left(1 - \varepsilon^2 \left(\frac{\partial R}{\partial z} \right)^2 \right) (\varepsilon^2 \frac{\partial v_r}{\partial z} + \frac{\partial v_z}{\partial r}) \right)$$

$$= \varepsilon \left(\frac{\partial R(z)}{\partial z} \frac{\partial \sigma(T)}{\partial r} + \frac{\partial \sigma(T)}{\partial z} \right) \sqrt{1 + \varepsilon^2 \left(\frac{\partial R(z)}{\partial z} \right)^2} \quad \text{on } r = R(z); \quad (138)$$

$$\varepsilon \mathbf{Ca} \mu(T) \left(\left(1 + \varepsilon^2 \left(\frac{\partial R(z)}{\partial z} \right)^2 \right) p - 2 \left(\frac{\partial v_r}{\partial r} + \varepsilon^2 \left(\frac{\partial R(z)}{\partial z} \right)^2 \frac{\partial v_z}{\partial z} \right.$$

$$\left. - \varepsilon^2 \frac{\partial R(z)}{\partial z} \frac{\partial v_r}{\partial z} - \frac{\partial R(z)}{\partial z} \frac{\partial v_z}{\partial r} \right) \right) = \frac{\sigma(T)}{R(z)} \sqrt{1 + \varepsilon^2 \left(\frac{\partial R(z)}{\partial z} \right)^2}$$

$$- \varepsilon^2 \sigma(T) \frac{\partial^2 R(z)}{\partial z^2} / \sqrt{1 + \varepsilon^2 \left(\frac{\partial R(z)}{\partial z} \right)^2} \quad \text{on } r = R(z). \quad (139)$$

Finally, the heat transfer to the environment is given by Newton's cooling law

$$\frac{\partial T}{\partial r} - \varepsilon^2 \frac{\partial R}{\partial z}\frac{\partial T}{\partial z} = -\mathbf{Bi}\frac{h}{\lambda(T)}T\sqrt{1+\varepsilon^2\left(\frac{\partial R(z)}{\partial z}\right)^2}\quad\text{on }r=R(z); \qquad (140)$$

Before starting the two-scale expansion it is essential to relate with ε the magnitude of all non–dimensional coefficients entering the system. The experimental data from [13] imply that

$$\mathbf{Re}\sim\varepsilon,\quad \mathbf{Bo}\sim 1,\quad \mathbf{Ca}\sim\frac{1}{\varepsilon},\quad \mathbf{Pe}\sim\frac{1}{\varepsilon},\quad \mathbf{Bi}\sim\varepsilon\quad\text{as}\quad\varepsilon\to 0. \qquad (141)$$

In order to perform the asymptotic analysis of our equations, we expand all unknown functions with respect to ε, i.e. for an arbitrary function $f=f(r,z)$ we set $f=f^{(0)}+\varepsilon^2\mathbf{Pe}\,f^{(1)}+\varepsilon^2 f^{(2)}+\dots$.

Let $\Omega_0=\{\,0\le r<R^{(0)}(z,t),0<z<1\,\}$. After inserting the expansions for unknowns into the system (133)–(140), we find out that the zero order terms in the system (133)–(136) are:

$$\frac{\partial v_r^{(0)}}{\partial r}+\frac{v_r^{(0)}}{r}+\frac{\partial v_z^{(0)}}{\partial z}=-v_z^{(0)}\partial_z\log\rho(T^{(0)})-v_r^{(0)}\partial_r\log\rho(T^{(0)})$$
$$-\partial_t\rho(T^{(0)})\quad\text{in }\Omega_0; \qquad (142)$$

$$\frac{\partial}{\partial r}\left(\partial_t v_z^{(0)}+v_z^{(0)}\partial_z\log\rho(T^{(0)})+v_r^{(0)}\partial_r\log\rho(T^{(0)})\right)=-\frac{\partial p^{(0)}}{\partial r}+\frac{\partial^2 v_r^{(0)}}{\partial r^2}+\frac{1}{r}\frac{\partial v_r^{(0)}}{\partial r}$$
$$-\frac{v_r^{(0)}}{r^2}+\frac{\partial\log\mu(T^{(0)})}{\partial r}\left(2\frac{\partial v_r^{(0)}}{\partial r}-p\right)+\frac{\partial\log\mu(T^{(0)})}{\partial z}\frac{\partial v_z^{(0)}}{\partial r}\quad\text{in }\Omega_0; \qquad (143)$$

$$0=\frac{\partial^2 v_z^{(0)}}{\partial r^2}+\frac{1}{r}\frac{\partial v_z^{(0)}}{\partial r}+\frac{\partial\log\mu(T^{(0)})}{\partial r}\frac{\partial v_z^{(0)}}{\partial r}\quad\text{in }\Omega_0; \qquad (144)$$

$$0=\frac{1}{r}\frac{\partial T^{(0)}}{\partial r}+\frac{\partial^2 T^{(0)}}{\partial r^2}\quad\text{in }\Omega_0, \qquad (145)$$

with the boundary conditions

$$v_r^{(0)}=\partial_t R^{(0)}+v_z^{(0)}\frac{\partial R^{(0)}}{\partial z}\quad\text{on }r=R^{(0)}(z); \qquad (146)$$

$$\frac{\partial v_z^{(0)}}{\partial r}=0\quad\text{on }r=R^{(0)}(z,t); \qquad (147)$$

$$p^{(0)}=\frac{1}{\varepsilon\mathbf{Ca}}\frac{\sigma(T^{(0)})}{\mu(T^{(0)})R^{(0)}}+2\left(\frac{\partial v_r^{(0)}}{\partial r}-\partial_z R^{(0)}\frac{\partial v_z^{(0)}}{\partial r}\right)\quad\text{on }r=R^{(0)}(z,t); \qquad (148)$$

$$\frac{\partial T^{(0)}}{\partial r}=0\quad\text{on }r=R^{(0)}(z,t). \qquad (149)$$

We note that it is advantageous to have $\log \mu$, in the differential equations, since, contrary to μ, its variation is not dramatic. Using (145) we obtain $T^{(0)} = T^{(0)}(z,t)$. Next, (144) yields $v_z^{(0)} = v_z^{(0)}(z,t)$.

Radial component of the velocity is calculated using the (142). We integrate it and get

$$\frac{\partial v_z^{(0)}}{\partial r} = 0 = \frac{\partial T^{(0)}}{\partial r}, \; v_r^{(0)}(r,z) = -\frac{r}{2}\left(\frac{\partial v_z^{(0)}}{\partial z} + \partial_z \log\rho(T^{(0)})v_z^{(0)} + \partial_t \log\rho(T^{(0)})\right) \tag{150}$$

Consequently, (143) reads $\dfrac{\partial p^{(0)}}{\partial r} = 0$ and, after using the boundary value (148), we obtain

$$p^{(0)} = \frac{1}{\varepsilon \mathbf{Ca}} \frac{\sigma(T^{(0)})}{\mu(T^{(0)})R^{(0)}} + 2\frac{\partial v_r^{(0)}}{\partial r} = \frac{1}{\varepsilon \mathbf{Ca}} \frac{\sigma(T^{(0)})}{\mu(T^{(0)})R^{(0)}}$$
$$- \frac{\partial v_z^{(0)}}{\partial z} - \partial_z \log\rho(T^{(0)})v_z^{(0)} - \partial_t \log\rho(T^{(0)})). \tag{151}$$

The kinematic condition (146) now transforms to

$$0 = \frac{\partial}{\partial t}\left(\rho(T^{(0)})(R^{(0)})^2\right) + \frac{\partial}{\partial z}\left(\rho(T^{(0)})v_z^{(0)}(R^{(0)})^2\right). \tag{152}$$

At the order $\mathscr{O}(\varepsilon^2 \mathbf{Pe})$, we have the following boundary value problem for the temperature correction $T^{(1)}$:

$$\rho(T^{(0)})c_p(T^{(0)})\left(\frac{\partial T^{(0)}}{\partial t} + v_z^{(0)}\frac{\partial T^{(0)}}{\partial z}\right) = \frac{1}{r}\frac{\partial}{\partial r}(r\lambda(T^{(0)})\frac{\partial T^{(1)}}{\partial r}) \text{ in } \Omega_0, \tag{153}$$

$$\varepsilon^2 \mathbf{Pe}\lambda(T^{(0)})\frac{\partial T^{(1)}}{\partial r} + \mathbf{Bi}\frac{(v_z^{(0)}R^{(0)})^m T^{(0)}}{\varepsilon^2 R^{(0)}} = 0, \text{ on } r = R^{(0)}(z), \tag{154}$$

where m is the exponent from Kase–Matsuo's formula. The Neumann problem (153)–(154) for $T^{(1)}$ with respect to r has solution if and only if the usual compatibility condition from Fredholm's alternative is satisfied. It reads

$$\rho(T^{(0)})(R^{(0)})^2 c_p(T^{(0)})\left(\frac{\partial T^{(0)}}{\partial t} + v_z^{(0)}\frac{\partial T^{(0)}}{\partial z}\right) + \frac{2\mathbf{Bi}}{\varepsilon^2 \mathbf{Pe}}\left(v_z^{(0)}R^{(0)}\right)^m T^{(0)} = 0 \text{ on } (0,1). \tag{155}$$

Next, at the order $\mathscr{O}(\varepsilon^2)$, (135) and condition (138) give

$$\mathbf{Re}\frac{\rho(T^{(0)})}{\mu(T^{(0)})}\left(\partial_t v_z^{(0)} + v_z^{(0)}\partial_z v_z^{(0)}\right) + \frac{\partial p^{(0)}}{\partial z} + \frac{\partial \log\mu(T^{(0)})}{\partial z}\left(p^{(0)} - 2\frac{\partial v_z^{(0)}}{\partial z}\right)$$

$$
= \frac{\partial^2 v_z^{(0)}}{\partial z^2} + \frac{\partial^2 v_z^{(2)}}{\partial r^2} + \frac{1}{r}\frac{\partial v_z^{(2)}}{\partial r} + \mathbf{Bo}\,\frac{\rho(T^{(0)})}{\mu(T^{(0)})}
$$

$$
- \frac{\partial}{\partial z}\left(v_z^{(0)}\partial_z \log \rho(T^{(0)}) + \partial_t \log \rho(T^{(0)})\right) \quad \text{in } \Omega_0. \tag{156}
$$

$$
2\frac{\partial R^{(0)}(z)}{\partial z}\left(\frac{\partial v_r^{(0)}}{\partial r} - \frac{\partial v_z^{(0)}}{\partial z}\right) + \frac{\partial v_r^{(0)}}{\partial z} + \frac{\partial v_z^{(2)}}{\partial r}
$$

$$
= \frac{1}{\varepsilon \mathbf{Ca}\,\mu(T^{(0)})}\frac{\partial \sigma(T^{(0)})}{\partial z} \quad \text{on } r = R^{(0)}(z,t). \tag{157}
$$

Now (150) yields

$$
\partial_z\left(\mu(T^0)(p^0 - 2\partial_z v_z^0)\right) = \partial_z\left(\frac{1}{\varepsilon \mathbf{Ca}}\frac{\sigma(T^{(0)})}{R^{(0)}} - 3\mu(T^0)\frac{\partial v_z^{(0)}}{\partial z}\right.
$$

$$
\left. -\mu(T^0)\partial_z \log \rho(T^{(0)})v_z^{(0)} - \mu(T^0)\partial_t \log \rho(T^{(0)})\right) \tag{158}
$$

and integration of the (156) yields

$$
\partial_r v_z^{(2)}|_{r=R^{(0)}} = \frac{R^{(0)}}{2}\left(\mathbf{Re}\,\frac{\rho(T^{(0)})}{\mu(T^{(0)})}\left(\partial_t v_z^{(0)} + v_z^{(0)}\partial_z v_z^{(0)}\right) - \mathbf{Bo}\,\frac{\rho(T^{(0)})}{\mu(T^{(0)})}\right.
$$

$$
- \frac{1}{\mu(T^{(0)})}\frac{\partial}{\partial z}\left(\mu(T^{(0)})\left(3\frac{\partial v_z^{(0)}}{\partial z} + \partial_z \log \rho(T^{(0)})v_z^{(0)}\right.\right.
$$

$$
\left.\left.\left. + \partial_t \log \rho(T^{(0)})\right)\right) + \frac{1}{\mu(T^{(0)})}\partial_z\left(\frac{1}{\varepsilon \mathbf{Ca}}\frac{\sigma(T^{(0)})}{R^{(0)}}\right)\right). \tag{159}
$$

On the other hand, after inserting (150) into the boundary condition (157) we get

$$
\partial_r v_z^{(2)}|_{r=R^{(0)}} = \frac{1}{\varepsilon \mathbf{Ca}}\frac{\partial_z \sigma(T^{(0)})}{\mu(T^{(0)})} + \frac{R^{(0)}}{2}\left(\partial_{zz}v_z^0 + \partial_z\left(v_z^0 \partial_z \log \rho(T^{(0)}) + \partial_t \log \rho(T^{(0)})\right)\right)
$$

$$
+ \partial_z R^{(0)}\left(3\frac{\partial v_z^{(0)}}{\partial z} + v_z^0 \partial_z \log \rho(T^{(0)}) + \partial_t \log \rho(T^{(0)})\right). \tag{160}
$$

After comparing (159)–(160), we obtain the effective momentum equation:

$$
\frac{\partial}{\partial z}\left(3\mu(T^{(0)})(R^{(0)})^2\frac{\partial v_z^{(0)}}{\partial z} + \mu(T^{(0)})(R^{(0)})^2\left(v_z^{(0)}\partial_z \log \rho(T^{(0)}) + \partial_t \log \rho(T^{(0)})\right)\right.
$$

$$
\left. + \frac{1}{\varepsilon \mathbf{Ca}}\sigma(T^{(0)})R^{(0)}\right) = \rho(T^{(0)})(R^{(0)})^2\left(\mathbf{Re}\left(\partial_t v_z^{(0)} + v_z^{(0)}\partial_z v_z^{(0)}\right) - \mathbf{Bo}\right). \tag{161}
$$

Although $\mathbf{Re} \sim \varepsilon$, we retain the inertia term, just for sake of completeness. One can imagine situations corresponding to large Reynolds numbers. As conclusion we summarize our results in dimensional form:

Proposition 5.1. *Let* $v_{eff} = v_E v_z^{(0)}$ *be the effective axial velocity,* $R_{eff} = R_E R^{(0)}$ *the effective fiber radius and* $T_{eff} = T_E T^{(0)}$ *the effective temperature. Let us suppose that the quantities* Q_0 *(the mass flow),* R_f *(the final fiber radius),* V_f *(the pulling velocity),* F_L *(the traction force) and* T_E *(the extrusion temperature) are given positive constants. Then all other relevant physical quantities are determined by* $\{v_{eff}, R_{eff}, T_{eff}\}$ *and given by:*

effective radial velocity:

$$v_r^{eff}(r,z) = -\frac{r}{2}\left(\partial_z v_{eff}(z) + v_{eff}\partial_z \log\rho\left(T_{eff}(z)\right) + \partial_t \log\rho\left(T_{eff}(z)\right)\right) \quad (162)$$

effective pressure: $\quad p_{eff}(z) = \dfrac{\sigma\left(T_{eff}(z)\right)}{R_{eff}(z)} - \mu\left(T_{eff}(z)\right)\partial_z v_{eff}(z)$

$$-\frac{\mu\left(T_{eff}(z)\right)}{3}\left(v_{eff}(z)\partial_z \log\rho\left(T_{eff}(z)\right) + \partial_t \log\rho\left(T_{eff}(z)\right)\right); \quad (163)$$

effective axial stress:

$$\Sigma_{eff}(r,z)\mathbf{e}_z = \mu\left(T_{eff}(z)\right)\left(3\partial_z v_{eff}(z) + \partial_t \log\rho\left(T_{eff}(z)\right)\right.$$

$$+v_{eff}(z)\partial_z \log\rho\left(T_{eff}(z)\right)\mathbf{e}_z - \frac{\sigma\left(T_{eff}(z)\right)}{R_{eff}(z)}\mathbf{e}_z + \mu\left(T_{eff}(z)\right)\left(3\partial_z v_{eff}(z)\right.$$

$$+\partial_t \log\rho\left(T_{eff}(z)\right) + v_{eff}(z)\partial_z \log\rho\left(T_{eff}(z)\right)\right)r\frac{\partial_z R_{eff}(z)}{R_{eff}(z)}\mathbf{e}_r + r\frac{\partial_z \sigma\left(T_{eff}(z)\right)}{R_{eff}(z)}\mathbf{e}_r. \quad (164)$$

effective traction : $\mathscr{F}_{eff} = \dfrac{\pi R_{eff}^2}{g}\left(\mu\left(T_{eff}(z)\right)\left(3\partial_z v_{eff}(z) + \partial_t \log\rho\left(T_{eff}(z)\right)\right.\right.$

$$\left.\left.+v_{eff}(z)\partial_z \log\rho\left(T_{eff}(z)\right)\right) - \frac{\sigma\left(T_{eff}(z)\right)}{R_{eff}(z)}\right). \quad (165)$$

Functions $\{v_{eff}, R_{eff}, T_{eff}\}$ *are given by the Cauchy problem*

$$\frac{\partial}{\partial t}\left(\rho\left(T_{eff}\right)\left(R_{eff}\right)^2\right) + \frac{\partial}{\partial z}\left(\rho\left(T_{eff}\right)\left(R_{eff}\right)^2 v_{eff}\right) = 0, \quad 0 < z < L; \quad (166)$$

$$\frac{\partial}{\partial z}\left\{\mu\left(T_{eff}(z)\right)\left(R_{eff}(z)\right)^2\left[3\frac{\partial v_{eff}}{\partial z} + \frac{D}{Dt}\log\rho\left(T_{eff}(z)\right)\right] + \sigma\left(T_{eff}(z)\right)R_{eff}(z)\right\}$$

$$= \left(\frac{D}{Dt}v_{eff} - g\right)\rho\left(T_{eff}(z)\right)\left(R_{eff}(z)\right)^2, \quad 0 < z < L; \quad (167)$$

$$c_p\left(T_{eff}\right)\rho\left(T_{eff}\right)R_{eff}^2\frac{DT_{eff}}{Dt}$$

$$= -C\lambda_\infty\left(\frac{2\rho_\infty v_{eff}R_{eff}}{\mu_\infty}\right)^m\left(T_{eff} - T_\infty\right), \quad 0 < z < L; \quad (168)$$

$$R_{eff}(L) = R_f, \quad v_{eff}(L) = V_f, \quad T_{eff}(L) = T_E, \quad (169)$$

where $\dfrac{D}{Dt} = \dfrac{\partial}{\partial t} + v_{\mathit{eff}}\dfrac{\partial}{\partial z}$ *denotes the total derivative. Finally, we have*

$$v_z(r,z) = v_{\mathit{eff}}(z) + \mathscr{O}(\varepsilon^2 \mathbf{Pe}\,); \qquad v_r(r,z) = v_r^{\mathit{eff}}(r,z) + \mathscr{O}(\varepsilon^2 \mathbf{Pe}\,);$$
$$R(r,z) = R_{\mathit{eff}} + \mathscr{O}(\varepsilon^2 \mathbf{Pe}\,); \qquad p(r,z) = p_{\mathit{eff}}(z) + \mathscr{O}(\varepsilon^2 \mathbf{Pe}\,);$$
$$\Sigma(r,z) = \Sigma_{\mathit{eff}}(r,z) + \mathscr{O}(\varepsilon^2 \mathbf{Pe}\,).$$

Remark 5.2. It is important to recall that in (167) the term $\dfrac{D}{Dt}v_{\mathit{eff}}$ is negligible if compared to g. Such a term, namely the fluid acceleration, is present in (167) because we decided to retain inertia although $\mathscr{O}(\varepsilon)$.

5.2 Solvability of the Boundary Value Problems for the Stationary Effective Equations

Clearly, the values at the extrusion boundary could be replaced by the values at the interface S_E between the stages (c) and (d) of the fiber drawing process.

In the industrial simulations it makes sense to solve the full 3D Navier–Stokes system in the stage (c), to solve the (166)–(168) corresponding to the stage (d) and to couple them at the interface S_E. Coupling at the interface requires construction of the boundary layer.

Following ideas from [24], a typical iterative procedure for the Navier–Stokes equations with free boundary is the following: for a given free boundary we solve the Navier–Stokes equations with normal stress given at the lateral boundary. Then we update position of the free boundary using the kinematic free boundary condition. Iterations are repeated until the stabilization. Such procedure requires solving the (166)–(168) with the boundary conditions (169) replaced by

$$v_{\mathit{eff}}(L) = V_f, \; v_{\mathit{eff}}(0) = v_E, \; T_{\mathit{eff}}(0) = T_E. \tag{170}$$

In this section we present the study of the stationary version of the boundary value problem (166)–(168), (170), which was undertaken in [5]. We simply drop the index *eff* and set $Q = Q_0/\pi$.

In the absence of the gravity and inertia effects, with constant density and with the heat transfer coefficient depending only on the temperature, the problem was solved in the reference [14]. Using the temperature as variable, it was possible to write an explicit solution for the radius and prove existence and uniqueness. In the general situation, the approach from [14] is not possible any more. Nevertheless, their change of the unknown function will be useful in our existence proof. We prove an existence result under the following physical properties on the coefficients:

(H1) Functions μ, ρ and $\dfrac{\sigma}{\rho^{1/3}}(T)$ are defined on $\mathbb{R}$, bounded from above and from below by positive constants and decreasing. We suppose them infinitely differentiable. $\rho_f = \min_{T_\infty \leq T \leq T_E} \rho(T)$.

(H2) $0 < v_E = v|_{z=0} < V_f = v|_{z=L}$ and $T_E > T_\infty$.

(H3) c_p is an infinitely differentiable strictly positive function.

We introduce the new unknown w by

$$w = \log \frac{V_f \rho_f^{1/3}}{v \rho^{1/3}(T)}. \tag{171}$$

Let $G = V_f \rho_g^{1/3}$ and $C_1 = \dfrac{C\lambda_\infty}{Q} \left(\dfrac{2\rho_\infty \sqrt{G}}{\mu_\infty} \right)^m$, m being the Kase–Matsuo exponent, see(128). Then the boundary value problem (166)–(168), (170) transforms to

$$\frac{\partial}{\partial z} \left(-3 \frac{\mu(T)}{\rho(T)} \frac{\partial w}{\partial z} + \frac{1}{\sqrt{QG}} \frac{\sigma(T)}{\rho^{1/3}(T)} e^{w/2} \right)$$
$$= -\frac{g}{G} \rho^{1/3}(T) e^w - G \rho^{-1/3} e^{-w} \partial_z w - \frac{5G}{6} \rho^{-4/3} e^{-w} \partial_z \rho, \quad 0 < z < L; \tag{172}$$

$$c_p(T) \frac{\partial T}{\partial z} + C_1 \rho^{-2m/3}(T) e^{-mw/2}(T - T_\infty) = 0, \quad 0 < z < L; \tag{173}$$

$$w(0) = w_0 = \log \frac{V_f \rho_f^{1/3}}{v_E \rho^{1/3}} > 0, \quad w(L) = 0, \quad T(0) = T_E. \tag{174}$$

We will obtain existence of C^∞-solutions to problem (172)–(174), such that $w \leq w_0$. Then the velocity v is calculated using the formula $v = V_f \left(\dfrac{\rho_f}{\rho(T)} \right)^{1/3} e^{-w}$. It is a C^∞-function and satisfies $v(z) \geq v_E$ on $[0, L]$. We note as well that $\rho_f = \rho(T(L))$ is not given. For simplicity, we suppose w_0 known. Otherwise, we should do one more fixed point calculation for ρ_f, which does not pose problems.

Definition 5.3. The corresponding variational formulation for problem (172)–(174) is: Find functions $w \in H^1(0, L)$ and $T \in H^1(0, L)$, $\partial_z T \leq 0$, such that the boundary conditions (174) are satisfied and

$$\int_0^L 3 \frac{\mu(T)}{\rho(T)} \frac{\partial w}{\partial z} \frac{\partial \varphi}{\partial z} \, dz - \int_0^L \frac{1}{\sqrt{QG}} \frac{\sigma(T)}{\rho^{1/3}(T)} e^{\min\{w, w_0\}/2} \frac{\partial \varphi}{\partial z} \, dz$$
$$+ \int_0^L \frac{g}{G} \rho^{1/3}(T) e^{\min\{w, w_0\}} \varphi \, dz + \int_0^L G \rho^{-1/3}(T) e^{-w} \frac{\partial w}{\partial z} \varphi \, dz$$
$$+ \int_0^L \frac{5G}{6} \rho^{-4/3}(T) e^{-w} \partial_z \rho \, \varphi \, dz = 0, \quad \forall \varphi \in H_0^1(0, L). \tag{175}$$

$$\frac{\partial T}{\partial z} = -\frac{C_1}{c_p(T)} \rho^{-2m/3}(T) e^{-m \min\{w, w_0\}/2}(T - T_\infty), \quad 0 < z < L. \tag{176}$$

Proposition 5.4. *(see [5]). Let* μ, σ *and* ρ *satisfy* **(H1)–(H2)** *and let* $\{w, T\}$ *be a variational solution to (174), (175) and (176). Then we have* $w \leq w_0$ *and* $T_E \geq T \geq T_\infty$.

Corollary 5.5. *(see [5]). Under hypothesis* **(H1)–(H2)**, *any variational solution* $\{w, T\}$ *to (174), (175) and (176) solves equations (172)–(173).*

Theorem 5.6. *(see [5]). Let us suppose hypotheses* **(H1)–(H3)** *hold true. Let* $\tilde{T}$ *be the solution for*

$$\frac{\partial \tilde{T}}{\partial z} = -\frac{C_1}{c_p(\tilde{T})} \rho^{-2m/3}(\tilde{T}) e^{-mw_0/2}(\tilde{T} - T_\infty), \ 0 < z < L; \ \tilde{T}(0) = T_E. \tag{177}$$

Let $\kappa = \max_{T_\infty \leq T \leq T_E} |\partial_T \rho(T)|$ *and let* $\mathscr{A} = \int_0^L \frac{dz}{\mu(\tilde{T}(z))}$. *Then there is* $\delta_0 > 0$ *such that for* $\kappa \mathscr{A} < \delta_0$, *problem (172)–(174) admits a solution* $\{w, T\} \in C^\infty[0, L]^2$, *such that* $\partial_z T \leq 0$ *and* $w(z) \leq w_0$.

Remark 5.7. We see that in the case of constant density, the necessary condition from the Theorem is always fulfilled. Furthermore, since viscosity takes large values as temperature decreases, $\mathscr{A}$ is a very small quantity. Consequently, the Theorem covers all situations of practical interest. Proof uses Brouwer's fixed point theorem.

Appendix 1

Let us consider the state

$$\mathscr{S} = \{\mathbf{v} = 0, \vartheta = 1, \rho = 1, P = 0\},$$

corresponding to the fluid at rest subject only to a uniform pressure and to isothermal conditions. We now study the linear stability of $\mathscr{S}$, in absence of any body forces[10] and external heat sources, in two cases:

1. The system dynamics is governed by

$$\begin{cases} \rho(\vartheta) \operatorname{div} \mathbf{v} = \alpha \dfrac{D\vartheta}{Dt}, \\[3mm] \rho(\vartheta) \dfrac{D\mathbf{v}}{Dt} = \dfrac{1}{\mathbf{Re}} \left[-\nabla P + \Delta \mathbf{v} + \dfrac{1}{3} \nabla(\operatorname{div} \mathbf{v}) \right], \\[3mm] \rho(\vartheta) \dfrac{D\vartheta}{Dt} = \dfrac{1}{\mathbf{Pe}} \Delta \vartheta + \dfrac{1 + \vartheta(T_w - 1)}{(T_w - 1)^2} \dfrac{\alpha \mathbf{Ec}}{\mathbf{Re}} \dfrac{DP}{Dt} + \dfrac{2}{(T_w - 1)} \dfrac{\mathbf{Ec}}{\mathbf{Re}} |\hat{D}(\mathbf{v})|^2, \end{cases} \tag{178}$$

---

[10] We consider $g = 0$, and, as a consequence, $p = P$ from (31).

where, for the sake of simplicity, we have considered β, c_{p1}, μ and λ constant, i.e. $\beta = \beta_R$, $c_{p1} = c_{pR}$, $\mu = \mu_R$ and $\lambda = \lambda_R$.

2. The system dynamics is governed by

$$
\begin{cases}
\rho(\vartheta)\,\operatorname{div}\mathbf{v} = \alpha\dfrac{D\vartheta}{Dt}, \\[2mm]
\rho(\vartheta)\dfrac{D\mathbf{v}}{Dt} = \dfrac{1}{\mathbf{Re}}\left[-\nabla P + \Delta\mathbf{v} + \dfrac{1}{3}\nabla(\operatorname{div}\mathbf{v})\right], \\[2mm]
\rho(\vartheta)\dfrac{D\vartheta}{Dt} = \dfrac{1}{\mathbf{Pe}}\Delta\vartheta,
\end{cases}
\tag{179}
$$

corresponding to $\mathbf{Ec} = 0$, i.e. to the fact that the kinetic energy of the fluid is really negligible when compared to its thermal energy.

Let us first linearize the equations considering small (dimensionless) perturbations of the state $\mathscr{S}$, namely

$$
\mathbf{v} = \mathbf{v}^*(\mathbf{x},t), \quad \vartheta = 1 + \vartheta^*(\mathbf{x},t), \quad \rho = 1 + \rho^*(\mathbf{x},t), \quad P = P^*(\mathbf{x},t).
$$

In particular, we have also

$$
\rho = 1 - \alpha\vartheta^*, \quad \Rightarrow \quad \rho^* = -\alpha\vartheta^*.
$$

Case 1. Dynamics governed by (178). We will see, as put in evidence in [2], that, in such a case, the rest state is linearly unstable.
The linearized form of system (178) is

$$
\begin{cases}
\operatorname{div}\mathbf{v}^* = \alpha\dfrac{\partial\vartheta^*}{\partial t}, \\[2mm]
\dfrac{\partial\mathbf{v}^*}{\partial t} = \dfrac{1}{\mathbf{Re}}\left[-\nabla P^* + \Delta\mathbf{v}^* + \dfrac{1}{3}\nabla(\operatorname{div}\mathbf{v}^*)\right], \\[2mm]
\dfrac{\partial\vartheta^*}{\partial t} = \dfrac{1}{\mathbf{Pe}}\Delta\vartheta^* + \beta\dfrac{\partial P^*}{\partial t},
\end{cases}
\tag{180}
$$

where

$$
\beta = \frac{T_w}{(T_w - 1)^2}\frac{\alpha\mathbf{Ec}}{\mathbf{Re}}.
$$

We now take the Fourier transform[11] with respect to the spatial variable $\mathbf{x}$ of the dependent variable $\mathbf{v}^*$, P^* and ϑ^*, namely

---

[11] The Fourier transform, with respect to the spatial variable $\mathbf{x}$, of a function $f(\mathbf{x},t)$ is defined by

$$
\hat{f}(\mathbf{k},t) = \frac{1}{(2\pi)^{3/2}}\int_{\mathbb{R}^3} f(\mathbf{x},t)\exp(-i\mathbf{k}\cdot\mathbf{x})\,d\mathbf{x}.
$$

$$\text{F.T.}$$
$$\mathbf{v}^*(\mathbf{x},t) \longrightarrow \widehat{\mathbf{v}}(\mathbf{k},t),$$
$$\vartheta^*(\mathbf{x},t) \longrightarrow \widehat{\vartheta}(\mathbf{k},t),$$
$$P^*(\mathbf{x},t) \longrightarrow \widehat{P}(\mathbf{k},t),$$

getting

$$
\begin{cases}
i\mathbf{k}\cdot\widehat{\mathbf{v}} = \alpha\dfrac{\partial\widehat{\vartheta}}{\partial t}, \\[2mm]
\dfrac{\partial\widehat{\mathbf{v}}}{\partial t} = -\dfrac{i\mathbf{k}}{\mathrm{Re}}\widehat{P} - \dfrac{k^2}{\mathrm{Re}}\widehat{\mathbf{v}} - \dfrac{1}{3\mathrm{Re}}\mathbf{k}\,(\mathbf{k}\cdot\widehat{\mathbf{v}}), \\[2mm]
\dfrac{\partial\widehat{\vartheta}}{\partial t} = -\dfrac{k^2}{\mathrm{Pe}}\widehat{\vartheta} + \beta\dfrac{\partial\widehat{P}}{\partial t},
\end{cases}
\tag{181}
$$

where $k = |\mathbf{k}|$. Decomposing $\widehat{\mathbf{v}}$ as follows

$$\widehat{\mathbf{v}} = \widehat{\mathbf{v}}_{\|} + \widehat{\mathbf{v}}_{\perp},$$

where

$$
\begin{cases}
\widehat{\mathbf{v}}_{\|} = \left(\widehat{\mathbf{v}}\cdot\dfrac{\mathbf{k}}{k}\right)\dfrac{\mathbf{k}}{k} = \widehat{v}_{\|}\dfrac{\mathbf{k}}{k}, \\[2mm]
\widehat{\mathbf{v}}_{\perp} = \widehat{\mathbf{v}} - \widehat{\mathbf{v}}_{\|},
\end{cases}
$$

system (181) rewrites

$$
\begin{cases}
\dfrac{\partial\widehat{\mathbf{v}}_{\perp}}{\partial t} = -\dfrac{k^2}{\mathrm{Re}}\widehat{\mathbf{v}}_{\perp}, \\[2mm]
\dfrac{\partial\widehat{v}_{\|}}{\partial t} = -\dfrac{ik}{\mathrm{Re}}\widehat{P} - \dfrac{4}{3\mathrm{Re}}k^2\widehat{v}_{\|}, \\[2mm]
\dfrac{\partial\widehat{\vartheta}}{\partial t} = i\dfrac{k}{K_\rho}\widehat{v}_{\|}, \\[2mm]
\dfrac{\partial\widehat{P}}{\partial t} = \dfrac{k^2}{\beta\mathrm{Pe}}\widehat{\vartheta} + i\dfrac{k}{\alpha\beta}\widehat{v}_{\|}.
\end{cases}
\tag{182}
$$

So, we immediately realize that $\widehat{\mathbf{v}}_{\perp}$ vanishes as $t \to \infty$. Therefore we are left with

$$
\frac{\partial}{\partial t}\begin{pmatrix}\widehat{\vartheta}\\ \widehat{v}_{\|}\\ \widehat{P}\end{pmatrix} = \mathbb{A}\begin{pmatrix}\widehat{\vartheta}\\ \widehat{v}_{\|}\\ \widehat{P}\end{pmatrix}, \quad \mathbb{A} = \begin{pmatrix} 0 & i\dfrac{k}{\alpha} & 0 \\[2mm] 0 & -\dfrac{4}{3\mathrm{Re}}k^2 & -\dfrac{ik}{\mathrm{Re}} \\[2mm] \dfrac{k^2}{\beta\mathrm{Pe}} & i\dfrac{k}{\alpha\beta} & 0 \end{pmatrix}.
$$

We have to evaluate the eigenvalues of the matrix $\mathbb{A}$. Denoting them by λ_i, $i = 1, 2, 3$ we have

$$\det \mathbb{A} = \lambda_1 \lambda_2 \lambda_2 = \frac{k^4}{\beta \, \mathbf{Pe} \, \mathbf{Re}} \geq 0.$$

Now, since the secular equation is a third degree equation in λ whose coefficient are real, at least one eigenvalue is real. The above relations therefore implies that there exists at least a positive eigenvalue. We thus conclude that the state $\mathscr{S}$ is instable.

Case 2. Dynamics governed by (179).

Considering, as before, the Fourier transform of the linearized form of system (179) we obtain

$$\begin{cases} \dfrac{\partial \widehat{\mathbf{v}}_\perp}{\partial t} = -\dfrac{k^2}{\mathbf{Re}} \widehat{\mathbf{v}}_\perp, \\[2ex] \dfrac{\partial \widehat{v}_\parallel}{\partial t} = -\dfrac{ik}{\mathbf{Re}} \widehat{P} - \dfrac{4}{3\mathbf{Re}} k^2 \widehat{v}_\parallel, \\[2ex] \dfrac{\partial \widehat{\vartheta}}{\partial t} = i\dfrac{k}{\alpha} \widehat{v}_\parallel, \\[2ex] \dfrac{\partial \widehat{\vartheta}}{\partial t} = -\dfrac{k^2}{\mathbf{Pe}} \widehat{\vartheta}. \end{cases} \tag{183}$$

Form $(183)_4$ and $(183)_1$ we deduce that $\widehat{\vartheta}$ and $\widehat{\mathbf{v}}_\perp$ vanishes as $t \to \infty$. In particular, $\widehat{\vartheta} = \widehat{\vartheta}_o \exp\left(-\dfrac{k^2}{\mathbf{Pe}} t\right)$ and

$$\frac{\partial \widehat{\vartheta}}{\partial t} = -\frac{\widehat{\vartheta}_o k^2}{\mathbf{Pe}} \exp\left(-\frac{k^2}{\mathbf{Pe}} t\right), \quad \Rightarrow \quad \frac{\partial \widehat{\vartheta}}{\partial t} \to 0, \text{ as } t \to \infty.$$

So, by $(183)_3$, we have that also $\widehat{v}_\parallel$ vanishes as $t \to \infty$. Finally, $(183)_2$ implies that $\widehat{P} \to 0$, as $t \to \infty$. We therefore conclude that the state $\mathscr{S}$ is linearly stable.

Appendix 2

Referring to Fig. 5, let us consider the simplest version of the Matovich–Pearson model for the stage (d). We have

$$\begin{cases} \dfrac{\partial}{\partial x} (v_1 \mathbf{A}) = 0, \\[2ex] \dfrac{\partial}{\partial x} \left(\mathbf{A} \dfrac{\partial v_1}{\partial x}\right) = 0, \end{cases} \tag{184}$$

where, as in formula (127), $\mathbf{A} = \mathbf{A}(x)$ is the fiber cross section area, $\mathbf{A} = \pi R^2$.

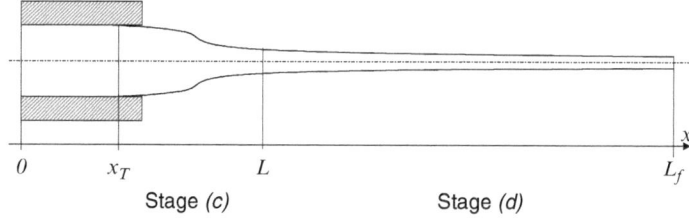

0 x_T L L_f x

Stage (c)   Stage (d)

Fig. 5 A schematic of stages (c) and (d). The spinneret is located at $x = L_f$. $x = 0$ is the die inflow surface. Picture not in scale

As boundary conditions, following [23], we consider

$$v_1 \left(x = L_f \right) = V_{sp}, \quad \text{and} \quad \Phi \left(x = L_f \right) = \Phi_{sp} = \frac{F_{sp}}{\pi R_{sp}^2}, \tag{185}$$

where Φ denotes the longitudinal stress

$$\Phi = 3\mu_f \frac{\partial v_1}{\partial x},$$

with μ_f fluid viscosity (considered constant along stage (d)). Solving (184), (185) we obtain

$$v_1 \left(x \right) = V_{sp} \exp \left\{ -\frac{\Phi_{sp}}{3\mu_f V_{sp}} \left(L_f - x \right) \right\}, \tag{186}$$

$$R \left(x \right) = R_{sp} \exp \left\{ \frac{\Phi_{sp}}{6\mu_f V_{sp}} \left(L_f - x \right) \right\}. \tag{187}$$

Hence the explicit expressions of V_L and Φ_L, in terms of the terminal quantities are the following

$$V_L = V_{sp} \exp \left\{ -\frac{\Phi_{sp}}{3\mu_f V_{sp}} \left(L_f - L \right) \right\}, \quad \Phi_L = -\Phi_{sp} \exp \left\{ -\frac{\Phi_{sp}}{3\mu_f V_{sp}} \left(L_f - L \right) \right\}. \tag{188}$$

As we have seen in Sect. 5.1, the Matovich–Pearson model can be generalized considering the residual spatial dependence of viscosity, still neglecting inertia, surface tension and density variation. In that framework, according to (127) and (167) (see also [7]), $(184)_2$ modifies to

$$\frac{\partial}{\partial x} \left(\mu \left(x \right) \mathbf{A} \left(x \right) \frac{\partial v_1}{\partial x} \right) = 0.$$

The expressions replacing (186) and (187) are

$$v_1(x) = V_{sp} \exp\left\{ -\frac{\Phi_{sp}}{3V_{sp}} \int_x^{L_f} \frac{dx'}{\mu(x')} \right\}, \tag{189}$$

$$R(x) = R_{sp} \exp\left\{ \frac{\Phi_{sp}}{6V_{sp}} \int_x^{L_f} \frac{dx'}{\mu(x')} \right\}. \tag{190}$$

Thus, coming back to Sect. 4.1, namely to the definition of L, on the basis of (189), (190) we may confirm the consistency of our approach, taking $x = L$ and observing that $\frac{\Phi_{sp}}{6V_{sp}} \int_L^{L_f} \frac{dx'}{\mu(x')} = \mathcal{O}(1)$ if $\int_L^{L_f} \frac{dx'}{\mu(x')} = \frac{L_f - L}{\overline{\mu}}$. Since $L_f - L \approx 10\,\mathrm{m}$, this requires $\overline{\mu} = \mathcal{O}\left(10^7\,\mathrm{Pa\,s}\right)$, consistently with the thermal field experimentally known in stage (d).

References

1. Antontsev, S.N., Kazhikhov, A.V., Monakhov, V.N.: Boundary Value Problems in Mechanics of Nonhomogeneous Fluids, North-Holland, Amsterdam (1990)
2. Bechtel, S.E., Forest, M.G., Rooney, F.J., Wang, Q.: Thermal expansion models of viscous fluids based on limits of free energy. Phys. Fluids **15**, 2681–2693 (2003)
3. Bechtel, S.E., Rooney, F.J., Forest, M.G: Internal constraint theories for the thermal expansion of viscous fluids. Int. J. Eng. Sci. **42**, 43–64 (2004)
4. Beirao da Veiga, H.: An L^p-Theory for the n-Dimensional, stationary, compressible Navier-Stokes Equations, and the Incompressible Limit for Compressible Limit for Compressible Fluids. The Equilibrium Solutions. Commun. Math. Phys. **109**, 229–248 (1987)
5. Clopeau, T., Farina, A., Fasano, A., Mikelić, A. : Asymptotic equations for the terminal phase of glass fiber drawing and their analysis, accepted for publication in Nonlinear Analysis TMA: Real World Applications, 2009, http://dx.doi.org/10.1016/j.nonrwa.2008.09.017.
6. Desjardins, B., Grenier, E., Lions, P.-L., Masmoudi, N.: Incompressible limit for solutions of the isentropic Navier-Stokes equations with Dirichlet boundary conditions. J. Math. Pures Appl. **78**, 461–471 (1999)
7. Dewynne, J.N., Ockendon, J.R., Wilmott, P.: On a mathematical model for fibre tapering, SIAM J. Appl. Math. **49**, 983–990 (1989)
8. Diaz, J.I., Galiano, G.: Existence and uniqueness of solutions of the Boussinesq system with nonlinear thermal diffusion. Topol. Methods Nonlin. Anal. **11**, 59–82 (1998)
9. Farina, A., Fasano, A., Mikelić, A.: On the equations governing the flow of mechanically incompressible, but thermally expansible, viscous fluids, M^{3}AS : Math. Models Methods Appl. Sci. **18**, 813–858 (2008)
10. Feireisl, E., Novotny, A.: The low Mach number limit for the full Navier-Stokes-Fourier system. Arch. Ration. Mech. Anal. **186**, 77–107 (2007)
11. Gallavotti, G.: Foundations of Fluid Dynamics. Springer, Berlin (2002)
12. Green, A.E., Naghdi, P.M., Trapp, J.A., Thermodynamics of a continuum with internal constraints. Int. J. Eng. Sci. **8**, 891–908 (1970)
13. Gupta, G., Schultz, W.W.: Non-isothermal flows of Newtonian slender glass fibers. Int. J. Non-Linear Mech. **33**, 151–163 (1998)
14. Hagen, T.: On the Effects of Spinline Cooling and Surface Tension in Fiber Spinnning, ZAMM Z. Angew. Math. Mech. **82**, 545–558 (2002)
15. Hoff, D.: The zero Mach number limit of compressible flows. Comm. Math. Phys. **192**, 543–554 (1998)
16. Kagei, Y., Růžička, M., Thäter, G.: Natural Convection with Dissipative Heating. Commun. Math. Phys. **214**, 287–313 (2000)

17. Kase, S., Matsuo, T.: Studies of melt spinning I. Fundamental Equations on the Dynamics of Melt Spinning, J. Polym. Sci. A **3**, 2541–2554 (1965)
18. Kase, S., Matsuo, T.: Studies of melt spinning. II, Steady state and transient solutions of fundamental equations compared with experimental results. J. Polym. Sci. **11**, 251–287 (1967)
19. Lions, P.L., Masmoudi, N.: Incompressible limit for a viscous compressible fluid. J. Math. Pures Appl. **77**, 585–627 (1998)
20. Matovich, M.A., Pearson, J.R.A.: Spinning a molten threadline-steady state isothermal viscous flows. Ind. Engrg. Chem. Fundam. **8**, 512–520 (1969)
21. von der Ohe, R.: Simulation of glass fiber forming processes. Thesis PhD (2005) – AUC Imprint: ISBN: 8791200253. – Aalborg : Department of Production, Aalborg University (2005)
22. Rajagopal, K.R., Růžička, M., Srinivasa, A.R.: On the Oberbeck-Boussinesq approximation. Math. Models Methods Appl. Sci. **6**, 1157–1167 (1996)
23. Renardy, M.: Draw resonance revisited, SIAM J. Appl. Math. **66**, 1261–1269 (2006)
24. Renardy, M.: An existence theorem for a free surface flow problem with open boundaries. Commun. Partial Differ. Equ. **17**, 1387–1405 (1992)
25. Schlichting, H., Gersten, K.: Boundary-Layer Theory, 8th edn. Springer, Heidelberg (2000)
26. Shelby, J.E.: Introduction to Glass Science and Technology, 2nd edn. RSCP Publishing, London (2005)
27. Temam, R.: Infinite-Dimensional Dynamical systems in Mechanics and Physics. Springer, New York (1988)
28. Zeytounian, R.Kh.: Modélisation asymptotique en mécanique des fluides newtoniens. Springer, New York (1994)
29. Zeytounian, R.Kh.: The Bénard-Marangoni thermocapillary instability problem. Phys. Uspekhis **41**, 241–267 (1998)
30. Zeytounian, R.Kh.: Joseph Boussinesq and his approximation: A contemporary view. C.R. Mécanique **331**, 575–586 (2003)

List of Participants

1. Aionicesei Elena
   Univ. of Maribor, Slovenia
   `aionicesei@yahoo.com`
2. Antonietti Paola Francesca
   Politecnico of Milano, Italy
   `Paola.Antonietti@mate.polimi.it`
3. Benedetti Irene
   Univ. of Firenze, Italy
   `bendetti@math.unifi.it`
4. Borsi Iacopo
   Univ. of Firenze, Italy
   `borsi@math.unifi.it`
5. Butt Azhar Iqbal Kashif
   TU Kaiserslautern, Germany
   `butt@mathematik.uni-kl.de`
6. Ceseri Maurizio
   Univ. of Firenze, Italy
   `ceseri@math.unifi.it`
7. Curkovic Andrijana
   Univ. of Split, Croatia
   `aradovcic@gmail.com`
8. Dierich Frank
   Freiberg Univ. of Mining and Technology, Germany
   `frankdierich@web.de`
9. Doschoris Michael
   Univ. of Patras, Greece
   `mdoscho@chemeng.upatras.gr`
10. Ebert Svend
    Freiberg Univ. of Mining and Technology, Germany
    `svendebert@web.de`
11. Faienza Loredana
    Univ. of Firenze, Italy
    `loryfaienza@libero.it`

A. Fasano (ed.), *Mathematical Models in the Manufacturing of Glass*,
Lecture Notes in Mathematics 2010, DOI 10.1007/978-3-642-15967-1,
© Springer-Verlag Berlin Heidelberg 2011

12. Fusi Lorenzo
    Univ. of Firenze, Italy
    `fusi@math.unifi.it`
13. Geyer Anna
    Univ. of Vienna, Austria
    `anna.geyer@gmx.at`
14. Hadjloizi Demetra
    Univ. Patras, Greece
    `dhadjiloizi@rea.chemeng.upatras.gr`
15. Lebedyanskaya Elena
    Voronezh State agricolture Univ., Russia
    `leblen@mail.ru`
16. Nagwanshi Rekha
    Vikram Univ. Ujjain MP, India
    `sarojnagwanshi@gmail.com`
17. Niedziela Maciej
    Univ. of Zielona Gora, Poland
    `m.niedziela@wmie.uz.zgora.pl`
18. Nouri Fatma Zohra
    Univ. Badji Mokhtar, France
    `fz_nouri@yahoo.fr`
19. Olech Michal
    The Univ. of Wroclaw, Poland
    `olech@math.uni.wroc.pl`
20. Pazanin Igor
    Univ. of Zagreb, Croatia
    `pazanin@math.hr`
21. Peter Cristian
    Univ. of Vienna, Austria
    `rantanplan2@gmx.at`
22. Ricci Riccardo
    Univ. of Firenze, Italy
    `ricci@math.unifi.it`
23. Rosso Fabio
    Univ. of Firenze, Italy
    `rosso@math.unifi.it`
24. Senger Benno
    Technische Univ. Bergakandemie Freiberg, German
    `benno.senger@student.tu-freiberg.de`
25. Serpa Cristina
    University of Lisbon, Portugal
    `cristinaserpa@hotmail.com`
26. Speranza Alessandro
    Univ. of Firenze, Italy
    `alessandro.speranza@math.unifi.it`

27. Starcevic Maja
    Univ. of Zagreb, Croatia
    `mstarcev@math.hr`
28. Sulkowski Tomasz
    Univ. of Zielona Gora, Poland
    `t.sulkowski@wmie.uz.zgora.pl`
29. Togobytska Nataliya
    Weierstrass Inst. for Appl Anals and Stoch, Germany
    `togobyts@wias-berlin.de`
30. Turbin Mikhail
    Voronezh State Univ., Russia
    `mrmike@math.vsu.ru`
31. Verani Marco
    Politecnico of Milano, Italy
    `marco.verani@polimi.it`
32. Vorotnikov Dmitry
    Voronezh State Univ., Russia
    `mitvorot@math.vsu.ru`
33. Zubkov Vladimir
    Univ. of Limerick, Ireland
    `vladimir.zubkov@ul.ie`
34. Zvyagin Andrey
    Voronezh State Univ., Russia
    `zvyagin.a@mail.ru`

Lecture Notes in Mathematics

For information about earlier volumes
please contact your bookseller or Springer
LNM Online archive: springerlink.com

I. From Classical Probability to Quantum Stochastic Calculus. Editors: M. Schürmann, U. Franz (2005)

Vol. 1866: O.E. Barndorff-Nielsen, U. Franz, R. Gohm, B. Kümmerer, S. Thorbjønsen, Quantum Independent Increment Processes II. Structure of Quantum Lévy Processes, Classical Probability, and Physics. Editors: M. Schürmann, U. Franz, (2005)

Vol. 1867: J. Sneyd (Ed.), Tutorials in Mathematical Biosciences II. Mathematical Modeling of Calcium Dynamics and Signal Transduction. (2005)

Vol. 1868: J. Jorgenson, S. Lang, $Pos_n(R)$ and Eisenstein Series. (2005)

Vol. 1869: A. Dembo, T. Funaki, Lectures on Probability Theory and Statistics. Ecole d'Eté de Probabilités de Saint-Flour XXXIII-2003. Editor: J. Picard (2005)

Vol. 1870: V.I. Gurariy, W. Lusky, Geometry of Müntz Spaces and Related Questions. (2005)

Vol. 1871: P. Constantin, G. Gallavotti, A.V. Kazhikhov, Y. Meyer, S. Ukai, Mathematical Foundation of Turbulent Viscous Flows, Martina Franca, Italy, 2003. Editors: M. Cannone, T. Miyakawa (2006)

Vol. 1872: A. Friedman (Ed.), Tutorials in Mathematical Biosciences III. Cell Cycle, Proliferation, and Cancer (2006)

Vol. 1873: R. Mansuy, M. Yor, Random Times and Enlargements of Filtrations in a Brownian Setting (2006)

Vol. 1874: M. Yor, M. Émery (Eds.), In Memoriam Paul-André Meyer - Séminaire de Probabilités XXXIX (2006)

Vol. 1875: J. Pitman, Combinatorial Stochastic Processes. Ecole d'Eté de Probabilités de Saint-Flour XXXII-2002. Editor: J. Picard (2006)

Vol. 1876: H. Herrlich, Axiom of Choice (2006)

Vol. 1877: J. Steuding, Value Distributions of L-Functions (2007)

Vol. 1878: R. Cerf, The Wulff Crystal in Ising and Percolation Models, Ecole d'Eté de Probabilités de Saint-Flour XXXIV-2004. Editor: Jean Picard (2006)

Vol. 1879: G. Slade, The Lace Expansion and its Applications, Ecole d'Eté de Probabilités de Saint-Flour XXXIV-2004. Editor: Jean Picard (2006)

Vol. 1880: S. Attal, A. Joye, C.-A. Pillet, Open Quantum Systems I, The Hamiltonian Approach (2006)

Vol. 1881: S. Attal, A. Joye, C.-A. Pillet, Open Quantum Systems II, The Markovian Approach (2006)

Vol. 1882: S. Attal, A. Joye, C.-A. Pillet, Open Quantum Systems III, Recent Developments (2006)

Vol. 1883: W. Van Assche, F. Marcellàn (Eds.), Orthogonal Polynomials and Special Functions, Computation and Application (2006)

Vol. 1884: N. Hayashi, E.I. Kaikina, P.I. Naumkin, I.A. Shishmarev, Asymptotics for Dissipative Nonlinear Equations (2006)

Vol. 1885: A. Telcs, The Art of Random Walks (2006)

Vol. 1886: S. Takamura, Splitting Deformations of Degenerations of Complex Curves (2006)

Vol. 1887: K. Habermann, L. Habermann, Introduction to Symplectic Dirac Operators (2006)

Vol. 1888: J. van der Hoeven, Transseries and Real Differential Algebra (2006)

Vol. 1889: G. Osipenko, Dynamical Systems, Graphs, and Algorithms (2006)

Vol. 1890: M. Bunge, J. Funk, Singular Coverings of Toposes (2006)

Vol. 1891: J.B. Friedlander, D.R. Heath-Brown, H. Iwaniec, J. Kaczorowski, Analytic Number Theory, Cetraro, Italy, 2002. Editors: A. Perelli, C. Viola (2006)

Vol. 1892: A. Baddeley, I. Bárány, R. Schneider, W. Weil, Stochastic Geometry, Martina Franca, Italy, 2004. Editor: W. Weil (2007)

Vol. 1893: H. Hanßmann, Local and Semi-Local Bifurcations in Hamiltonian Dynamical Systems, Results and Examples (2007)

Vol. 1894: C.W. Groetsch, Stable Approximate Evaluation of Unbounded Operators (2007)

Vol. 1895: L. Molnár, Selected Preserver Problems on Algebraic Structures of Linear Operators and on Function Spaces (2007)

Vol. 1896: P. Massart, Concentration Inequalities and Model Selection, Ecole d'Été de Probabilités de Saint-Flour XXXIII-2003. Editor: J. Picard (2007)

Vol. 1897: R. Doney, Fluctuation Theory for Lévy Processes, Ecole d'Été de Probabilités de Saint-Flour XXXV-2005. Editor: J. Picard (2007)

Vol. 1898: H.R. Beyer, Beyond Partial Differential Equations, On linear and Quasi-Linear Abstract Hyperbolic Evolution Equations (2007)

Vol. 1899: Séminaire de Probabilités XL. Editors: C. Donati-Martin, M. Émery, A. Rouault, C. Stricker (2007)

Vol. 1900: E. Bolthausen, A. Bovier (Eds.), Spin Glasses (2007)

Vol. 1901: O. Wittenberg, Intersections de deux quadriques et pinceaux de courbes de genre 1, Intersections of Two Quadrics and Pencils of Curves of Genus 1 (2007)

Vol. 1902: A. Isaev, Lectures on the Automorphism Groups of Kobayashi-Hyperbolic Manifolds (2007)

Vol. 1903: G. Kresin, V. Maz'ya, Sharp Real-Part Theorems (2007)

Vol. 1904: P. Giesl, Construction of Global Lyapunov Functions Using Radial Basis Functions (2007)

Vol. 1905: C. Prévôt, M. Röckner, A Concise Course on Stochastic Partial Differential Equations (2007)

Vol. 1906: T. Schuster, The Method of Approximate Inverse: Theory and Applications (2007)

Vol. 1907: M. Rasmussen, Attractivity and Bifurcation for Nonautonomous Dynamical Systems (2007)

Vol. 1908: T.J. Lyons, M. Caruana, T. Lévy, Differential Equations Driven by Rough Paths, Ecole d'Été de Probabilités de Saint-Flour XXXIV-2004 (2007)

Vol. 1909: H. Akiyoshi, M. Sakuma, M. Wada, Y. Yamashita, Punctured Torus Groups and 2-Bridge Knot Groups (I) (2007)

Vol. 1910: V.D. Milman, G. Schechtman (Eds.), Geometric Aspects of Functional Analysis. Israel Seminar 2004-2005 (2007)

Vol. 1911: A. Bressan, D. Serre, M. Williams, K. Zumbrun, Hyperbolic Systems of Balance Laws. Cetraro, Italy 2003. Editor: P. Marcati (2007)

Vol. 1912: V. Berinde, Iterative Approximation of Fixed Points (2007)

Vol. 1913: J.E. Marsden, G. Misiołek, J.-P. Ortega, M. Perlmutter, T.S. Ratiu, Hamiltonian Reduction by Stages (2007)

Vol. 1914: G. Kutyniok, Affine Density in Wavelet Analysis (2007)

Vol. 1915: T. Bıyıkoğlu, J. Leydold, P.F. Stadler, Laplacian Eigenvectors of Graphs. Perron-Frobenius and Faber-Krahn Type Theorems (2007)

Vol. 1916: C. Villani, F. Rezakhanlou, Entropy Methods for the Boltzmann Equation. Editors: F. Golse, S. Olla (2008)

Vol. 1917: I. Veselić, Existence and Regularity Properties of the Integrated Density of States of Random Schrödinger (2008)

Vol. 1918: B. Roberts, R. Schmidt, Local Newforms for GSp(4) (2007)

Vol. 1919: R.A. Carmona, I. Ekeland, A. Kohatsu-Higa, J.-M. Lasry, P.-L. Lions, H. Pham, E. Taflin, Paris-Princeton Lectures on Mathematical Finance 2004. Editors: R.A. Carmona, E. Çinlar, I. Ekeland, E. Jouini, J.A. Scheinkman, N. Touzi (2007)

Vol. 1920: S.N. Evans, Probability and Real Trees. Ecole d'Été de Probabilités de Saint-Flour XXXV-2005 (2008)

Vol. 1921: J.P. Tian, Evolution Algebras and their Applications (2008)

Vol. 1922: A. Friedman (Ed.), Tutorials in Mathematical BioSciences IV. Evolution and Ecology (2008)

Vol. 1923: J.P.N. Bishwal, Parameter Estimation in Stochastic Differential Equations (2008)

Vol. 1924: M. Wilson, Littlewood-Paley Theory and Exponential-Square Integrability (2008)

Vol. 1925: M. du Sautoy, L. Woodward, Zeta Functions of Groups and Rings (2008)

Vol. 1926: L. Barreira, V. Claudia, Stability of Nonautonomous Differential Equations (2008)

Vol. 1927: L. Ambrosio, L. Caffarelli, M.G. Crandall, L.C. Evans, N. Fusco, Calculus of Variations and Non-Linear Partial Differential Equations. Cetraro, Italy 2005. Editors: B. Dacorogna, P. Marcellini (2008)

Vol. 1928: J. Jonsson, Simplicial Complexes of Graphs (2008)

Vol. 1929: Y. Mishura, Stochastic Calculus for Fractional Brownian Motion and Related Processes (2008)

Vol. 1930: J.M. Urbano, The Method of Intrinsic Scaling. A Systematic Approach to Regularity for Degenerate and Singular PDEs (2008)

Vol. 1931: M. Cowling, E. Frenkel, M. Kashiwara, A. Valette, D.A. Vogan, Jr., N.R. Wallach, Representation Theory and Complex Analysis. Venice, Italy 2004. Editors: E.C. Tarabusi, A. D'Agnolo, M. Picardello (2008)

Vol. 1932: A.A. Agrachev, A.S. Morse, E.D. Sontag, H.J. Sussmann, V.I. Utkin, Nonlinear and Optimal Control Theory. Cetraro, Italy 2004. Editors: P. Nistri, G. Stefani (2008)

Vol. 1933: M. Petkovic, Point Estimation of Root Finding Methods (2008)

Vol. 1934: C. Donati-Martin, M. Émery, A. Rouault, C. Stricker (Eds.), Séminaire de Probabilités XLI (2008)

Vol. 1935: A. Unterberger, Alternative Pseudodifferential Analysis (2008)

Vol. 1936: P. Magal, S. Ruan (Eds.), Structured Population Models in Biology and Epidemiology (2008)

Vol. 1937: G. Capriz, P. Giovine, P.M. Mariano (Eds.), Mathematical Models of Granular Matter (2008)

Vol. 1938: D. Auroux, F. Catanese, M. Manetti, P. Seidel, B. Siebert, I. Smith, G. Tian, Symplectic 4-Manifolds and Algebraic Surfaces. Cetraro, Italy 2003. Editors: F. Catanese, G. Tian (2008)

Vol. 1939: D. Boffi, F. Brezzi, L. Demkowicz, R.G. Durán, R.S. Falk, M. Fortin, Mixed Finite Elements, Compatibility Conditions, and Applications. Cetraro, Italy 2006. Editors: D. Boffi, L. Gastaldi (2008)

Vol. 1940: J. Banasiak, V. Capasso, M.A.J. Chaplain, M. Lachowicz, J. Miękisz, Multiscale Problems in the Life Sciences. From Microscopic to Macroscopic. Będlewo, Poland 2006. Editors: V. Capasso, M. Lachowicz (2008)

Vol. 1941: S.M.J. Haran, Arithmetical Investigations. Representation Theory, Orthogonal Polynomials, and Quantum Interpolations (2008)

Vol. 1942: S. Albeverio, F. Flandoli, Y.G. Sinai, SPDE in Hydrodynamic. Recent Progress and Prospects. Cetraro, Italy 2005. Editors: G. Da Prato, M. Röckner (2008)

Vol. 1943: L.L. Bonilla (Ed.), Inverse Problems and Imaging. Martina Franca, Italy 2002 (2008)

Vol. 1944: A. Di Bartolo, G. Falcone, P. Plaumann, K. Strambach, Algebraic Groups and Lie Groups with Few Factors (2008)

Vol. 1945: F. Brauer, P. van den Driessche, J. Wu (Eds.), Mathematical Epidemiology (2008)

Vol. 1946: G. Allaire, A. Arnold, P. Degond, T.Y. Hou, Quantum Transport. Modelling, Analysis and Asymptotics. Cetraro, Italy 2006. Editors: N.B. Abdallah, G. Frosali (2008)

Vol. 1947: D. Abramovich, M. Mariño, M. Thaddeus, R. Vakil, Enumerative Invariants in Algebraic Geometry and String Theory. Cetraro, Italy 2005. Editors: K. Behrend, M. Manetti (2008)

Vol. 1948: F. Cao, J-L. Lisani, J-M. Morel, P. Musé, F. Sur, A Theory of Shape Identification (2008)

Vol. 1949: H.G. Feichtinger, B. Helffer, M.P. Lamoureux, N. Lerner, J. Toft, Pseudo-Differential Operators. Quantization and Signals. Cetraro, Italy 2006. Editors: L. Rodino, M.W. Wong (2008)

Vol. 1950: M. Bramson, Stability of Queueing Networks, Ecole d'Eté de Probabilités de Saint-Flour XXXVI-2006 (2008)

Vol. 1951: A. Moltó, J. Orihuela, S. Troyanski, M. Valdivia, A Non Linear Transfer Technique for Renorming (2009)

Vol. 1952: R. Mikhailov, I.B.S. Passi, Lower Central and Dimension Series of Groups (2009)

Vol. 1953: K. Arwini, C.T.J. Dodson, Information Geometry (2008)

Vol. 1954: P. Biane, L. Bouten, F. Cipriani, N. Konno, N. Privault, Q. Xu, Quantum Potential Theory. Editors: U. Franz, M. Schuermann (2008)

Vol. 1955: M. Bernot, V. Caselles, J.-M. Morel, Optimal Transportation Networks (2008)

Vol. 1956: C.H. Chu, Matrix Convolution Operators on Groups (2008)

Vol. 1957: A. Guionnet, On Random Matrices: Macroscopic Asymptotics, Ecole d'Eté de Probabilités de Saint-Flour XXXVI-2006 (2009)

Vol. 1958: M.C. Olsson, Compactifying Moduli Spaces for Abelian Varieties (2008)

Vol. 1959: Y. Nakkajima, A. Shiho, Weight Filtrations on Log Crystalline Cohomologies of Families of Open Smooth Varieties (2008)

Vol. 1960: J. Lipman, M. Hashimoto, Foundations of Grothendieck Duality for Diagrams of Schemes (2009)

Vol. 1961: G. Buttazzo, A. Pratelli, S. Solimini, E. Stepanov, Optimal Urban Networks via Mass Transportation (2009)

Vol. 1962: R. Dalang, D. Khoshnevisan, C. Mueller, D. Nualart, Y. Xiao, A Minicourse on Stochastic Partial Differential Equations (2009)

Vol. 1963: W. Siegert, Local Lyapunov Exponents (2009)

Vol. 1964: W. Roth, Operator-valued Measures and Integrals for Cone-valued Functions and Integrals for Cone-valued Functions (2009)

Vol. 1965: C. Chidume, Geometric Properties of Banach Spaces and Nonlinear Iterations (2009)

Vol. 1966: D. Deng, Y. Han, Harmonic Analysis on Spaces of Homogeneous Type (2009)

Vol. 1967: B. Fresse, Modules over Operads and Functors (2009)

Vol. 1968: R. Weissauer, Endoscopy for GSP(4) and the Cohomology of Siegel Modular Threefolds (2009)

Vol. 1969: B. Roynette, M. Yor, Penalising Brownian Paths (2009)

Vol. 1970: M. Biskup, A. Bovier, F. den Hollander, D. Ioffe, F. Martinelli, K. Netočný, F. Toninelli, Methods of Contemporary Mathematical Statistical Physics. Editor: R. Kotecký (2009)

Vol. 1971: L. Saint-Raymond, Hydrodynamic Limits of the Boltzmann Equation (2009)

Vol. 1972: T. Mochizuki, Donaldson Type Invariants for Algebraic Surfaces (2009)

Vol. 1973: M.A. Berger, L.H. Kauffmann, B. Khesin, H.K. Moffatt, R.L. Ricca, De W. Sumners, Lectures on Topological Fluid Mechanics. Cetraro, Italy 2001. Editor: R.L. Ricca (2009)

Vol. 1974: F. den Hollander, Random Polymers: École d'Été de Probabilités de Saint-Flour XXXVII – 2007 (2009)

Vol. 1975: J.C. Rohde, Cyclic Coverings, Calabi-Yau Manifolds and Complex Multiplication (2009)

Vol. 1976: N. Ginoux, The Dirac Spectrum (2009)

Vol. 1977: M.J. Gursky, E. Lanconelli, A. Malchiodi, G. Tarantello, X.-J. Wang, P.C. Yang, Geometric Analysis and PDEs. Cetraro, Italy 2001. Editors: A. Ambrosetti, S.-Y.A. Chang, A. Malchiodi (2009)

Vol. 1978: M. Qian, J.-S. Xie, S. Zhu, Smooth Ergodic Theory for Endomorphisms (2009)

Vol. 1979: C. Donati-Martin, M. Émery, A. Rouault, C. Stricker (Eds.), Séminaire de Probablitiés XLII (2009)

Vol. 1980: P. Graczyk, A. Stos (Eds.), Potential Analysis of Stable Processes and its Extensions (2009)

Vol. 1981: M. Chlouveraki, Blocks and Families for Cyclotomic Hecke Algebras (2009)

Vol. 1982: N. Privault, Stochastic Analysis in Discrete and Continuous Settings. With Normal Martingales (2009)

Vol. 1983: H. Ammari (Ed.), Mathematical Modeling in Biomedical Imaging I. Electrical and Ultrasound Tomographies, Anomaly Detection, and Brain Imaging (2009)

Vol. 1984: V. Caselles, P. Monasse, Geometric Description of Images as Topographic Maps (2010)

Vol. 1985: T. Linß, Layer-Adapted Meshes for Reaction-Convection-Diffusion Problems (2010)

Vol. 1986: J.-P. Antoine, C. Trapani, Partial Inner Product Spaces. Theory and Applications (2009)

Vol. 1987: J.-P. Brasselet, J. Seade, T. Suwa, Vector Fields on Singular Varieties (2010)

Vol. 1988: M. Broué, Introduction to Complex Reflection Groups and Their Braid Groups (2010)

Vol. 1989: I.M. Bomze, V. Demyanov, Nonlinear Optimization. Cetraro, Italy 2007. Editors: G. di Pillo, F. Schoen (2010)

Vol. 1990: S. Bouc, Biset Functors for Finite Groups (2010)

Vol. 1991: F. Gazzola, H.-C. Grunau, G. Sweers, Polyharmonic Boundary Value Problems (2010)

Vol. 1992: A. Parmeggiani, Spectral Theory of Non-Commutative Harmonic Oscillators: An Introduction (2010)

Vol. 1993: P. Dodos, Banach Spaces and Descriptive Set Theory: Selected Topics (2010)

Vol. 1994: A. Baricz, Generalized Bessel Functions of the First Kind (2010)

Vol. 1995: A.Y. Khapalov, Controllability of Partial Differential Equations Governed by Multiplicative Controls (2010)

Vol. 1996: T. Lorenz, Mutational Analysis. A Joint Framework for Cauchy Problems In and Beyond Vector Spaces (2010)

Vol. 1997: M. Banagl, Intersection Spaces, Spatial Homology Truncation, and String Theory (2010)

Vol. 1998: M. Abate, E. Bedford, M. Brunella, T.-C. Dinh, D. Schleicher, N. Sibony, Holomorphic Dynamical Systems. Editors: G. Gentili, J. Guenot, G. Patrizio (2010)

Vol. 1999: H. Schoutens, The Use of Ultraproducts in Commutative Algebra (2010)

Vol. 2000: H. Yserentant, Regularity and Approximability of Electronic Wave Functions (2010)

Vol. 2001: T. Duquesne, O. Reichmann, K.-i. Sato, C. Schwab, Lévy Matters I. Editors: O.E. Barndorff-Nielson, J. Bertoin, J. Jacod, C. Klüppelberg (2010)

Vol. 2002: C. Pötzsche, Geometric Theory of Discrete Nonautonomous Dynamical Systems (2010)

Vol. 2003: A. Cousin, S. Crépey, O. Guéant, D. Hobson, M. Jeanblanc, J.-M. Lasry, J.-P. Laurent, P.-L. Lions, P. Tankov, Paris-Princeton Lectures on Mathematical Finance 2010. Editors: R.A. Carmona, E. Cinlar, I. Ekeland, E. Jouini, J.A. Scheinkman, N. Touzi (2010)

Vol. 2004: K. Diethelm, The Analysis of Fractional Differential Equations (2010)

Vol. 2005: W. Yuan, W. Sickel, D. Yang, Morrey and Campanato Meet Besov, Lizorkin and Triebel (2011)

Vol. 2006: C. Donati-Martin, A. Lejay, W. Rouault (Eds.), Séminaire de Probabilités XLIII (2011)

Vol. 2007: G. Gromadzki, F.J. Cirre, J.M. Gamboa, E. Bujalance, Symmetries of Compact Riemann Surfaces (2010)

Vol. 2008: P.F. Baum,G. Cortiñas, R. Meyer, R. Sánchez-García, M. Schlichting, B. Toën, Topics in Algebraic and Topological K-Theory (2011)

Vol. 2009: J.-L. Colliot-Thélène, P.S. Dyer, P. Vojta, Arithmetic Geometry. Cetraro, Italy 2007. Editors: P. Corvaja, C. Gasbarri (2011)

Vol. 2010: A. Farina, A. Klar, R.M.M. Mattheij, A. Mikelić, N. Siedow, Mathematical Models in the Manufacturing of Glass. Montecatini Terme, Italy 2008. Editor: A. Fasano (2011)

Recent Reprints and New Editions

Vol. 1702: J. Ma, J. Yong, Forward-Backward Stochastic Differential Equations and their Applications. 1999 – Corr. 3rd printing (2007)

Vol. 830: J.A. Green, Polynomial Representations of GL_n, with an Appendix on Schensted Correspondence and Littelmann Paths by K. Erdmann, J.A. Green and M. Schoker 1980 – 2nd corr. and augmented edition (2007)

Vol. 1693: S. Simons, From Hahn-Banach to Monotonicity (Minimax and Monotonicity 1998) – 2nd exp. edition (2008)

Vol. 470: R.E. Bowen, Equilibrium States and the Ergodic Theory of Anosov Diffeomorphisms. With a preface by D. Ruelle. Edited by J.-R. Chazottes. 1975 – 2nd rev. edition (2008)

Vol. 523: S.A. Albeverio, R.J. Høegh-Krohn, S. Mazzucchi, Mathematical Theory of Feynman Path Integral. 1976 – 2nd corr. and enlarged edition (2008)

Vol. 1764: A. Cannas da Silva, Lectures on Symplectic Geometry 2001 – Corr. 2nd printing (2008)

4. Manuscripts should in general be submitted in English. Final manuscripts should contain at least 100 pages of mathematical text and should always include
   - a general table of contents;
   - an informative introduction, with adequate motivation and perhaps some historical remarks: it should be accessible to a reader not intimately familiar with the topic treated;
   - a global subject index: as a rule this is genuinely helpful for the reader.

   Lecture Notes volumes are, as a rule, printed digitally from the authors' files. We strongly recommend that all contributions in a volume be written in the same LaTeX version, preferably LaTeX2e. To ensure best results, authors are asked to use the LaTeX2e style files available from Springer's web-server at

   ftp://ftp.springer.de/pub/tex/latex/svmonot1/ (for monographs) and
   ftp://ftp.springer.de/pub/tex/latex/svmultt1/ (for summer schools/tutorials).

   Additional technical instructions are available on request from: lnm@springer.com.

5. Careful preparation of the manuscripts will help keep production time short besides ensuring satisfactory appearance of the finished book in print and online. After acceptance of the manuscript authors will be asked to prepare the final LaTeX source files and also the corresponding dvi-, pdf- or zipped ps-file. The LaTeX source files are essential for producing the full-text online version of the book. For the existing online volumes of LNM see: http://www.springerlink.com/openurl.asp?genre=journal&issn=0075-8434.

   The actual production of a Lecture Notes volume takes approximately 12 weeks.

6. Volume editors receive a total of 50 free copies of their volume to be shared with the authors, but no royalties. They and the authors are entitled to a discount of 33.3% on the price of Springer books purchased for their personal use, if ordering directly from Springer.

7. Commitment to publish is made by letter of intent rather than by signing a formal contract. Springer-Verlag secures the copyright for each volume. Authors are free to reuse material contained in their LNM volumes in later publications: a brief written (or e-mail) request for formal permission is sufficient.

Addresses:

Professor J.-M. Morel, CMLA,
École Normale Supérieure de Cachan,
61 Avenue du Président Wilson,
94235 Cachan Cedex, France
E-mail: Jean-Michel.Morel@cmla.ens-cachan.fr

Professor F. Takens, Mathematisch Instituut,
Rijksuniversiteit Groningen, Postbus 800,
9700 AV Groningen, The Netherlands
E-mail: F.Takens@rug.nl

Professor B. Teissier,
Institut Mathématique de Jussieu,
UMR 7586 du CNRS,
Équipe "Géométrie et Dynamique",
175 rue du Chevaleret,
75013 Paris, France
E-mail: teissier@math.jussieu.fr

For the "Mathematical Biosciences Subseries" of LNM:

Professor P.K. Maini, Center for Mathematical Biology,
Mathematical Institute, 24-29 St Giles,
Oxford OX1 3LP, UK
E-mail: maini@maths.ox.ac.uk

Springer, Mathematics Editorial I, Tiergartenstr. 17,
69121 Heidelberg, Germany,
Tel.: +49 (6221) 487-8259
Fax: +49 (6221) 4876-8259
E-mail: lnm@springer.com